Comparison Methods for Stochastic Models and Risks

Comparison Methods for Stochastic Models and Risks

Alfred Müller
University of Karlsruhe, Germany

Dietrich Stoyan
Freiberg University of Mining and Technology, Germany

JOHN WILEY & SONS, LTD

Copyright © 2002   John Wiley & Sons, Ltd
                      Baffins Lane, Chichester,
                      West Sussex, PO19 1UD, England

                      National   01243 779777
                      International (+44) 1243 779777

e-mail (for orders and customer service enquiries): cs-books@wiley.co.uk

Visit our Home Page on http://www.wiley.co.uk
            or http://www.wiley.com

Other Wiley Editorial Offices

John Wiley & Sons, Inc., 605 Third Avenue,
New York, NY 10158-0012, USA

WILEY-VCH Verlag GmbH
Pappelallee 3, D-69469 Weinheim, Germany

John Wiley & Sons Australia, Ltd
33 Park Road, Milton, Queensland 4064, Australia

John Wiley & Sons (Canada) Ltd, 22 Worcester Road
Rexdale, Ontario, M9W 1L1, Canada

John Wiley & Sons (Asia) Pte Ltd, 2 Clementi Loop #02-01,
Jin Xing Distripark, Singapore 129809

British Library Cataloguing in Publication Data

A catalogue record for this book is available from the British Library

ISBN 0-471-49446-1

Set in LATEX, printed from PostScript files supplied by the author.

Contents

Preface

Stochastic orders are now a well-established topic of research, which is still in intensive development and offers many open problems. They lead to powerful approximation methods and bounds in situations where realistic stochastic models are too complex for rigorous treatment; the aim of mathematical research is then to find well-adapted orders which lead to close bounds and good approximations. Stochastic orders are also helpful in situations where fundamental model distributions are only known partially. In economics they are valuable tools in the theory of individual decisions under risk, where a decision maker has to compare actions leading to different uncertain payments. Important fields of application are also queueing theory, reliability theory, statistical physics, epidemiology, insurance mathematics, and others. The authors are sure that there are still many unknown possibilities for further fruitful applications and theoretical work; the potential of comparison methods based on stochastic orders is by no means exhausted. Therefore the present book is methodologically oriented.

The book aims to give the state-of-the-art in this field at the beginning of the 21st century. Originally, it was planned as a continuation or even second edition of Stoyan (1983), because that book has been sold out for years and has served as a standard reference until now. However, in the last 15 to 20 years so many new ideas and concepts have arisen that a nearly completely new exposition became necessary.

The most important new features of this book compared with Stoyan (1983) are

(a) the unified treatment of the theory of integral stochastic orders presented in Chapter 2, which has led to a satisfactory treatment of the generators of such orders (of which D. S. had dreamt in the 1970s);

(b) an extensive treatment of multivariate orders and characterizations of dependence structures, as presented in Chapter 3;

(c) the enormous widening of the field of applications of stochastic orders, in particular its use in economics and actuarial sciences.

Because the authors aim to present these main results but nevertheless want to limit the volume of their book, they decided to omit in the new book some of the topics in the forerunner. This applies to statistical problems

and to continuity or robustness results; for the latter topic the book by
Rachev (1991) is a modern reference. Furthermore, there are no lists of open
problems, though it was a good idea to publish them in 1983, since this
has led to solutions of some of them. While Stoyan (1983) still could aim
to present the whole knowledge of the area, this is impossible today. Notice
that already the bibliography on stochastic orders published by Mosler and
Scarsini (1993) contains more than 200 pages of references to articles on this
topic. Nevertheless, the authors hope to present the most important and
promising ideas in the context of stochastic orders. This, however, implies
that not all proofs can be given; the authors hope that their selection makes
sense and opens the way to understanding the theory also for novices and
leads the reader to the periodical literature and to research in the subject.
The aim is to give complete proofs to all basic results, so that a beginner can
read the book without consulting other references. This especially applies to
Chapter 1, which presents the theory of univariate stochastic orders. However,
some proofs of deeper results are omitted since graduate level mathematics is
required.

As in the former book the treatment is systematic and detailed, the selection
of the examples is made so that a broad spectrum of applications is covered.
As Stoyan (1983), the present book is based on a so-called *Habilitationsschrift*
(now that of A. M.), which is a monograph leading to the second German
doctoral degree. By the way, the authors have known each other since 1995,
when D.S. become aware of the Ph.D. thesis of A.M.

The book begins with a systematic description of univariate stochastic
orders. The first one is the usual stochastic order $\leq_{st}$, which is based on
the comparison of distribution functions. Then several orders sharper than
$\leq_{st}$ are considered, in particular hazard rate order and likelihood ratio
order. The other important family of stochastic orders treated in Chapter
1 is formed by the convex orders and their relatives such as $\leq_{rand}$. They
enable the comparison of distributions of different variability. Moreover, it is
demonstrated how stochastic orders can be used to characterize notions of
aging for lifetime distributions.

Chapter 2 gives the theory of integral stochastic orders from a very general
viewpoint. The problem is to find the largest set of functions g satisfying

$$X \preceq Y \quad \text{implies} \quad Eg(X) \leq Eg(Y)$$

for arbitrary random variables X and Y with values in a general space S,
where the stochastic order $\preceq$ is defined by

$$X \preceq Y \quad \text{if} \quad Ef(X) \leq Ef(Y)$$

for all functions f in some set $\mathcal{F}$. Such a stochastic order is called integral
stochastic order and $\mathcal{F}$ is one of its generators. It is also of interest to find a
possibly small generator, which makes it easy to check $X \preceq Y$. It is shown how

interesting properties of such orders are inherited from its generators. This chapter also presents the famous Strassen theorem, in its general form and in two forms in which it can be applied in the context of the usual stochastic order and the convex order.

Chapter 3 is devoted to multivariate stochastic orders. It starts with possible generalizations of the usual stochastic order $\leq_{st}$, followed by a part on variability orders like convex order and its generalizations. Then an extensive survey describes stochastic orders comparing the strength of dependence between the components of a random vector. Here the supermodular order and the directionally convex order are important examples. Also the related concepts of positive dependence like 'association' or 'total positivity' are discussed. It is a still not finished task to study systematically comparison criteria for parametric families of multivariate distributions; the present book concentrates on normal distributions.

Chapter 4 offers general principles for deriving comparison and monotonicity results for stochastic models. In the simplest case, such a result says that system characteristics of two models with comparable component characteristics are comparable, too, with respect to a suitable stochastic order. Powerful methods for the proof of such relations are the coupling method and the mapping method. While the coupling method, which is in many applications based on the deep Strassen theorem, is the stronger method, the mapping method is more elementary and can also be applied in the context of some stochastic orders which are not suitable for the coupling method.

The comparison properties of stochastic processes (on the real line) are the topic of Chapter 5. After the presentation of general principles, the case of Markov chains is thoroughly studied, in terms of their transition operators. Then some examples of non-Markov processes are discussed and, finally, point processes.

Chapter 6 presents the classical applications of stochastic orders in queueing theory, which lead in particular to comparison results for waiting times with respect to the convex order. Furthermore, some examples for the application of dependence orders are given.

Chapter 7 presents a multifaceted collection of comparison results for various stochastic models and processes. It begins with some problems from reliability theory, some classical and some close to modern research. The next topic is comparison of geometric point processes and random sets. Then the application of stochastic orders to some models of statistical physics is demonstrated, namely the Ising model and Gibbs processes. These results were originally obtained by physicists, who have developed their own terminology. The authors would be happy if this book could diminish a little the gap between applied probability and statistical physics.

Applications in economics and actuarial sciences are considered in Chapter 8. In these two fields, too, many of the concepts have been developed independently under different names. The integral stochastic orders investigated in Chapter 2, for instance, are well-established in economics under the name stochastic dominance rules. After the presentation of the basics of that theory, some financial applications are given, including consistency of mean-risk measures and some portfolio optimization problems. Finally, some actuarial applications are considered, with an emphasis on demonstrating the effect of dependencies between individual claims on the riskiness of portfolios.

The authors are grateful to many colleagues who supported their work for this book. Nicole Bäuerle, Reinhard Bergmann, Hans-Otto Georgii, Günter Last, Karl Mosler, Stefan Napel, Tomasz Rolski, Marco Scarsini, Helga Stoyan, Ryszard Szekli, and Elke Thönnes and an anonymous referee were so kind to read parts of the manuscript and made valuable suggestions for improvement. Other colleagues helped by discussing special topics or by answering questions and sending still unpublished papers. Here Peter M. Alberti, Xiuli Chao, James A. Fill, Serguei Foss, Jerzy Kamburowski, Antonietta Mira, Moshe Shaked, Richard Tweedie, and Armin Uhlmann have to be mentioned. Thanks also go to Gunter Döge, Nora Gürtler, Maria Robakowski, Martin Sonntag, Helga Stoyan and André Tscheschel for providing assistance on technical problems, and to Michaela Tassler for typing parts of the manuscript. Finally, the authors thank the publishers for a good cooperation, in particular Robert Calver, who helped in many ways.

Alfred Müller and Dietrich Stoyan
Karlsruhe and Freiberg, January, 07, 2002.

1

Univariate Stochastic Orders

1.1 Introduction

Stochastic order relations are special cases of partial order relations. Therefore it is natural to start with the definition of a partial order.

Definition 1.1.1. A binary relation $\preceq$ on an arbitrary set S is called a *(partial) order* if it satisfies the three following conditions:

1. *Reflexivity*: $x \preceq x$ for all $x \in S$;
2. *Transitivity*: if $x \preceq y$ and $y \preceq z$ then $x \preceq z$;
3. *Antisymmetry*: if $x \preceq y$ and $y \preceq x$ then $x = y$.

It is sometimes convenient to write $y \succeq x$ as an equivalent form of $x \preceq y$.

Now consider the case that S is the set (or a suitable subset) of all distribution functions of real-valued random variables. A partial order on this set is called a *stochastic order*. If X is a random variable on the reals we write P_X for its distribution and F_X for its distribution function, i.e.

$$F_X(t) = P_X((-\infty, t]) = P(X \le t)$$

for all real t. Often it is convenient not to distinguish between an order relation for distribution functions and the corresponding relations for distributions and random variables. Therefore the following convention will be used throughout this book.

Convention: If the random variables X and Y have distributions P_X and P_Y and distribution functions F_X and F_Y with $F_X \preceq F_Y$, then the notation $P_X \preceq P_Y$ and $X \preceq Y$ will be used, whenever this is convenient. In some places we will not strictly discriminate between distribution functions and distributions; the same character will be used.

Note that there can be different random variables with the same distribution, so that the relation $\preceq$ is antisymmetric as a relation among distributions, but not antisymmetric as a relation for random variables.

1.2 Usual Stochastic Order

The most natural candidate for a stochastic order is that of pointwise comparison of the distribution functions. If $F_X(t) \geq F_Y(t)$ for all real t, then X assumes small values with higher probability than Y does, and hence X assumes large values with smaller probability than Y does. This leads to the following definition for an order relation that compares the location of random variables.

Definition 1.2.1. The random variable X is said to be smaller than the random variable Y with respect to *usual stochastic order* (written $X \leq_{st} Y$), if

$$F_X(t) \geq F_Y(t) \quad \text{for all real } t,$$

or equivalently, if

$$\bar{F}_X(t) \leq \bar{F}_Y(t) \quad \text{for all real } t,$$

where $\bar{F}_X(t)$ denotes the so-called *survival function* of X, given by

$$\bar{F}_X(t) = P(X > t).$$

Remark 1.2.2. This stochastic order relation is frequently just called *the stochastic order*. To distinguish it from the many other stochastic orders used in this book, we will use the name *usual stochastic order* as Shaked and Shanthikumar (1994) also do; Szekli (1995) denotes it as *strong stochastic order*. It has a long history. Mann and Whitney (1947) and Lehmann (1955) used it in their work on statistical problems. Karlin (1960) introduced it to the operations research literature when considering inventory problems. In the economic literature it is known as *first order stochastic dominance*, abbreviated as $\leq_{FSD}$ (see also Section 8.1.2). Here, Quirk and Saposnik (1962) and Fishburn (1970) are important early references.

Remark 1.2.3. At first sight it might seem to be counterintuitive to say that $F_X \leq_{st} F_Y$, if $F_X(t) \geq F_Y(t)$ for all real t. On the other hand, it is clear that we want to call Y stochastically larger than X if it assumes *large* values with higher probability. The distribution function, however, describes the probability of assuming small values, hence the reversal of the inequality sign. In the given form, $\leq_{st}$ can be considered as a generalization of the order $\leq$ on the real axis, since for real numbers a and b

$$a \leq b \quad \text{implies} \quad a \leq_{st} b$$

where on the right-hand side a and b are considered as degenerated random variables. In other words,

$$a \leq b \quad \text{implies} \quad \delta_a \leq_{st} \delta_b$$

for the corresponding one-point distributions.

The following theorem shows that $\leq_{st}$ is closely related to the pointwise comparison of random variables.

Theorem 1.2.4. *For random variables X and Y with distribution functions F_X and F_Y the following statements are equivalent:*

(i) $X \leq_{st} Y$;

(ii) *there is a probability space $(\Omega, \mathcal{A}, P)$ and random variables $\hat{X}$ and $\hat{Y}$ on it with the distribution functions F_X and F_Y such that $\hat{X}(\omega) \leq \hat{Y}(\omega)$ for all $\omega \in \Omega$.*

Proof. (i) $\Rightarrow$ (ii): Denote by

$$F^{-1}(u) = \inf\{x : F(x) \geq u\} \text{ for } 0 < u < 1$$

the inverse distribution function (also called quantile function) corresponding to the distribution function F. Let U be a random variable uniformly distributed on $(0,1)$, and define $\hat{X} = F_X^{-1}(U)$ and $\hat{Y} = F_Y^{-1}(U)$. Then $\hat{X}$ has distribution function F_X and $\hat{Y}$ has distribution function F_Y. According to (i) $F_X \geq F_Y$ and therefore $F_X^{-1} \leq F_Y^{-1}$. Thus $\hat{X} \leq \hat{Y}$ with certainty. (ii) $\Rightarrow$ (i) is obvious. $\qquad\qquad\square$

Remark 1.2.5. A natural candidate for a partial order that compares the size of random variables would be the relation $X \preceq_{a.s.} Y$, which holds if and only if $X(\omega) \leq Y(\omega)$ for P–almost all ω. This order relation, however, does not only depend on the distributions. This can easily be seen from the observation that $X \preceq_{a.s.} X$ always holds, but $X \preceq_{a.s.} Y$ does not hold if X and Y are independent and identically distributed with the same non-degenerated distribution. Therefore the definition of $\leq_{st}$ used here is much more useful, and as Theorem 1.2.4 reveals, it is the strongest order $\preceq$ for distributions with the property that $X \leq Y$ a.s. implies $F_X \preceq F_Y$.

Stochastic order can also be characterized by the *relative inverse distribution function* $\Phi_{F,G}$ defined as

$$\Phi_{F,G}(t) = G^{-1}(F(t)). \qquad\qquad (1.2.1)$$

The relative inverse distribution function is strongly related to the *quantile–quantile plot* (or shortly *Q–Q plot*), which is a well-known tool for graphical analysis of statistical data. The Q–Q plot is obtained by plotting the quantiles $G^{-1}(p)$ and $F^{-1}(p)$ against each other for all $0 < p < 1$. For continuous distributions the Q–Q plot and the graph of the relative inverse distribution function $\phi_{F,G}$ are identical. In general, however, the points of the Q–Q plot form only a subset of the graph of $\phi_{F,G}$. This applies especially in the practically important case of empirical distributions.

The following theorem provides a simple graphical procedure for checking stochastic order between two given distribution functions F and G. Informally it states that $G \leq_{st} F$ holds if and only if the Q–Q plot of G^{-1} against F^{-1} lies below the 45° line.

Theorem 1.2.6. *Let F and G be arbitrary distribution functions. Then $G \leq_{st} F$ if and only if*

$$\Phi_{F,G}(x) = G^{-1}(F(x)) \leq x \quad \text{for all } x.$$

Proof. First notice that by definition $G^{-1}(G(x)) \leq x$ and $G(G^{-1}(x)) \geq x$ for all x. Therefore, if $G \leq_{st} F$ then $F(x) \leq G(x)$ for all x and hence

$$G^{-1}(F(x)) \leq G^{-1}(G(x)) \leq x \quad \text{for all real } x.$$

Conversely, if $x \geq G^{-1}(F(x))$ for all x then

$$G(x) \geq G(G^{-1}(F(x))) \geq F(x) \quad \text{for all } x.$$

$\square$

If F and G are continuous then $G \leq_{st} F$ is also equivalent to $F^{-1}(G(x)) \geq x$ for all real x, i.e. the Q–Q plot of F^{-1} against G^{-1} lies above the 45° line. For discrete distributions this is still true for the Q–Q plot, for the relative inverse distribution function $F^{-1}(G(x))$, however, this can be wrong. (As a simple counterexample consider equal two-point distributions $F = G$ with atoms at 0 and 1 with equal probability $1/2$.) Therefore the statement in Szekli (1995), theorem C on p.7, is wrong in the case of discrete distributions. The correct statement is that in Theorem 1.2.6.

Plotting the distribution functions $G(x)$ and $F(x)$ against each other yields the so called *probability–probability plot* (or shortly *P–P plot*). For continuous distributions this is the graph of the function

$$\Psi_{F,G}(t) = G(F^{-1}(t)) \quad \text{for } 0 < t < 1. \tag{1.2.2}$$

For this function an analogous result to Theorem 1.2.6 can be shown. In particular, the following characterization of the usual stochastic order holds.

Theorem 1.2.7. *$F \leq_{st} G$ holds if and only if the corresponding P–P plot lies below the 45° line.*

Now an important characterization in terms of expectations of increasing functions will be given. As always in this book the notion *increasing* will be used in the weak sense, i.e. a function $f : \mathbb{R} \to \mathbb{R}$ is called increasing, if $x \leq y$ implies $f(x) \leq f(y)$.

Theorem 1.2.8. *The following statements are equivalent:*

(i) $X \leq_{st} Y$;

(ii) *the inequality*

$$Ef(X) \leq Ef(Y) \tag{1.2.3}$$

holds for all increasing functions f, for which both expectations exist.

Moreover, if for a given f inequality (1.2.3) holds for all X and Y with $X \leq_{st} Y$, then f must be increasing.

Proof. (i) $\Rightarrow$ (ii): According to Theorem 1.2.4 it can be assumed without loss of generality that $X \leq Y$ almost surely. Thus, if f is increasing, then $f(X) \leq f(Y)$ a.s. too, and hence $Ef(X) \leq Ef(Y)$ follows from the monotonicity of expectation.

(ii) $\Rightarrow$ (i): This follows immediately from the observation that $P(X > t) = Ef_t(X)$ for the indicator function

$$f_t(x) = \mathbf{1}_{(t,\infty)}(x) = \begin{cases} 1 & \text{for } x > t \\ 0 & \text{otherwise} \end{cases}$$

which is of course increasing.

To show the last statement, assume that f is not increasing, so that there are $x \leq y$ with $f(x) > f(y)$, and let $P(X = x) = P(Y = y) = 1$. Then $X \leq_{st} Y$, but $Ef(X) > Ef(Y)$. $\square$

It turns out that the stochastic order $\leq_{st}$ has many interesting properties. As a first result, it will be shown that it implies ordering of expectations and that different distributions with the same expectation cannot be ordered with respect to $\leq_{st}$.

Theorem 1.2.9. *Let X and Y be random variables with finite expectations.*
a) If $X \leq_{st} Y$ then $EX \leq EY$;
b) if $X \leq_{st} Y$ and $EX = EY$ then X and Y have the same distribution.

Proof. a) follows from Theorem 1.2.8 by using $f(x) = x$.
b) The well-known representation

$$EX = \int_0^\infty [1 - F_X(t)]\, \mathrm{d}t - \int_{-\infty}^0 F_X(t)\, \mathrm{d}t$$

yields

$$EY - EX = \int_{-\infty}^\infty [F_X(t) - F_Y(t)]\, \mathrm{d}t. \tag{1.2.4}$$

If $X \leq_{st} Y$ and $EX = EY$, then the left-hand-side of (1.2.4) vanishes and the right-hand side is an integral of a nonnegative right-continuous function. This is only possible if this function vanishes, too. Hence $F_X = F_Y$. $\square$

Remark 1.2.10. For mathematically less trained people there is a more natural order than the usual stochastic order, sometimes called the 'engineer's order', which is simply based on comparison of means. The random variable X is called *smaller* than Y *in mean* (written $X \leq_\mu Y$) if

$$EX \leq EY.$$

Theorem 1.2.9 a) thus states that usual stochastic order implies the mean order.

Remark 1.2.11. Theorem 1.2.9 b) has been shown in Scarsini and Shaked (1990). It has important consequences for order restricted statistical inference. If one has two samples and wants to test the hypothesis $F = G$ against the alternative $F \leq_{st} G$ then a simple consistent test can be based on the difference of the means. Better tests can be based on the function $\max\{F(x) - G(x), 0\}$ since this function vanishes under the hypothesis, but is strictly positive for some real x under the alternative. For a detailed account of order restricted statistical inference refer to Robertson, Wright, and Dykstra (1988).

Part a) of the previous theorem can be generalized to the comparison of higher moments. The following results follow easily from Theorem 1.2.4.

Theorem 1.2.12. *If $X \leq_{st} Y$, then $EX^n \leq EY^n$ for $n = 1, 3, 5, \ldots$, whenever the expectations exist. If, moreover, X and Y are non-negative then $EX^n \leq EY^n$ holds for all $n = 1, 2, \ldots$.*

Theorem 1.2.13. *If $X \leq_{st} Y$ then $f(X) \leq_{st} f(Y)$ for all increasing functions f.*

Theorem 1.2.14. *The stochastic order relation $\leq_{st}$ is closed with respect to weak convergence, i.e. if $F_n \leq_{st} G_n$ holds for all n and the sequences (F_n) and (G_n) converge weakly to F and G then $F \leq_{st} G$.*

Proof. Since $F_n(t) \geq G_n(t)$ for all real t, weak convergence implies $F(t) \geq G(t)$, if t is a continuity point of both F and G. Since the set of discontinuity points of F and G is at most countable, the continuity points are dense in $\mathbb{R}$. Hence $F(t) \geq G(t)$ must hold for all real t, since distribution functions are right-continuous.   □

Another important property of a stochastic order relation is closure under mixture. It will be shown next that the usual stochastic order has this property.

Theorem 1.2.15. *Usual stochastic order is closed under mixtures, i.e. if X, Y and Θ are random variables such that $[X|\Theta = \theta] \leq_{st} [Y|\Theta = \theta]$ for all θ in the support of Θ, then $X \leq_{st} Y$.*

Proof. This is an immediate consequence of Theorem 1.2.8, since for any increasing function f

$$Ef(X) = E_\Theta E[f(X)|\Theta] \leq E_\Theta E[f(Y)|\Theta] = Ef(Y).$$

$\square$

Theorem 1.2.16. *Let $X_1, \ldots, X_n$ and $Y_1, \ldots, Y_n$ be independent random variables with $X_i \leq_{st} Y_i$ for $i = 1, \ldots, n$, and assume that $\psi : \mathbb{R}^n \to \mathbb{R}$ is an increasing function. Then*

$$\psi(X_1, \ldots, X_n) \leq_{st} \psi(Y_1, \ldots, Y_n).$$

Proof. The proof uses Theorem 1.2.8 and proceeds by induction. For $n = 1$ one has to show that $Ef(\psi(X_1)) \leq Ef(\psi(Y_1))$ for all increasing functions f. This follows from the fact that the composition of two increasing functions is again increasing. Hence assume that the assertion holds for $n - 1$. Define $g(x) = Ef(\psi(X_1, \ldots, X_{n-1}, x))$ and $h(x) = Ef(\psi(Y_1, \ldots, Y_{n-1}, x))$ for all real x. According to the induction hypothesis, $g(x) \leq h(x)$ holds for all $x \in \mathbb{R}$ and both g and h are increasing functions, since f and ψ are increasing. This yields

$$Ef(\psi(X_1, \ldots, X_n)) = Eg(X_n) \leq Eh(X_n) \leq Eh(Y_n) = Ef(\psi(Y_1, \ldots, Y_n))$$

and thus $\psi(X_1, \ldots, X_n) \leq_{st} \psi(Y_1, \ldots, Y_n)$. $\square$

This theorem is one of the bases of the mapping method in Section 4.3.2.

Choosing in Theorem 1.2.16 the special case $\psi(x_1, \ldots, x_n) = \sum_{i=1}^n x_i$ yields that $\leq_{st}$ is preserved under convolution, i.e. under addition of independent random variables.

Theorem 1.2.17. *Let $X_1, \ldots, X_n$ and $Y_1, \ldots, Y_n$ be independent random variables with $X_i \leq_{st} Y_i$ for $i = 1, \ldots, n$. Then*

$$X_1 + \cdots + X_n \leq_{st} Y_1 + \cdots + Y_n.$$

Other functions of interest are $\psi(x_1, \ldots, x_n) = \max_{i=1}^n x_i$ and $\psi(x_1, \ldots, x_n) = \min_{i=1}^n x_i$, as well as any other order statistic. For a different proof of the following corollary in a more general context we refer to Theorem 1.7.11.

Corollary 1.2.18. *Under the assumptions of Theorem 1.2.16 let $X_{(1:n)} \leq X_{(2:n)} \leq \cdots \leq X_{(n:n)}$ and $Y_{(1:n)} \leq Y_{(2:n)} \leq \cdots \leq Y_{(n:n)}$ be the order statistics corresponding to $X_1, \ldots, X_n$ and $Y_1, \ldots, Y_n$, respectively. Then*

$$X_{(i:n)} \leq_{st} Y_{(i:n)} \quad for\ i = 1, \ldots, n.$$

1.3   Hazard Rate Order

There are many situations where stronger concepts than the usual stochastic order are necessary or useful. Consider the situation that somebody wants to buy a car, and that she can choose between two types with different random lifetimes X and Y. Of course, if $X \leq_{st} Y$ and the price is the same then she will choose the second type. But now assume that she wants to buy a used car which is one year old, so that the remaining lifetimes are X' and Y', where $P(X' > t) = P(X > 1 + t | X > 1)$ and similarly for Y'. Is the second type still the better one, i.e. is $X' \leq_{st} Y'$? The following example shows that this is not necessarily the case.

Example 1.3.1. ($X \leq_{st} Y$ is not preserved under aging) Assume that X is uniformly distributed on $(0, 3)$ and that Y has the density

$$f(x) = \begin{cases} \frac{1}{6} & , \ 0 < x \leq 1 \\ \frac{1}{2} & , \ 1 < x \leq 2 \\ \frac{1}{3} & , \ 2 < x < 3 \end{cases}.$$

Then $X \leq_{st} Y$ holds. But X' is uniformly distributed on $(0, 2)$ and Y' has the density

$$f'(x) = \begin{cases} \frac{3}{5} & , \ 0 < x \leq 1 \\ \frac{2}{5} & , \ 1 < x < 2 \end{cases}$$

so that $X' \geq_{st} Y'$!

Thus the following question arises: which stronger assumptions are necessary to ensure that usual stochastic order holds for all used cars of any age t, i.e. when is $[X | X > t] \leq_{st} [Y | Y > t]$ for all t?

Inserting the definition of $\leq_{st}$, this inequality can be rewritten as

$$P(X > s + t | X > t) \leq P(Y > s + t | Y > t) \quad \text{for all } s \geq 0 \text{ and all } t. \quad (1.3.1)$$

This is obviously equivalent to the statement

$$\frac{\bar{F}_Y(t)}{\bar{F}_X(t)} \leq \frac{\bar{F}_Y(s + t)}{\bar{F}_X(s + t)} \quad \text{for all } s \geq 0 \text{ and all } t,$$

where $\bar{F}_X(t) = 1 - F_X(t)$. This is a motivation for the following definition.

Definition 1.3.2. The random variable X is said to be smaller than the random variable Y with respect to the *hazard rate order* (written $X \leq_{hr} Y$), if the function

$$t \mapsto \frac{\bar{F}_Y(t)}{\bar{F}_X(t)}$$

is increasing.

The name of this order is due to the fact that, assuming the existence of continuous densities and hence of the hazard rates, there is an equivalent characterization in terms of pointwise comparison of the so-called *hazard rates*.

The hazard rate (also called *failure rate*) $r_X(t)$ is the infinitesimal rate of failure at time t, i.e.

$$r_X(t) = \lim_{\varepsilon \to 0} \frac{P(X \le t + \varepsilon | X > t)}{\varepsilon} = \frac{f_X(t)}{\bar{F}_X(t)} = -\frac{d}{dt} \ln(\bar{F}_X(t))$$

for $\bar{F}_X(t) > 0$.

Theorem 1.3.3. *If X and Y have continuous densities, then $X \le_{hr} Y$ is equivalent to $r_X(t) \ge r_Y(t)$ for all real t.*

Proof. $\bar{F}_Y(t)/\bar{F}_X(t)$ is increasing if and only if

$$\ln\left(\frac{\bar{F}_Y(t)}{\bar{F}_X(t)}\right) = \ln(\bar{F}_Y(t)) - \ln(\bar{F}_X(t))$$

is increasing. Since $r_X(t) = -d/dt \ln(\bar{F}_X(t))$, the result follows from the fact that a differentiable function is increasing if and only if its derivative is non-negative. □

It would be more natural to use the characterization of Theorem 1.3.3 as a definition, but Definition 1.3.2 has the advantage that it does not rely on the existence of densities. It applies also to discrete distributions and even to mixtures of discrete and continuous distributions.

The next result shows that hazard rate order has a nice graphical interpretation in terms of P–P plots. Recall that a subset A of an Euclidean space is called *star-shaped* with respect to a point $\mathbf{s}$, if $\mathbf{x} \in A$ implies that A contains the whole line segment between $\mathbf{x}$ and $\mathbf{s}$. A real function f is called star-shaped with respect to (a, b), if its epigraph is star-shaped with respect to (a, b). This is equivalent to the statement that

$$\frac{f(x) - b}{x - a} \quad \text{is increasing in } x.$$

Theorem 1.3.4. *$F \le_{hr} G$ holds if and only if the P–P plot is star-shaped with respect to $(1, 1)$, i.e.*

$$\frac{G(F^{-1}(x)) - 1}{x - 1} \quad \text{is increasing in } x.$$

Proof. Since F^{-1} is increasing this follows immediately by inserting $t = F^{-1}(x)$ in the definition of hazard rate order. □

Theorem 1.3.5. *The hazard rate order is closed with respect to weak convergence.*

Proof. Let (F_n) and (G_n) be sequences of distribution functions, converging weakly to distribution functions F and G respectively. Moreover, assume that $F_n \leq_{hr} G_n$ for all n. Then the h_n given by $h_n(t) = \bar{G}_n(t)/\bar{F}_n(t)$ form a sequence of right-continuous increasing functions, which converges to the function $h(t) = \bar{G}(t)/\bar{F}(t)$ in all t where both F and G are continuous. Since these points are dense and h is right-continuous, h must also be increasing. $\square$

As the continuous distribution functions are dense in the set of all distribution functions, the previous result shows that it is always sufficient to consider distributions with densities when showing results about hazard rate order. In fact, it reveals that hazard rate order could be defined alternatively as follows. First define it for continuous distributions via the characterization in Theorem 1.3.3, and then take the closure with respect to weak convergence to define it for arbitrary distributions.

Unfortunately, the hazard rate order is neither closed with respect to convolution nor with respect to mixtures, as can be seen from the following example.

Example 1.3.6. ($X \leq_{hr} Y$ does not imply $X + Z \leq_{hr} Y + Z$.) Let X be uniformly distributed on the interval $(0,3)$ and Y uniformly distributed on $(0,4)$. Then $X \leq_{hr} Y$. Assume that Z is independent of X and Y with $P(Z = 0) = P(Z = 3) = 1/2$. Then

$$\frac{\bar{F}_{Y+Z}(3)}{\bar{F}_{X+Z}(3)} = \frac{5}{4} > \frac{9}{8} = \frac{\bar{F}_{Y+Z}(4)}{\bar{F}_{X+Z}(4)}.$$

Hence $X + Z \not\leq_{hr} Y + Z$. The same setting can also be considered as a counterexample that hazard rate order is not closed under mixtures, since the distributions of $X + Z$ and $Y + Z$ are mixtures of the distributions of X and $X + 3$, and Y and $Y + 3$, respectively, and it holds $X \leq_{hr} Y$ as well as $X + 3 \leq_{hr} Y + 3$.

Shanthikumar and Yao (1991) showed that closure with respect to convolution holds, if all involved random variables have an *increasing failure rate* (IFR) distribution. This means that the failure rate function r is increasing, or equivalently, that $\bar{F}$ is log-concave, i.e. $\log \bar{F}$ is concave, see Section 1.8.

The next result shows that hazard rate order is invariant with respect to any monotone transformation of the scale.

Theorem 1.3.7. *Assume that $X \leq_{hr} Y$ and that $g : \mathbb{R} \to \mathbb{R}$ is increasing. Then $g(X) \leq_{hr} g(Y)$.*

Proof. This follows immediately from the identity $\bar{F}_{g(X)}(t) = \bar{F}_X(g^{-1}(t))$. $\square$

Theorem 1.3.8. $X \leq_{hr} Y$ implies $X \leq_{st} Y$.

Proof. As $\lim_{t \to -\infty} \bar{F}_Y(t)/\bar{F}_X(t) = 1$, $X \leq_{hr} Y$ implies $\bar{F}_Y(t)/\bar{F}_X(t) \geq 1$ and hence $F_X(t) \geq F_Y(t)$ for all real t. □

It has already been demonstrated in Example 1.3.1 that the implication of Theorem 1.3.8 is strict.

Sometimes it is useful to consider the stochastic order that is obtained when the survival function is replaced by the distribution function in the definition of hazard rate order. This yields the so called reversed hazard rate order.

Definition 1.3.9. The random variable X is said to be smaller than the random variable Y with respect to the *reversed hazard rate order* (written $X \leq_{rh} Y$), if the function

$$t \mapsto \frac{F_Y(t)}{F_X(t)}$$

is increasing.

The reversed hazard rate order shares many properties with the usual hazard rate order. In fact, there is a strong duality between $\leq_{hr}$ and $\leq_{rh}$. This is due to the following result that was found by Nanda and Shaked (2002).

Theorem 1.3.10. *Let g be a continuous strictly decreasing function. Then $X \leq_{hr} Y$ if and only if $g(X) \geq_{rh} g(Y)$.*

Proof. If g is continuous and strictly decreasing then the same holds for g^{-1}. The assertion thus follows from the identity

$$F_{g(X)}(t) = P(g(X) \leq t) = P(X \geq g^{-1}(t)) = \bar{F}_X(g^{-1}(t)).$$

□

The most important consequences of this theorem are collected below.

Theorem 1.3.11. *$F \leq_{rh} G$ holds if and only if the P–P plot is star-shaped with respect to $(0,0)$, i.e.*

$$\frac{G(F^{-1}(x))}{x} \quad \text{is increasing in } x.$$

Theorem 1.3.12. *The reversed hazard rate order is closed with respect to weak convergence.*

Theorem 1.3.13. *Assume that $X \leq_{rh} Y$ and that $g : \mathbb{R} \to \mathbb{R}$ is increasing. Then $g(X) \leq_{rh} g(Y)$.*

Theorem 1.3.14. *$X \leq_{rh} Y$ implies $X \leq_{st} Y$.*

Theorem 1.3.15. *$X \leq_{rh} Y$ holds if and only if*

$$[X|X \leq t] \leq_{st} [Y|Y \leq t] \quad \text{for all real } t.$$

1.4 Likelihood Ratio Order

An interesting feature of the hazard rate order is that $X \leq_{hr} Y$ holds if and only if $[X|X > t] \leq_{st} [Y|Y > t]$ for all real t. This is important for analyzing lifetime distributions. There are other situations, however, where one would like to have $[X|X \in A] \leq_{st} [Y|Y \in A]$ for all possible events A. It will turn out that this requirement leads to the so called *likelihood ratio order*.

Definition 1.4.1. The random variable Y is said to be larger than X in *likelihood ratio order* (written $X \leq_{lr} Y$), if X and Y have densities with respect to some dominating measure μ such that

$$f_X(t)f_Y(s) \leq f_X(s)f_Y(t) \quad \text{for all } s \leq t. \tag{1.4.1}$$

Remark 1.4.2. a) We have chosen a definition with an arbitrary dominating measure μ, so that the definition includes the case of densities of continuous distributions (where μ is Lebesgue measure), the case of discrete distributions (where μ is counting measure) as well as more general situations like mixtures of these two.

b) Equation (1.4.1) simply states that the ratio f_Y/f_X is increasing, but written in a form that avoids the difficulty of declaring separately what to do if the numerator or denominator or both of them vanish.

c) Alternatively the definition can be stated as follows: For *all* dominating measures μ and all corresponding densities f_X and f_Y

$$f_X(t)f_Y(s) \leq f_X(s)f_Y(t) \quad \text{for all } s \leq t,$$

μ-a.s. on the union of the supports of X and Y.

Likelihood ratio order can easily be characterized in terms of P–P plots.

Theorem 1.4.3. $X \leq_{lr} Y$ *holds if and only if the P–P plot (i.e.* $F_Y(F_X^{-1}(t))$) *is convex.*

Proof. We only give the proof for the continuous case. There differentiation yields that $F_Y(F_X^{-1}(t))$ is convex if and only if

$$\frac{f_Y(F_X^{-1}(t))}{f_X(F_X^{-1}(t))} \quad \text{is increasing,}$$

hence if and only if $X \leq_{lr} Y$. $\qquad\square$

The likelihood ratio order is stronger than the usual stochastic order.

Theorem 1.4.4. *If* $X \leq_{lr} Y$ *then* $X \leq_{st} Y$.

Proof. This is a consequence of Theorem 1.4.5 below, but we give a simple direct proof here, too. Since f_Y/f_X is increasing, there is some t_0 such that $f_Y(t)/f_X(t) \leq 1$ for $t < t_0$ and $f_Y(t)/f_X(t) \geq 1$ for $t > t_0$. Hence $f_Y(t) \leq f_X(t)$ for $t < t_0$ and $f_Y(t) \geq f_X(t)$ for $t > t_0$. Thus $F_Y(t) - F_X(t)$ is decreasing for $t < t_0$ and increasing for $t > t_0$. As

$$\lim_{t \to -\infty} (F_Y(t) - F_X(t)) = \lim_{t \to \infty} (F_Y(t) - F_X(t)) = 0,$$

this implies $F_Y(t) \leq F_X(t)$ for all t. $\qquad\qquad\square$

Likelihood ratio order is even stronger than the hazard rate orders.

Theorem 1.4.5. $X \leq_{lr} Y$ *implies* $X \leq_{hr} Y$ *and* $X \leq_{rh} Y$.

Proof. This follows from Theorem 1.4.3, taking into account Theorem 1.3.4 and Theorem 1.3.11, respectively. Notice that a convex function is star-shaped with respect to any point on its graph. $\qquad\qquad\square$

An alternative proof of Theorem 1.4.5 could be derived from the next result, which characterizes the likelihood ratio order in terms of conditional distributions.

Theorem 1.4.6. *The following statements are equivalent:*

(i) $X \leq_{lr} Y$;

(ii) *for all intervals* $U = [a, b]$ *and* $V = [c, d]$ *with* $a < b < c < d$

$$P(X \in V)P(Y \in U) \leq P(X \in U)P(Y \in V); \qquad (1.4.2)$$

(iii) $[X|a \leq X \leq b] \leq_{st} [Y|a \leq Y \leq b]$ *for all* $a < b$ *with* $P(a \leq X \leq b) > 0$ *and* $P(a \leq Y \leq b) > 0$;

(iv) $[X|X \in A] \leq_{st} [Y|Y \in A]$ *for all events* A *with* $P(X \in A) > 0$ *and* $P(Y \in A) > 0$;

(v) $[X|X \in A] \leq_{lr} [Y|Y \in A]$ *for all events* A *with* $P(X \in A) > 0$ *and* $P(Y \in A) > 0$.

Proof. (i) $\Rightarrow$ (v): The density of $[X|X \in A]$ is given by

$$f_{[X|X \in A]}(t) = \frac{f_X(t)}{P(X \in A)}.$$

Therefore the claim follows immediately from the definition of the likelihood ratio order.

(v) $\Rightarrow$ (iv): This is an immediate consequence of Theorem 1.4.4.

(iv) $\Rightarrow$ (iii) is trivial.

(iii) $\Rightarrow$ (ii): If (iii) holds then

$$\frac{P(b \leq X \leq c)}{P(a \leq X \leq c)} \leq \frac{P(b \leq Y \leq c)}{P(a \leq Y \leq c)} \qquad \text{for } a < b < c.$$

This can only be true if

$$\frac{P(a \leq X \leq b)}{P(b \leq X \leq c)} \geq \frac{P(a \leq Y \leq b)}{P(b \leq Y \leq c)} \qquad \text{for } a < b < c,$$

or equivalently, if

$$\frac{P(a \leq X \leq b)}{P(a \leq Y \leq b)} \geq \frac{P(b \leq X \leq c)}{P(b \leq Y \leq c)} \qquad \text{for } a < b < c.$$

Similarly,

$$\frac{P(b \leq X \leq c)}{P(b \leq Y \leq c)} \geq \frac{P(c \leq X \leq d)}{P(c \leq Y \leq d)} \qquad \text{for } b < c < d,$$

and hence

$$\frac{P(a \leq X \leq b)}{P(a \leq Y \leq b)} \geq \frac{P(c \leq X \leq d)}{P(c \leq Y \leq d)},$$

whenever $a < b < c < d$.

(ii) $\Rightarrow$ (i): Let μ be any dominating measure. Inequality (1.4.2) implies

$$\frac{P(X \in V)}{\mu(V)} \frac{P(Y \in U)}{\mu(U)} \leq \frac{P(X \in U)}{\mu(U)} \frac{P(Y \in V)}{\mu(V)}.$$

Now we choose $U = [s - \varepsilon, s + \varepsilon]$ and $V = [t - \varepsilon, t + \varepsilon]$ where $s < t$ and $0 < \varepsilon < (t - s)/2$. Taking the limit $\varepsilon \to 0$ yields

$$f_X(t) f_Y(s) \leq f_X(s) f_Y(t)$$

μ-a.s. and hence $X \leq_{lr} Y$. □

Remark 1.4.7. The equivalence of (i) and (iv) in the previous theorem is due to Whitt (1980a). He introduced the name *uniform conditional stochastic order* for the stochastic order relation $\leq_{ucso}$ defined as

$$X \leq_{ucso} Y \quad \text{if } [X|X \in A] \leq_{st} [Y|Y \in A]$$

for all events A with $P(X \in A) > 0$ and $P(Y \in A) > 0$. Theorem 1.4.6 thus shows that likelihood ratio order is equivalent to uniform conditional stochastic order.

The likelihood ratio order is, like the hazard rate order, preserved under monotone transformations.

Theorem 1.4.8. *Assume that $X \leq_{lr} Y$ and that g is an increasing function. Then $g(X) \leq_{lr} g(Y)$.*

Proof. As $P(a \leq g(X) \leq b) = P(g^{-1}(a) \leq X \leq g^{-1}(b))$, the result is a consequence of characterization (ii) in Theorem 1.4.6. □

In the next theorem it is shown that likelihood ratio order is closed with respect to weak convergence. The proof uses some deeper facts from measure theory.

Theorem 1.4.9. *The likelihood ratio order $\leq_{lr}$ is closed with respect to weak convergence.*

Proof. (Müller (1997b)) Let (P_n) and (Q_n) be sequences of probability measures converging weakly to P and Q, and assume $P_n \leq_{lr} Q_n$ for all n. By characterization (ii) in Theorem 1.4.6

$$P_n(I_y(h)) \, Q_n(I_x(h)) \leq P_n(I_x(h)) \, Q_n(I_y(h)) \text{ for all } x < y, \qquad (1.4.3)$$

where $I_x(h)$ denotes the open interval $(x - h, x + h)$. Now let μ be a σ-finite measure with $\mu(I_x(h)) > 0$ for all x and h which dominates P and Q. (Such a measure exists, as the Lebesgue measure can be added to an arbitrary dominating measure without affecting the dominance property.)

By the portmanteau theorem (see Billingsley (1999), Theorem 2.1), inequality (1.4.3) implies

$$P(I_y(h)) \, Q(I_x(h)) \leq P(I_x(h)) \, Q(I_y(h)),$$

if the intervals are μ-continuity sets.

Next we construct a sequence (h_n) with $h_n \downarrow 0$ such that the intervals $I_x(h_n)$ are μ-continuity sets for μ-almost all $x \in \mathbb{R}$. Denote by M the countable set of atoms of μ. Then the set $M - M = \{x - y : x, y \in M\}$ is also countable. Now take $h_n \in (M - M)^c$ with $h_n \downarrow 0$. Then $I_x(h_n)$ is a μ-continuity set for all $x \in M$. Furthermore, the set of all $x \in M^c$ for which $I_x(h_n)$ is not a μ-continuity set, is at most countable. Hence $I_x(h_n)$ is a μ-continuity set for μ-almost all real x.
Consequently,

$$\frac{P(I_y(h_n))}{\mu(I_y(h_n))} \, \frac{Q(I_x(h_n))}{\mu(I_x(h_n))} \leq \frac{P(I_x(h_n))}{\mu(I_x(h_n))} \, \frac{Q(I_y(h_n))}{\mu(I_y(h_n))}$$

for μ-almost all $x < y$.
Now the general differentiation theorem for measures (Wheeden and Zygmund (1977), theorem 10.49 and corollary 10.50) yields

$$f(y) \, g(x) \leq f(x) \, g(y) \text{ for } \mu\text{-almost all } x < y,$$

where f and g are μ-densities of P and Q. Hence $P \leq_{lr} Q$. $\qquad \square$

1.5 Convex Orders

1.5.1 *Fundamental Properties*

The previous sections have dealt with stochastic orders which compare the size of random variables. Very often, however, also the variability of a random variable is of interest, since it describes the riskiness of an uncertain outcome. If two random variables X and Y with the same mean describe the returns of two risky investments, then every risk-averse decision maker will choose that with lower variability. Therefore variability orderings are of special interest in the context of decision making under risk, see Chapter 8.

It will turn out that there is a natural connection between variability of random variables and stochastic orders based on convex functions. Recall that a real function f is called *convex*, if

$$f(\alpha x + (1 - \alpha)y) \leq \alpha f(x) + (1 - \alpha)f(y)$$

for all x and y and all $0 < \alpha < 1$, and that f is *concave*, if $-f$ is convex.

Definition 1.5.1. Let X and Y be random variables with finite means. Then we say that

(i) X is less than Y in *convex order* (written $X \leq_{cx} Y$), if $Ef(X) \leq Ef(Y)$ for all real convex functions f such that the expectations exist;

(ii) X is less than Y in *increasing convex order* (written $X \leq_{icx} Y$), if $Ef(X) \leq Ef(Y)$ for all increasing convex functions f such that the expectations exist;

(iii) X is less than Y in *increasing concave order* (written $X \leq_{icv} Y$), if $Ef(X) \leq Ef(Y)$ for all increasing concave functions f such that the expectations exist.

Remark 1.5.2. These orders are also known under various other names. In Ross (1983) $\leq_{icx}$ is called *stochastically more variable*, denoted as $\leq_v$. Stoyan (1983) called it *smaller in mean residual life* and denoted it by $\leq_c$. In the literature of actuarial sciences it is known as *stop-loss order* $\leq_{sl}$, see Section 8.3. The convex order is known in that field as *stop-loss order with equal means*, denoted by $\leq_{sl,=}$.

The order $\leq_{icv}$ occurs naturally in the theory of decisions under risk. There it is usually called *second order stochastic dominance*, denoted by $\leq_{SSD}$, see Section 8.1.2.

Throughout this section we will adhere to the assumption that all occurring random variables have a finite mean. If the mean of X does not exist and f is a convex or concave function, then $Ef(X)$ can only be finite if f is constant; for any non-trivial convex function $Ef(X)$ is either infinite or does not exist. Therefore Definition 1.5.1 only makes sense for random variables with finite mean. Indeed, any attempt to generalize the idea of a variability order to random variables with non-existing means leads to serious difficulties, see, for example, Elton and Hill (1992).

Since it is easy to see that $X \leq_{icx} Y$ if and only if $-Y \leq_{icv} -X$, it is sufficient to investigate one of these two dual concepts. Therefore $\leq_{icx}$ will be considered here, and the corresponding results for $\leq_{icv}$ will be omitted. Moreover, convex order and increasing convex order are also closely related, as the following result reveals.

Theorem 1.5.3. *The following statements are equivalent:*

(i) $X \leq_{cx} Y$;

(ii) $X \leq_{icx} Y$ *and* $EX = EY$.

Proof. (i) $\Rightarrow$ (ii): It is obvious that $X \leq_{cx} Y$ implies $X \leq_{icx} Y$. Moreover, the functions $f(x) = x$ and $f(x) = -x$ are both convex. Hence $X \leq_{cx} Y$ implies $EX \leq EY$ and $-EX \leq -EY$, thus $EX = EY$.

(ii) $\Rightarrow$ (i): Let f be an arbitrary convex function. Assume for the moment that there is some finite α such that

$$x \mapsto f(x) + \alpha x \quad \text{is increasing.} \tag{1.5.1}$$

Then $Ef(X) + \alpha EX \leq Ef(Y) + \alpha EY$ and hence $Ef(X) \leq Ef(Y)$, as $EX = EY$. If such an α does not exist, then approximate f monotonically by

$$f_n(x) = \begin{cases} f(x) & \text{for } x \geq -n, \\ f(-n) + f'_+(-n)(x+n) & \text{otherwise.} \end{cases}$$

where f'_+ denotes the right derivative that always exists because of convexity. All these functions are convex and fulfill (1.5.1) with $\alpha = -f'_+(-n)$. Hence the assertion follows from the monotone convergence theorem. $\square$

As an immediate consequence of Theorem 1.5.3 and the definition of the orders we get the following result concerning the comparison of moments.

Corollary 1.5.4. *a) If* $X \leq_{cx} Y$, *then*

$$EX^n \leq EY^n \text{ and } E(X - EX)^n \leq E(Y - EY)^n \quad \text{for } n = 2, 4, 6, \ldots.$$

In particular, $X \leq_{cx} Y$ *implies* $Var(X) \leq Var(Y)$.
b) If, moreover, X *and* Y *are non-negative random variables, then*

$$EX^n \leq EY^n \quad \text{for all } n \in \mathbb{N}.$$

Next it will be shown that convex and increasing convex order are closed with respect to convolutions.

Theorem 1.5.5. *a) If* $X \leq_{cx} Y$ *and* Z *is independent of* X *and* Y, *then* $X + Z \leq_{cx} Y + Z$.
b) If $X \leq_{icx} Y$ *and* Z *is independent of* X *and* Y, *then* $X + Z \leq_{icx} Y + Z$.

Proof. a) Let f be a convex function. Define $g(z) = Ef(X + z)$ and $h(z) = Ef(Y + z)$. As $x \mapsto f(x + z)$ is convex for all real z, $X \leq_{cx} Y$ implies $g(z) \leq h(z)$ for all $z \in \mathbb{R}$. Therefore

$$Ef(X + Z) = Eg(Z) \leq Eh(Z) = Ef(Y + Z).$$

The proof of b) is similar. □

Like the usual stochastic order, convex and increasing convex order are closed under mixtures.

Theorem 1.5.6. *If X, Y and Θ are random variables such that $[X|\Theta = \theta] \leq_{cx}$ ($\leq_{icx}$) $[Y|\Theta = \theta]$ for all θ in the support of Θ, then $X \leq_{cx} Y$ ($X \leq_{icx} Y$).*

Proof. Under the given assumption for any (increasing) convex function f

$$Ef(X) = E_\Theta E[f(X)|\Theta = \theta] \leq E_\Theta E[f(Y)|\Theta = \theta] = Ef(Y).$$

□

It is impossible to verify directly the inequality $Ef(X) \leq Ef(Y)$ for *all* convex functions or *all* increasing convex functions. Thus there is a need for simpler characterizations of these orders which are easier to check. The next results show that it is sufficient to consider only a small subclass of the set of all convex functions, the so-called *wedge functions* $\phi_t(x) = (x - t)_+ = \max\{x - t, 0\}$.

Theorem 1.5.7. *The following statements are equivalent:*

(i) $X \leq_{icx} Y$;

(ii) $E(X - t)_+ \leq E(Y - t)_+$ *for all real t.*

Proof. The implication (i) $\Rightarrow$ (ii) is trivial. Therefore assume that $E(X-t)_+ \leq E(Y - t)_+$ for all $t \in \mathbb{R}$, and let f be an arbitrary increasing convex function. We have to consider three cases.

1. Assume for the moment that $\lim_{t\to-\infty} f(t) = 0$. It is well known that f then is the maximum of a countable set $\{\ell_1, \ell_2, \dots\}$ of increasing linear functions; take, for example, the lines of support in all rational points. Now define

$$f_n(t) = \max\{0, \ell_1(t), \dots, \ell_n(t)\}.$$

Then f_n converges to f from below, and each f_n is piecewise linear with a finite number of kinks. Therefore f_n can be written as

$$f_n(x) = \sum_{i=1}^{n} a_{in}(x - b_{in})_+$$

for some constants $a_{in} \geq 0$ and $b_{in} \in \mathbb{R}$. Hence

$$Ef_n(X) = \sum_{i=1}^{n} a_{in} E(X - b_{in})_+ \leq \sum_{i=1}^{n} a_{in} E(Y - b_{in})_+ = Ef_n(Y).$$

Applying the monotone convergence theorem implies $Ef(X) \leq Ef(Y)$.

2. If $\lim_{t \to -\infty} f(t) = \alpha \in \mathbb{R}$, then the problem can be reduced to case 1 by considering the function $f - \alpha$.

3. Let $\lim_{t \to -\infty} f(t) = -\infty$. Then $f_n(x) = \max\{f(x), -n\}$ fulfills the assumptions of case 2 for all n, and f_n converges to f monotonically. Hence the assertion follows from the monotone convergence theorem.   □

The previous theorem shows that the increasing convex order can be characterized with the help of the function $\pi_X(t) = E(X - t)_+$, $t \in \mathbb{R}$. It is therefore not surprising that this function will play an important role in the investigation of convex orders. Integration by parts shows that

$$\pi_X(t) = E(X - t)_+ = \int_t^{\infty} \bar{F}_X(z)\mathrm{d}z.$$

Therefore the function π_X is sometimes called *integrated survival function*. In actuarial sciences this function is known as *stop-loss transform*, since $\pi_X(t)$ can be considered as the net premium of a stop-loss reinsurance contract; see Section 8.3 for more details on that topic.

As a first application of Theorem 1.5.7 the closure properties of increasing convex order with respect to different modes of convergence will be considered. Unfortunately it turns out that this order is not closed with respect to weak convergence, as the following example shows.

Example 1.5.8. ($\leq_{icx}$ is not closed with respect to weak convergence.) Let $P(X_n = 1) = 1$ for all $n \in \mathbb{N}$ and $P(Y_n = n) = 1/n = 1 - P(Y_n = 0)$. Then the sequences (X_n) and (Y_n) converge in distribution to X and Y respectively, where $P(X = 1) = P(Y = 0) = 1$. Moreover, it is easy to see that $X_n \leq_{cx} Y_n$ and hence $X_n \leq_{icx} Y_n$ for all n, but $X \nleq_{icx} Y$. In fact, obviously $X \geq_{icx} Y$. Notice that (Y_n) is an example of a sequence of random variables which converges in distribution but where the means do not converge, since $EY_n = 1$ for all n, whereas $EY = 0$.

Therefore a stronger definition of convergence is needed to obtain a positive result.

Theorem 1.5.9. *Let (X_n) and (Y_n) be sequences of random variables with $X_n \leq_{icx} Y_n$ for all $n \in \mathbb{N}$. If $X_n \to X$ and $Y_n \to Y$ in distribution, and if moreover $E(X_n)_+ \to EX_+$ and $E(Y_n)_+ \to EY_+$, then $X \leq_{icx} Y$.*

Proof. According to Theorem 1.5.7 it is sufficient to show $E(X - t)_+ \leq E(Y - t)_+$ for all real t. Therefore fix t. It follows from $X_n \leq_{icx} Y_n$

that $E(X_n - t)_+ \leq E(Y_n - t)_+$. It is $E(X_n - t)_+ = E(X_n)_+ - Eg_t(X_n)$, where $g_t(x) = \min\{x_+, t\}$ is a bounded and continuous function. Hence convergence in distribution implies $Eg_t(X_n) \to Eg_t(X)$. As $E(X_n)_+ \to EX_+$ by assumption, the result follows. $\square$

There are further results that can be derived from the characterization in Theorem 1.5.7. For their proofs it is helpful to collect some properties of the integrated survival function $\pi_X(t) = E(X - t)_+$.

Theorem 1.5.10. *a) The integrated survival function π_X of a random variable X with finite mean has the following properties:*

(i) π_X *is decreasing and convex;*

(ii) $\lim_{t \to \infty} \pi_X(t) = 0$ *and* $\lim_{t \to -\infty} (\pi_X(t) + t) = EX$.

b) To every function π which fulfills (i) and (ii) in a) there is a random variable X such that π is the integrated survival function of X. The distribution function of X is given by $F_X(t) = 1 + \pi'_+(t)$, where π'_+ is the right derivative of π.

Proof. a) The representation

$$\pi_X(t) = E(X - t)_+ = \int_t^\infty \bar{F}_X(z)dz$$

yields immediately that π_X is decreasing and convex. Moreover, the monotone convergence theorem implies

$$\lim_{t \to \infty} \pi_X(t) = \lim_{t \to \infty} E(\max\{X - t, 0\}) = 0$$

and

$$\lim_{t \to -\infty} (\pi_X(t) + t) = \lim_{t \to -\infty} E(\max\{X, t\}) = EX,$$

showing (ii).

b) If π is convex, then its right derivative π'_+ exists and is right-continuous and increasing. If $\lim_{t \to \infty} \pi(t) = 0$, then necessarily $\lim_{t \to \infty} \pi'_+(t) = 0$, and $\lim_{t \to -\infty} (\pi(t) + t)$ can only exist if $\lim_{t \to -\infty} \pi'_+(t) = -1$. Hence $1 + \pi'_+$ is a distribution function, and π is the corresponding integrated survival function. $\square$

The next result gives some equivalent characterizations for convergence of integrated survival functions.

Theorem 1.5.11. *Let X and X_n for $n = 1, 2, \ldots$ be random variables with finite means, and let π and π_n be their integrated survival functions. Then the following statements are equivalent:*

(i) $\pi_n \to \pi$ *pointwise.*

(ii) $X_n \to X$ *in distribution and* $E(X_n)_+ \to EX_+$.

Proof. (i) $\Rightarrow$ (ii): Assume that $\pi_n \to \pi$ pointwise. The convergence $E(X_n)_+ \to EX_+$ follows immediately from $E(X_n)_+ = \pi_n(0)$. To show the convergence in distribution, let F_n and F be the distribution functions of X_n and X respectively, and let x_0 be an arbitrary continuity point of F and hence also of $\bar{F} = 1 - F$. We have to show that $\lim_{n\to\infty} \bar{F}_n(x_0) = \bar{F}(x_0)$.

a) Suppose there is a subsequence $(\bar{F}_{n_k})$ of $(\bar{F}_n)$ with

$$\lim_{k\to\infty} \bar{F}_{n_k}(x_0) = b < a = \bar{F}(x_0).$$

Let $\varepsilon = (a - b)/2$. $\bar{F}$ is right-continuous and $\bar{F}_{n_k}$ is decreasing for all $k \in \mathbb{N}$. Hence, there is some $\delta > 0$ and some $k_0 \in \mathbb{N}$ such that for all $k \geq k_0$ and all $x \in [x_0, x_0 + \delta]$ the inequality $\bar{F}(x) - \bar{F}_{n_k}(x) > \varepsilon$ holds. Therefore

$$\pi_{n_k}(x_0) = \pi_{n_k}(x_0 + \delta) + \int_{x_0}^{x_0+\delta} \bar{F}_{n_k}(x)\,dx \qquad (1.5.2)$$

$$< \pi_{n_k}(x_0 + \delta) + \int_{x_0}^{x_0+\delta} (\bar{F}(x) - \varepsilon)\,dx.$$

Taking in (1.5.2) the limit $k \to \infty$ yields

$$\pi(x_0) \leq \pi(x_0 + \delta) + \pi(x_0) - \pi(x_0 + \delta) - \varepsilon\delta = \pi(x_0) - \varepsilon\delta,$$

a contradiction. Hence there is no such subsequence.

b) In the case

$$\lim_{k\to\infty} \bar{F}_{n_k}(x_0) > \bar{F}(x_0)$$

use the left continuity of $\bar{F}$ in x_0 to show that for $x \in [x_0 - \delta, x_0]$ the inequality $\bar{F}_{n_k}(x) - \bar{F}(x) > \varepsilon$ holds. This leads to a similar contradiction as in a).

(ii) $\Rightarrow$ (i): This follows immediately from

$$\pi_n(x) = E(X_n)_+ - \int_0^x \bar{F}_{X_n}(t)\,dt,$$

using the dominated convergence theorem. $\qquad\square$

Remark 1.5.12. Combining Theorem 1.5.11 with Theorem 1.5.7 yields an alternative proof of Theorem 1.5.9.

Now it can be shown that $\leq_{icx}$, $\leq_{cx}$ as well as $\leq_{st}$ can easily be characterized via integrated survival functions.

Theorem 1.5.13. *a)* $X \leq_{st} Y$ *if and only if* $t \mapsto \pi_Y(t) - \pi_X(t)$ *is decreasing.*
b) $X \leq_{icx} Y$ *if and only if* $\pi_X(t) \leq \pi_Y(t)$ *for all real* t.
c) $X \leq_{cx} Y$ *if and only if* $\pi_X(t) \leq \pi_Y(t)$ *for all real* t *and*

$$\lim_{t \to -\infty} (\pi_Y(t) - \pi_X(t)) = 0.$$

Proof. a) As

$$\pi_Y(t) - \pi_X(t) = \int_t^\infty [\bar{F}_Y(x) - \bar{F}_X(x)] \, dx,$$

the result follows from the definition of $\leq_{st}$.
b) This is the content of Theorem 1.5.7.
c) According to Theorem 1.5.10 a) (ii)

$$\lim_{t \to -\infty} (\pi_Y(t) - \pi_X(t)) = EY - EX.$$

Hence the assertion follows from Theorem 1.5.3, taking into account b). $\quad\square$

As an immediate consequence the following interesting separation result can be shown.

Theorem 1.5.14. *Let* X *and* Y *be random variables with* $X \leq_{icx} Y$. *Then there is a random variable* Z *such that*

$$X \leq_{st} Z \leq_{cx} Y.$$

Proof. Define $\pi_Z(t) = \max\{\pi_X(t), EY - t\}$. We will show that this is an integrated survival function and that a corresponding random variable Z fulfills $X \leq_{st} Z \leq_{cx} Y$.
(i) π_Z is the maximum of a convex and a linear function. Hence π_Z is convex. Furthermore it is decreasing with $\lim_{t \to \infty} \pi_Z(t) = 0$ and

$$\lim_{t \to -\infty} (\pi_Z(t) + t) = \lim_{t \to -\infty} \max\{\pi_X(t) + t, EY\} = \max\{EX, EY\} = EY.$$

$$(1.5.3)$$

Hence Theorem 1.5.10 implies that π_Z is a stop-loss transform of a random variable Z.
(ii)

$$\pi_Z(t) - \pi_X(t) = (EY - t - \pi_X(t))_+ = (EY - E(\max\{X, t\}))_+. \quad (1.5.4)$$

Hence $t \mapsto \pi_Z(t) - \pi_X(t)$ is decreasing. Thus Theorem 1.5.13 a) implies $X \leq_{st} Z$.
(iii) Jensen's inequality implies

$$EY - t \leq (EY - t)_+ \leq E(Y - t)_+ = \pi_Y(t).$$

Therefore $X \leq_{icx} Y$ implies $\pi_Z(t) = \max\{\pi_X(t), EY - t\} \leq \pi_Y(t)$. Now taking into account (1.5.3), Theorem 1.5.13 c) yields $Z \leq_{cx} Y$. $\quad\square$

Remark 1.5.15. For non-negative random variables Theorem 1.5.14 has been stated in Kaas and Heerwaarden (1992). For that case a proof similar to the one given here can be found in Müller (1996). The result can also be found in Makowski (1994), who gives a constructive proof for distributions with finite support and rational probabilities. For the general case he refers to the theorem of Strassen (1965), which we will consider in the next subsection.

1.5.2 Sufficient Conditions and Strassen's Theorem

A famous result of Strassen (1965) states that $X \leq_{cx} Y$ holds if and only if there are random variables $\hat{X}$ and $\hat{Y}$ with the same distributions as X and Y, such that $E[\hat{Y}|\hat{X}] = \hat{X}$. Strassen has shown this result in a much more general context, which will be considered in Section 2.6. The usual proof of this general result as it can be found in some textbooks (see, for example, Szekli (1995)) is based on some deep theorems from functional analysis like the Hahn–Banach theorem and the Riesz representation theorem. Here an elementary and constructive proof of this result for convex order of distributions on the real line is given which in fact can be translated into an algorithm for finding the joint distribution of $(\hat{X}, \hat{Y})$ in the case of distributions with finite support.

As a preparation for this proof we need some results about sufficient conditions for convex order, which are of interest in their own.

There are several sufficient conditions for convex order that are based on intuitive notions of higher variability of random variables. The first one is the so called *cut criterion* of Karlin and Novikoff (1963).

Definition 1.5.16. Let X and Y be random variables with distribution functions F_X and F_Y. Then X is said to be *less dangerous* than Y (written $X \leq_D Y$), if there is some $t_0 \in \mathbb{R}$ such that $F_X(t) \leq F_Y(t)$ for all $t < t_0$ and $F_X(t) \geq F_Y(t)$ for all $t \geq t_0$ and if in addition $EX \leq EY$.

Note that this definition has a simple interpretation in terms of P–P and Q–Q plots. Assuming continuity of the distribution functions, $F_X \leq_D F_Y$ holds if and only if $EX \leq EY$ and the Q–Q plot crosses the 45° line once, starting above it and ending below. Thus Y has the same distribution as $T(X)$, where T is a function crossing the diagonal once from below. This explains the names *cut criterion* or *crossing condition*. Similarly, the P–P plot has to cross the 45° line once, but starting below it and ending above.

The following theorem shows that dangerousness is a sufficient condition for increasing convex order. It can be traced back to Karlin and Novikoff (1963). In actuarial sciences this result has been rediscovered by Ohlin (1969).

Theorem 1.5.17. $X \leq_D Y$ *implies* $X \leq_{icx} Y$.

Proof. According to Theorem 1.5.7 it is sufficient to show that the difference

$\pi_Y - \pi_X$ of the integrated survival functions is non-negative. Since

$$\pi_Y(t) - \pi_X(t) = \int_t^\infty [F_X(x) - F_Y(x)]\, dx,$$

the assumption $X \leq_D Y$ implies that $\pi_Y - \pi_X$ is non-negative for $t \geq t_0$. Since

$$\lim_{t \to -\infty} (\pi_Y(t) - \pi_X(t)) = EY - EX,$$

it is

$$\pi_Y(t) - \pi_X(t) = EY - EX + \int_{-\infty}^t [F_Y(x) - F_X(x)]\, dx \geq 0$$

also for $t < t_0$. $\square$

Theorem 1.5.17 is a helpful tool to derive practical criteria for the comparison with respect to $\leq_{icx}$. As an important application it yields the following result about increasing convex ordering within location-scale families.

Theorem 1.5.18. *Let X be a random variable with $EX = 0$. If $0 \leq a \leq c$ and $b \leq d$, then $aX + b \leq_{icx} cX + d$.*

Proof. Since $F_{aX+b}(t) = F_X((t-b)/a)$ and $F_{cX+d}(t) = F_X((t-d)/b)$, it is $F_{aX+b}(t) \leq F_{cX+d}(t)$ if and only if $t \leq t_0 = (bc - ad)/(c - a)$. Moreover, $b \leq d$ implies $E(aX + b) \leq E(cX + d)$. Hence $aX + b \leq_D cX + d$ and thus the assertion follows from Theorem 1.5.17. $\square$

Notice that the relation $\leq_D$ is not transitive, since if F and G cross once, and G and H cross once, then F and H may cross twice. Therefore $\leq_{icx}$ is weaker than $\leq_D$. It can be shown, however, that $\leq_{icx}$ is in some sense the 'transitive closure' of $\leq_D$. In fact, the following result was shown in Müller (1996) for non-negative random variables, but the same method of proof can also be used for the general case.

Theorem 1.5.19. *$X \leq_{icx} Y$ if and only if there is a sequence (X_n) of random variables such that $X = X_1$, $X_n \leq_D X_{n+1}$ for all n, $X_n \to Y$ in distribution and $E(X_n)_+ \to EY_+$.*

Proof. The convex decreasing function π_Y is the supremum of a countable set of affine decreasing functions $A_1, A_2, \ldots$, which can be chosen, for example, as the support functions of π_Y in all rational points. Define $\pi_1 = \pi_X$ and recursively $\pi_{n+1} = \max\{\pi_n, A_n\}$. It is easy to see that all π_n fulfill conditions (i) and (ii) of Theorem 1.5.10. Hence (π_n) is a sequence of integrated survival functions converging to π_Y. According to Theorem 1.5.11 the proof is therefore complete, if it can be shown that the random variables X_n corresponding to π_n satisfy $X_n \leq_D X_{n+1}$.

If $A_n \leq \pi_n$, then this trivially holds. Therefore assume that there is an interval (a_n, b_n) such that

$$\pi_{n+1}(t) = \begin{cases} A_n(t) & \text{for } t \in (a_n, b_n) \\ \pi_n(t) & \text{otherwise} \end{cases}.$$

This means that π_{n+1} is linear on (a_n, b_n), and thus the corresponding distribution function F_{n+1} is constant there. Outside of (a_n, b_n) the functions π_{n+1} and π_n and hence also F_{n+1} and F_n coincide. As $EX_n = EX_{n+1}$ this implies $X_n \leq_D X_{n+1}$. □

The construction in the previous proof can also be used to give a constructive proof of the following theorem, which gives an almost sure characterization of convex order.

Theorem 1.5.20. (An Extension of Strassen's theorem) *For two random variables X and Y with finite means the following conditions are equivalent:*

(i) $X \leq_{cx} Y$;

(ii) *there are random variables $\hat{X}$ and $\hat{Y}$ with the same distributions as X and Y, such that $E[\hat{Y}|\hat{X}] = \hat{X}$ a.s., and in addition the conditional law $[\hat{Y}|\hat{X} = x]$ is stochastically increasing in x, i.e. $[\hat{Y}|\hat{X} = s] \leq_{st} [\hat{Y}|\hat{X} = t]$ for all $s < t$.*

Proof. (i) $\Rightarrow$ (ii): First choose $\hat{X} = X$. Then consider the construction of the sequence (π_n) in Theorem 1.5.19, and let P_n be the distributions corresponding to π_n. P_{n+1} is obtained from P_n by removing all mass from the interval (a_n, b_n) and moving it to the endpoints in such a way that the mean is preserved. Now define a Markov kernel Q_n as follows:

$$Q_n(x, \cdot) = \begin{cases} \delta_x; & x \notin (a_n, b_n) \\ \frac{x-a_n}{b_n-a_n}\delta_{b_n} + \frac{b_n-x}{b_n-a_n}\delta_{a_n}; & a_n \leq x \leq b_n \end{cases}, \qquad (1.5.5)$$

where δ_x is the one-point distribution with mass in x. It is obvious that $\int y Q_n(x, \mathrm{d}y) = x$. Moreover, $\int Q_n(x, A)P_n(\mathrm{d}x) = 0$ if $A \subset (a_n, b_n)$ and $\int Q_n(x, A)P_n(\mathrm{d}x) = P_n(A)$ if $A \subset (-\infty, a_n) \cup (b_n, \infty)$. Thus

$$\int Q_n(x, A)P_n(\mathrm{d}x) = P_{n+1}(A) \quad \text{for all measurable } A.$$

Hence Q_n is a Markov kernel with $\int y Q_n(x, \mathrm{d}y) = x$ that links P_n and P_{n+1}.

Thus there is a Markovian martingale (X_n) with $X_n \sim P_n$ for all $n \in \mathbb{N}$. Since

$$
\begin{aligned}
E|X_n| &= 2E(X_n)_+ - EX_n \\
&= 2\pi_n(0) - \lim_{t \to -\infty} (\pi_n(t) - t) \\
&\leq 2\pi_Y(0) - \lim_{t \to -\infty} (\pi_X(t) - t) \\
&= 2EY_+ - EX
\end{aligned}
$$

the sequence (X_n) is an L_1-bounded martingale, and hence it converges almost surely to a random variable $\hat{Y}$ with $E[\hat{Y}|\hat{X}] = \hat{X}$, and obviously $\hat{Y}$ has the integrated survival function π_Y and hence the same distribution as Y.

The proof of stochastic monotonicity is based on some results from Section 5.2. It is clear from the definition in (1.5.5) that Q_n fulfills the conditions of Corollary 5.2.4 and thus the corresponding operator is monotone (with respect to $\leq_{st}$). Now consider the two inhomogeneous Markov chains starting in $X_0' = s$ and $X_0'' = t$ with $s < t$, and both with transition kernels Q_n. Then it follows from (the obvious inhomogeneous version of) Corollary 5.2.12 that $X_n' \leq_{st} X_n''$ for all n. By construction the sequences (X_n') and (X_n'') converge a.s. to $[\hat{Y}|\hat{X} = s]$ and $[\hat{Y}|\hat{X} = t]$, respectively. Hence $[\hat{Y}|\hat{X} = x]$ is stochastically increasing in x.

(ii) $\Rightarrow$ (i) is a simple consequence of Jensen's inequality. $\qquad\square$

Corollary 1.5.21. *For two random variables X and Y with finite means the following conditions are equivalent:*

(i) $X \leq_{icx} Y$;

(ii) *there are random variables $\hat{X}$ and $\hat{Y}$ with the same distributions as X and Y, such that $E[\hat{Y}|\hat{X}] \geq \hat{X}$ a.s., and in addition the conditional law $[\hat{Y}|\hat{X} = x]$ is stochastically increasing in x.*

Proof. This follows immediately by combining Theorem 1.5.20 and Theorem 1.2.4, taking into consideration Theorem 1.5.14. $\qquad\square$

Remark 1.5.22. Theorem 1.5.20 (without the part on stochastic monotonicity) is a famous result which is usually attributed to Strassen (1965). In fact, he has shown a much more general result (see Theorem 2.6.6) for random elements in separable Banach spaces. For convex order of distributions on the real line this result is much older. In a simple version it had appeared already in Hardy, Littlewood, and Pólya (1934), section 2.19. Blackwell (1953) probably was the first to give a proof of the result stated here. He proved it first for distributions with finite support, using techniques from matrix algebra, and then he also used the martingale convergence theorem for the proof in the general case. The constructive proof given here, which uses the integrated survival function, and the fact that the transition kernel can be chosen to be stochastically increasing, is taken from Müller and Rüschendorf (2001).

It should be mentioned, however, that similar ideas have been used in the economic literature by Machina and Pratt (1997) and Müller (1998a). Müller and Rüschendorf (2001) give an application in optimal stopping theory where this additional property of stochastic monotonicity gives important further insight.

The value of Strassen's theorem will be demonstrated in the following two results. We are not aware of any different proof that avoids Strassen's theorem. Recall that the random variables $X_1, \ldots, X_n$ are said to be exchangeable if for any permutation p the random vector $(X_{p(1)}, \ldots, X_{p(n)})$ has the same distribution as $(X_1, \ldots, X_n)$.

Theorem 1.5.23. *Let $X_1, \ldots, X_n$ be exchangeable random variables and $f_1, \ldots, f_n$ measurable real functions. Define the function $\bar{f}$ by*

$$\bar{f}(x) = \frac{1}{n} \sum_{i=1}^{n} f_i(x).$$

Then

$$\sum_{i=1}^{n} \bar{f}(X_i) \leq_{cx} \sum_{i=1}^{n} f_i(X_i).$$

Proof. Let $\mathcal{P}$ be the set of all permutations of the numbers $1, \ldots, n$, and let π be a random permutation, i.e. π is a random variable uniformly distributed on $\mathcal{P}$. Then

$$E\left[\sum_{i=1}^{n} f_{\pi(i)}(X_i) | X_1, \ldots X_n\right] = \frac{1}{n!} \sum_{p \in \mathcal{P}} \left(f_{p(1)}(X_1) + \cdots + f_{p(n)}(X_n)\right)$$

$$= \sum_{i=1}^{n} \bar{f}(X_i),$$

so that

$$E\left[\sum_{i=1}^{n} f_{\pi(i)}(X_i) | \sum_{i=1}^{n} \bar{f}(X_i)\right] = \sum_{i=1}^{n} \bar{f}(X_i),$$

too. Since $\sum_{i=1}^{n} f_{\pi(i)}(X_i) = \sum_{i=1}^{n} f_i(X_i)$, the assertion follows from Theorem 1.5.20. □

Theorem 1.5.23 and its proof is taken from Denuit and Vermandele (1998). They apply this result in the context of the problem of finding an optimal reinsurance contract. Choosing $f_i(x) = x/(n-1)$ for $i = 1, \ldots, n-1$ and $f_n(x) \equiv 0$ yields the following corollary, which can already be found in Arnold and Villaseñor (1986).

Corollary 1.5.24. *Let $X_1, \ldots, X_n$ be exchangeable random variables. Then*

$$\frac{1}{k}\sum_{i=1}^{k} X_i \geq_{cx} \frac{1}{k+1}\sum_{i=1}^{k+1} X_i \quad for \ k = 1, \ldots, n-1.$$

The strong law of large numbers says that for a sequence of i.i.d. random variables with finite mean the sample mean converges to EX_1 almost surely. Corollary 1.5.24 tells us that this convergence is monotone in the sense that the variability of the sample mean decreases (in convex ordering sense) monotonically with the number of observations. In the context of martingale theory it is well known that the sample means (or more general any U-statistic) of exchangeable random variables form a reversed martingale, see for example, Chow and Teicher (1978), p. 228.

Now we turn back to the problem of giving sufficient conditions for convex order. We have already observed that the dangerousness relation has an intuitive interpretation via transforms of random variables. $X \leq_D Y$ holds if $Y =_{st} T(X)$, where $T(z) \geq z$ for all $z \geq t_0$ and $T(z) \leq z$ for all $z < t_0$. This means that the distribution of Y is obtained from that of X by moving the mass right of t_0 even more to the right and moving the mass left of t_0 even more to the left. One could also say that the distribution of Y is more spread out than that of X.

However, in the literature of economic decisions under risk the notion *mean preserving spread* is well established and describes a strongly related but more restrictive situation. Originally there existed different definitions for discrete and for continuous distributions (see the seminal paper of Rothschild and Stiglitz (1970)). However, they can be unified to the following definition, which is taken from Machina and Pratt (1997).

Definition 1.5.25. The distribution function G differs from the distribution function F by a *mean preserving spread* (written $F \leq_{MPS} G$), if they have the same finite mean and if there is an interval (a, b) such that G assigns no greater probability than F to any open subinterval of (a, b) and G assigns no smaller probability than F to any open interval either to the left or to the right of (a, b).

An equivalent definition, which is more formal but less intuitive, is the following one: G differs from F by a mean preserving spread if and only if there exist finite real numbers $a < b$ such that the function $G - F$ is increasing on $(-\infty, a)$ and on (b, ∞) and decreasing on (a, b).

This implies immediately that there is some point $t_0 \in [a, b]$ such that $G(x) \geq F(x)$ for $x < t_0$ and $G(x) \leq F(x)$ for $x > t_0$. Hence the following result holds.

Theorem 1.5.26. *If G differs from F by a mean preserving spread, then $F \leq_D G$ and hence $F \leq_{cx} G$.*

It is also clear that the notion of mean preserving spread is strictly stronger than dangerousness, since dangerousness requires only that $G - F$ is nonnegative on $(-\infty, t_0)$, whereas a mean preserving spread requires that $G - F$ is in addition unimodal in this region. As a simple example where $F \leq_D G$ holds but G does not differ from F by a mean preserving spread, one can choose for F the uniform distribution on $\{2, 4, 6, 8\}$ and for G the uniform distribution on $\{1, 3, 7, 9\}$.

Rothschild and Stiglitz (1970) have shown that for distributions F and G with finite support $F \leq_{cx} G$ holds if and only if G can be obtained from F by a sequence of mean preserving spreads. A close look at Theorem 1.5.19 shows that the sequence used there is in fact a sequence of mean preserving spreads. Thus the following result holds.

Theorem 1.5.27. *$X \leq_{cx} Y$ if and only if there is a sequence (X_n) of random variables such that $X = X_1$, $X_n \leq_{MPS} X_{n+1}$ for all n, $X_n \to Y$ in distribution and $EX_n \to EY$.*

For distributions with finite support the same result can be obtained for a notion of mean preserving spread strengthening that introduced by Rothschild and Stiglitz (1970). The following definition is stronger in two ways. First it requires that the spread is *local*, namely, that the mass removed from a point $x \in \mathbb{R}$ be shifted at most to the two points of support immediately to the left and to the right of x, and second, it requires that *all* mass has to be removed from x. The formal definition is as follows.

Definition 1.5.28. Let F and G be distribution functions of discrete distributions whose common support is a finite set of points $x_1 < x_2 < \cdots < x_n$ with probability mass functions f and g respectively. Then G is said to differ from F by a *local spread*, if there exists some $i \in \{2, \ldots, n-1\}$ such that $0 = g(x_i) \leq f(x_i)$, $g(x_{i+1}) \geq f(x_{i+1})$, $g(x_{i-1}) \geq f(x_{i-1})$ and $g(x_j) = f(x_j)$ for all $j \notin \{i-1, i, i+1\}$. A local spread is said to be *mean preserving* if F and G have the same mean. Write $F \leq_{ls} G$ if G is a mean preserving local spread of F.

The following theorem shows that though local mean preserving spread is a very particular concept, for discrete distributions it does the same job as a mean preserving spread in the sense of Rothschild and Stiglitz (1970).

Theorem 1.5.29. *Let F and G be distribution functions of discrete distributions with finite support. Then $F \leq_{cx} G$ holds if and only if there is a finite sequence $F_1, \ldots, F_k$ with $F = F_1$ and $F_k = G$, such that F_{i+1} differs from F_i by a mean preserving local spread for $i = 1, \ldots, k-1$.*

Proof. The if-part is clear. For the only-if part a construction similar to that in Theorem 1.5.19 will be used. Define the integrated distribution function $\phi_F(t) = \int_{-\infty}^{t} F(z)\, dz$ and similarly $\phi_G(t) = \int_{-\infty}^{t} G(z)\, dz$. It is easy to see that ϕ_F as well as ϕ_G is an increasing convex, piecewise affine function with $\phi_F(t) = 0$ for all $t \leq x_1$. It is easy to see that $F \leq_{cx} G$ holds if and only if $\phi_F(t) \leq \phi_G(t)$ for all t and $\lim_{t \to \infty} \phi_G(t) - \phi_F(t) = 0$. (The proof is similar to that of Theorem 1.5.13.)

Let us assume that $f(x_i) = g(x_i)$ for $i = 1, \ldots, \ell - 1$ for some ℓ. We will construct a sequence $F = F_1 \leq_{ls} F_2 \leq_{ls} \cdots \leq_{ls} F_m \leq_{ls} G$ such that $f_m(x_i) = g(x_i)$ for $i = 1, \ldots, \ell$. The assertion then follows by induction on ℓ. For this construction let h be the affine function the graph of which is the line through $(x_\ell, \phi_G(x_\ell))$ and $(x_{\ell+1}, \phi_G(x_{\ell+1}))$, and let $\xi = \inf\{t > x_\ell : h(t) = \phi_F(t)\}$. This ξ always exists and is finite since $h' < 1$ and $\phi'_F(t) = 1$ for $t > x_n$. If $\xi = x_\ell$ then $\phi_G(t) = \phi_F(t)$ for all $t \leq x_{\ell+1}$; hence $f(x_i) = g(x_i)$ for $i = 1, \ldots, \ell$ and thus we have nothing to show. Therefore let us assume that $\xi > x_\ell$. Then there is some m such that $x_{\ell+m-1} \leq \xi \leq x_{\ell+m}$. Let h_j be the affine function the graph of which is the line through $(x_\ell, \phi_F(x_\ell))$ and $(x_{\ell+j}, \phi_F(x_{\ell+j}))$, and define $\phi_{F_j} = \max\{\phi_F, h_j\}$, $j = 1, \ldots, m-1$, and let $\phi_{F_m} = \max\{\phi_F, h\}$. Then the corresponding distribution functions fulfill $F = F_1 \leq_{ls} F_2 \leq_{ls} \cdots \leq_{ls} F_m \leq_{ls} G$, and $\phi_G(t) = \phi_{F_m}(t)$ for all $t \leq x_{\ell+1}$, hence $f_m(x_i) = g(x_i)$ for $i = 1, \ldots, \ell$.   □

Notice that new points of support may intervene in the construction of the sequence $F_1, \ldots, F_k$ described in the proof of Theorem 1.5.29. As an example assume that F has point masses of $1/3$ in the points $-1, 0$ and 1 and that G has point masses $4/9, 1/9, 4/9$ in the same points. Then G is obtained from F by two mean preserving local spreads, but the distribution F_2 obtained after the first one has point masses of $4/9, 2/9$ and $3/9$ in the points $-1, 1/2$ and 1, respectively.

Next we will show that arbitrary distributions comparable with respect to $\leq_{cx}$ can be approximated by distributions fulfilling the conditions of Theorem 1.5.29. This is an extension of theorem 1b in Rothschild and Stiglitz (1970). Moreover, the new proof is simpler.

Theorem 1.5.30. *Let X and Y be two random variables such that $X \leq_{cx} Y$. Then there exist sequences (X_n) and (Y_n) of random variables with finite support such that $X_n \leq_{cx} Y_n$ for all n, $X_n \to X$ and $Y_n \to Y$ in distribution, and $EX_n = EX$ and $EY_n = EY$ for all n.*

Proof. Let π_X and π_Y be the integrated survival functions of X and Y. We will construct the integrated survival functions π_{X_n} and π_{Y_n} corresponding to X_n and Y_n, respectively. This can be done as follows. Since π_X and π_Y are increasing and convex, they are suprema of countable sets $\{\phi_1, \phi_2, \ldots\}$ respectively $\{\psi_1, \psi_2, \ldots\}$ of increasing affine functions. Now set

$$\pi_{X_n}(t) = \max\{0, t - EX, \phi_1(t), \phi_2(t), \ldots, \phi_n(t)\}$$

and

$$\pi_{Y_n}(t) = \max\{\pi_{X_n}(t), t - EY, \psi_1(t), \psi_2(t), \ldots, \psi_n(t)\}.$$

Then we obviously have $\pi_{X_n} \leq \pi_{Y_n}$, $\pi_{X_n} \to \pi_X$ and $\pi_{Y_n} \to \pi_Y$. Moreover, π_{X_n} and π_{Y_n} are obviously piecewise affine. Hence the assertion follows from Theorem 1.5.9 and 1.5.10. $\qquad\qquad\qquad\qquad\qquad\qquad\qquad\qquad\qquad\qquad\qquad\qquad\qquad\quad$ □

Theorem 1.5.29 and 1.5.30 are of some interest in their own, but we will see in Theorem 3.12.14 that they are also very useful tools in some proofs.

1.5.3   Majorization

For any n-dimensional vector $\mathbf{x} = (x_1, \dots, x_n)$ of reals, let

$$x_{[1]} \geq \dots \geq x_{[n]}$$

denote the components of $\mathbf{x}$ in decreasing order, and let

$$x_{(1)} \leq \dots \leq x_{(n)}$$

denote the components of $\mathbf{x}$ in increasing order. Moreover,

$$\mathbf{x}_\downarrow = (x_{[1]}, \dots, x_{[n]}) \quad \text{and } \mathbf{x}_\uparrow = (x_{(1)}, \dots, x_{(n)})$$

are the *decreasing rearrangement* and the *increasing rearrangement* of $\mathbf{x}$.

Definition 1.5.31. For $\mathbf{x}$ and $\mathbf{y}$ in $\mathbb{R}^n$,

$$\mathbf{x} \leq_M \mathbf{y} \quad \text{if} \quad \begin{cases} \sum_{i=1}^k x_{[i]} \leq \sum_{i=1}^k y_{[i]} & \text{for } k = 1, \dots, n-1 \\ \sum_{i=1}^n x_{[i]} = \sum_{i=1}^n y_{[i]}. \end{cases}$$

When $\mathbf{x} \leq_M \mathbf{y}$ then $\mathbf{x}$ is said to be *majorized* by $\mathbf{y}$.

Equivalently,

$$\mathbf{x} \leq_M \mathbf{y} \quad \text{if} \quad \begin{cases} \sum_{i=1}^k x_{(i)} \geq \sum_{i=1}^k y_{(i)} & \text{for } k = 1, \dots, n-1 \\ \sum_{i=1}^n x_{(i)} = \sum_{i=1}^n y_{(i)}. \end{cases}$$

This notation and terminology was introduced by Hardy et al. (1934). For a comprehensive treatment of the subject we refer to the monograph by Marshall and Olkin (1979), where most of the subsequent material is taken from.

Notice that this relation is only a preorder, since $\mathbf{x} \leq_M \mathbf{y}$ and $\mathbf{y} \leq_M \mathbf{x}$ holds whenever $\mathbf{x}_\downarrow = \mathbf{y}_\downarrow$, i.e. when $\mathbf{y}$ can be obtained from $\mathbf{x}$ by a permutation of the components. Majorization is a partial order, however, on the set of all $\mathbf{x}$ with $x_1 \geq \dots \geq x_n$.

Majorization is a tool to compare the dissimilarity within the components of vectors. One of its many applications is in economics, where it is used to compare the inequality of income distributions. Consider a population of n individuals, and let x_i be the wealth of individual i. One obtains $x_{(1)}, \dots, x_{(n)}$

by ordering the individuals from poorest to richest. Now define $S_0 = 0$ and let $S_k = \sum_{i=1}^{k} x_{(i)}$ be the total wealth of the k poorest individuals. Plotting the points $(k/n, S_k/S_n)$ and joining these points by line segments yields the so-called *Lorenz curve*, which is due to Lorenz (1905). If $\mathbf{y} = (y_1, \dots, y_n)$ denotes the vector of wealths of another population with the same total wealth

$$W = \sum_{i=1}^{n} x_i = \sum_{i=1}^{n} y_i$$

then the relation $\mathbf{x} \leq_M \mathbf{y}$ can be interpreted as $\mathbf{x}$ representing a more even distribution of wealth than $\mathbf{y}$, since in the population $\mathbf{y}$ the total wealth of the k richest persons is larger and the total wealth of the k poorest persons is smaller than the corresponding value for the population $\mathbf{x}$, for all $k = 1, \dots, n$. Since $\mathbf{x} \leq_M \mathbf{y}$ holds if and only if the corresponding Lorenz curves can be ordered pointwise, in the economic literature this relation is often called *Lorenz order*.

An important concept in the theory of majorization is the *principle of transfer*. This principle can easily be motivated in the context of income inequality. Any measure of income inequality should have the property that the income inequality within a population will be diminished, if a rich person j gives some money to a poor person k, but only so much that the ranking between the two persons is not modified; i.e. if there are j and k with $1 \leq j, k \leq n$ and $\varepsilon > 0$ such that

$$x_i = \begin{cases} y_i - \varepsilon & \text{for } i = j, \\ y_i + \varepsilon & \text{for } i = k, \\ y_i & \text{otherwise,} \end{cases} \qquad (1.5.6)$$

and

$$x_j \geq x_k. \qquad (1.5.7)$$

We will say that $\mathbf{x}$ is obtained from $\mathbf{y}$ by a *T-transform*, if $\mathbf{x}$ and $\mathbf{y}$ fulfill (1.5.6) and (1.5.7). Such transfers were first considered by Pigou (1912) and Dalton (1920). Therefore they are sometimes called *Pigou–Dalton transfer* in the economic literature.

There are quite similar applications also in statistical physics, see for example, Alberti and Uhlmann (1982). Consider n bodies each in thermal equilibrium and all of equal heat capacity. Let x_i be the heat content of body i. Then the situation is analogous to that with the wealth example. Here it is possible to connect some of the bodies for some time in order to allow some heat transfer. If there is no loss of heat, this will lead to a more even distribution of heat content. If all bodies are connected for a very long time, they all have the same heat content.

It is easy to see that $\mathbf{x} \leq_M \mathbf{y}$ holds if $\mathbf{x}$ is obtained from $\mathbf{y}$ by a T-transform, but the following theorem shows much more. A proof can be found in Marshall and Olkin (1979).

Theorem 1.5.32. *The following conditions are equivalent:*

(i) $\mathbf{x} \leq_M \mathbf{y}$;

(ii) $\mathbf{x}$ *can be obtained from* $\mathbf{y}$ *by a finite number of T-transforms.*

Another characterization of majorization can be given in terms of matrices $\mathbf{A}$ with $\mathbf{x} = \mathbf{A}\mathbf{y}$.

Definition 1.5.33. A non-negative $n \times n$ matrix $\mathbf{A} = (a_{ij})$ is called *doubly stochastic* if

$$\sum_{i=1}^{n} a_{ij} = 1 \quad \text{for } 1 \leq j \leq n$$

and

$$\sum_{j=1}^{n} a_{ij} = 1 \quad \text{for } 1 \leq i \leq n.$$

Theorem 1.5.34. *The following conditions are equivalent:*

(i) $\mathbf{x} \leq_M \mathbf{y}$;

(ii) *there is a doubly stochastic matrix* $\mathbf{A}$ *such that* $\mathbf{x} = \mathbf{A}\mathbf{y}$.

For a detailed proof we refer to Marshall and Olkin (1979), but we sketch the idea of the proof of the more interesting part (i) $\Rightarrow$ (ii). First show that there is a doubly stochastic matrix $\mathbf{A}$ such that $\mathbf{x} = \mathbf{A}\mathbf{y}$, if $\mathbf{x}$ is obtained from $\mathbf{y}$ by a T-transform. In this case $\mathbf{A}$ can be chosen as

$$\mathbf{A} = \lambda\mathbf{I} + (1 - \lambda)\mathbf{Q},$$

where $\mathbf{I}$ is the identity matrix, $\mathbf{Q}$ is the permutation matrix interchanging the coordinates j and k, and $\lambda = 1 - \varepsilon/(y_k - y_j)$. The general case follows from Theorem 1.5.32, as the product of doubly stochastic matrices is again doubly stochastic.

Schur (1923) introduced functions that are monotone with respect to majorization.

Definition 1.5.35. A function $f : \mathbb{R}^n \to \mathbb{R}$ is called *Schur-convex*, if

$$\mathbf{x} \leq_M \mathbf{y} \quad \text{implies } f(\mathbf{x}) \leq f(\mathbf{y}).$$

In our context the most important result about Schur-convex functions is the following, see Marshall and Olkin (1979), p. 64.

Theorem 1.5.36. *Let g be an arbitrary real function and define*

$$f(\mathbf{x}) = \sum_{i=1}^{n} g(x_i) \quad \text{for } \mathbf{x} \in \mathbb{R}^n.$$

Then f is Schur-convex if and only if g is convex.

This result can be used to show a relationship between majorization and convex ordering of random variables. Each n-dimensional vector $\mathbf{x} = (x_1, \ldots, x_n)$ can be identified with a random variable X which assigns probability $1/n$ to each x_i, if they are all different. When k components coincide, then probability k/n is assigned to this value.

Formally, let $\Omega = \{1, \ldots, n\}$ be endowed with its power set $\wp(\Omega)$ as σ-algebra, and let P be the uniform distribution on Ω, assigning probability $1/n$ to each ω. Then for given $\mathbf{x} = (x_1, \ldots, x_n)$ the random variable X is defined on $(\Omega, \wp(\Omega), P)$ as

$$X : \Omega \to \mathbb{R},$$
$$i \mapsto x_i.$$

If $\mathbf{x}$ describes the wealth of a population, then X can be considered as the wealth of an individual drawn at random.

Using this construction, we can derive the following corollary from Theorem 1.5.36.

Corollary 1.5.37. *For vectors $\mathbf{x}$ and $\mathbf{y}$ let X and Y denote the corresponding random variables as defined above. Then $\mathbf{x} \leq_M \mathbf{y}$ if and only if $X \leq_{cx} Y$.*

On the other hand, majorization can also be used to compare the inhomogeneity of probability distributions on a set of n fixed points, which shall be without loss of generality $\Omega = \{1, \ldots, n\}$. Let

$$\Sigma = \{\mathbf{p} = (p_1, \ldots, p_n) : \sum p_i = 1, p_i \geq 0\}$$

be the simplex of all probability mass functions on Ω. Then majorization on Σ is an order that compares the dissimilarity within probability mass functions.

This order relation for discrete distributions is also used in the context of statistical physics, see Alberti and Uhlmann (1982), or of quantum mechanics and quantum computing, see Thirring (1983) and Jonathan and Plenio (1999). In this context it makes sense to say that $\mathbf{p}$ is *more mixed* or *more chaotic* than $\mathbf{q}$ if $\mathbf{p} \leq_M \mathbf{q}$. It has also been considered in a series of papers by Hickey (1982, 1983, 1984, 1986). He calls $\mathbf{p}$ *more random* than $\mathbf{q}$. Using his terminology we give the following definition.

Definition 1.5.38. *Let X and Y be random variables with finite support and let $\mathbf{p} = (p_1, \ldots, p_n)$ and $\mathbf{q} = (q_1, \ldots, q_n)$ be their probability mass functions (defined on the union of their supports). X is said to be* less random than Y *(written $X \leq_{rand} Y$) if $\mathbf{q} \leq_M \mathbf{p}$.*

It is obvious that the relation $\leq_{rand}$ is reflexive and transitive. However, it is not antisymmetric. In fact, it is $X \leq_{rand} \phi(X)$ and $\phi(X) \leq_{rand} X$ whenever ϕ is one-to-one.

Notice that this stochastic order has a character different from all other orders considered so far. For all other orders it is important *where* the probability mass is placed and therefore topological and order structure of the underlying space (which always has been the real line up to now) are important. For the randomness order $\leq_{rand}$, however, things are totally different. Here it is only of interest *how much* probability mass is placed in each point, and therefore it does not rely on any topology or order in the space. In fact, it is obvious that for all one-to-one mappings ϕ

$$X \leq_{rand} Y \quad \text{implies} \quad \phi(X) \leq_{rand} \phi(Y). \tag{1.5.8}$$

The most important properties of this order are given in the next result. They reveal that this order indeed compares randomness. All of these properties can be found in Hickey (1983).

Theorem 1.5.39. *a) For any random variable X with finite support and any real function ϕ it holds*

$$\phi(X) \leq_{rand} X.$$

Especially, $E[Y|X] \leq_{rand} X$ for any random variable Y.

b) If X and Y are independent random variables with finite support then $X \leq_{rand} X + Y$.

c) In the set of all distributions with a given finite support the uniform distribution is the unique maximum and the one-point distributions are minimal elements with respect to $\leq_{rand}$.

Proof. a) Let $x_1, \ldots, x_n$ be the points of support of X ordered such that the probabilities $p_{[i]} = P(X = x_i)$ are given in decreasing order. Moreover, let $\mathbf{q} = (q_1, \ldots, q_n)$ be the probability mass vector of $\phi(X)$, where q_j may be zero for some j. Then

$$\sum_{i=1}^{k} p_{[i]} = \sum_{i=1}^{k} P(X = x_i) \leq \sum_{i=1}^{k} P(\phi(X) = \phi(x_i)) \leq \sum_{i=1}^{k} q_{[i]}.$$

b) Let $p_i = P(X = x_i)$ und $q_i = P(X + Y = z_i)$, $i = 1, \ldots, n$. Then $\mathbf{q} = \mathbf{A}\mathbf{p}$, where $a_{ij} = P(Y = z_i - x_j)$. It is easy to see that $\mathbf{A}$ is a doubly stochastic matrix. Hence the result follows from Theorem 1.5.34.

c) is obvious. $\qquad \square$

Note that the theory above can be generalized by introducing weights for the elements of Ω, see Marshall and Olkin (1979), ch. 14. In the case of this p-majorization the most chaotic state or the $\leq_{rand}$-maximum is not necessarily the uniform distribution. In terms of the heat transfer example this would mean that the bodies have different heat capacities. The notion of majorization can also be extended from vectors to matrices, see Marshall and Olkin (1979), ch. 15, Pečarić, Proschan, and Tong (1992), section 12.3, and Cohen, Kemperman, and Zbăganu (1998).

Moreover, the ordering $\leq_{rand}$ can easily be extended to discrete random variables with a countable support. Parts a) and b) of Theorem 1.5.39 remain true in that case. It is also possible to extend the definition of $\leq_{rand}$ to densities with respect to the Lebesgue measure on some general space. This can be done by extending the notion of majorization. We will describe this extension now, following Hardy et al. (1934), ch. 10; see also Marshall and Olkin (1979), p. 15.

Denote by ν the Lebesgue measure, and let f be a density, i.e. a non-negative function with integral $\int f(x)dx = 1$. Define first the 'decreasing rearrangement' f^* of f. For $y > 0$ let $m(y) = \nu\{x : f(x) \geq y\}$. Then $f^*(y) = \sup\{t : m(y) > t\}$, $y > 0$, is called the decreasing rearrangement of f. Now f is said to be majorized by g (written $f \leq_M g$) if

$$\int_0^t f^*(x)dx \leq \int_0^t g^*(x)dx \quad \text{for all } t > 0.$$

For this kind of majorization similar results as in the discrete case can be shown. Some of them are collected in the following theorem. For a proof see Ryff (1963).

Theorem 1.5.40. *a) For Lebesgue densities f and g on some arbitrary space the following conditions are equivalent:*

(i) *$f \leq_M g$;*

(ii) *$\int u(f(x))dx \geq \int u(g(x))dx$ for all continuous and convex u;*

(iii) *$\int (f(x) - a)_+ dx \geq \int (g(x) - a)_+ dx$ for all $a > 0$.*

b) If $f(x) = \int k(x,y)g(y)dy$ for some non-negative kernel k with

$$\int k(x,y)dx = \int k(x,y)dy = 1 \quad \text{for all } x \text{ and } y$$

then $f \leq_M g$.

With this continuous version of majorization it is possible to extend the definition of $\leq_{rand}$ to random variables with Lebesgue densities. In this case the following counterpart of Theorem 1.5.39 holds.

Theorem 1.5.41. *a) If X and Y are independent random variables with Lebesgue densities then $X \leq_{rand} X + Y$.*

b) In the set of all distributions with a given compact support the uniform distribution is the unique maximum with respect to $\leq_{rand}$.

Proof. a) Since $f_{X+Y}(t) = \int f_Y(t-x)f_X(x)\mathrm{d}x$ the assertion follows from Theorem 1.5.40 b).

b) is obvious. □

There is no counterpart to part a) of Theorem 1.5.39. It is not true in general that $\phi(X) \leq_{rand} X$. This only holds under additional assumptions on ϕ, see Theorem 1.7.8 for a result in this spirit. It should also be noted that $\leq_{rand}$ is a relation for densities, and *not* a relation of distributions. In general, different relations are obtained by considering majorization of densities with respect to different dominating measures, as we already mentioned in the discrete case. Therefore we only consider the case of Lebesgue densities. In contrast to the discrete version, this continuous version of $\leq_{rand}$ does depend on the structure of the space, which happens implicitly by the dominating measure. Therefore (1.5.8) is not true for continuous distributions.

A well known measure for *randomness* of probability functions (or density functions in the continuous case) is based on the *entropy* function

$$e(\mathbf{p}) = \sum_{i=1}^{n} -p_i \ln(p_i)$$

or

$$e(f) = \int -f(x)\ln(f(x))\mathrm{d}x,$$

where $0\ln(0) = 0$. This can be used to define *entropy ordering*

$$\mathbf{p} \leq_E \mathbf{q} \quad \text{if } e(\mathbf{p}) \leq e(\mathbf{q})$$

or

$$f \leq_E g \quad \text{if } e(f) \leq e(g)$$

in the continuous case. A detailed study of entropy and its applications in probability, information theory, physics and so on can be found in Cover and Thomas (1991).

It is easy to see that the function $f(x) = x\ln(x)$ is convex on $[0,1]$. Hence the following result can be deduced from Theorem 1.5.36 and Theorem 1.5.41, respectively.

Theorem 1.5.42. *If $X \leq_{rand} Y$ then $X \geq_E Y$.*

1.6 Higher Convexity Orders and Laplace Transform Order

A differentiable function is increasing if the first derivative is non-negative, and it is convex if the second derivative is non-negative. Therefore it seems natural to define stochastic orders by requiring $Ef(X) \leq Ef(Y)$ for functions with non-negative higher derivatives. Such order relations have been considered by Rolski and Stoyan (1974) and Rolski (1976) with applications in queueing

theory. Fishburn (1976, 1980) has introduced them in the literature on
decision theory, and in actuarial science they have recently been considered
by Denuit, Lefevre, and Shaked (1998). We will give a review of the most
important results in that field, using the terminology of Denuit et al. (1998).

An s times differentiable function is called s-convex, if the sth derivative
is non-negative. However, the differentiability assumption can be avoided by
working with differences instead of derivatives. Define for arbitrary $\varepsilon > 0$ the
difference operator

$$\Delta^\varepsilon f(x) = f(x + \varepsilon) - f(x).$$

The function f is increasing if and only if the differences $\Delta^\varepsilon f(x)$ are non-
negative for all $\varepsilon > 0$ and all real x. It is convex if and only if all second
differences $\Delta^{\varepsilon_1} \Delta^{\varepsilon_2} f(x)$ are non-negative for all $\varepsilon_1, \varepsilon_2 > 0$ and all real x.
Generalizing this to higher differences leads to the following definition.

Definition 1.6.1. Let s be a natural number. A real function f is called
s-convex, if

$$\Delta^{\varepsilon_1} \ldots \Delta^{\varepsilon_s} f(x) \geq 0$$

for all $\varepsilon_1, \ldots, \varepsilon_s > 0$ and all real x. f is called s-concave if $-f(-x)$ is s-convex.

These functions have been investigated first by Popoviciu in a series of
papers that have been summarized in Popoviciu (1944); see also Bullen (1971)
and Roberts and Varberg (1973). Many properties of s-convex functions can
be derived from the following alternative characterization. A function f is
increasing if and only if the determinant

$$\begin{vmatrix} 1 & 1 \\ f(x_0) & f(x_1) \end{vmatrix}$$

is nonnegative whenever $x_0 < x_1$. It is convex if and only if

$$\begin{vmatrix} 1 & 1 & 1 \\ x_0 & x_1 & x_2 \\ f(x_0) & f(x_1) & f(x_2) \end{vmatrix} \geq 0$$

whenever $x_0 < x_1 < x_2$, and similarly it is s-convex if and only if

$$\begin{vmatrix} 1 & 1 & \cdots & 1 \\ x_0 & x_1 & \cdots & x_s \\ \vdots & \vdots & \ddots & \vdots \\ x_0^{s-1} & x_1^{s-1} & \cdots & x_s^{s-1} \\ f(x_0) & f(x_1) & \cdots & f(x_s) \end{vmatrix} \geq 0$$

whenever $x_0 < x_1 < \cdots < x_s$.

By comparing expectations of s-convex or s-concave functions the following
stochastic orders can be defined.

Definition 1.6.2. a) The random variable X is said to be smaller than the random variable Y with respect to the *s-convex order* (written $X \leq_{s-cx} Y$), if $Ef(X) \leq Ef(Y)$ for all s-convex f for which the expectations exist.
b) X is smaller than Y in *s-increasing-convex order* (written $X \leq_{s-icx} Y$), if $Ef(X) \leq Ef(Y)$ for all functions f which are k-convex for all $1 \leq k \leq s$ and for which the expectations exist.
c) X is smaller than Y in *s-concave order* (written $X \leq_{s-cv} Y$), if $Ef(X) \leq Ef(Y)$ for all s-concave f for which the expectations exist.
d) X is smaller than Y in *s-increasing-concave order* (written $X \leq_{s-icv} Y$), if $Ef(X) \leq Ef(Y)$ for all functions f which are k-concave for all $1 \leq k \leq s$ and for which the expectations exist.

The ordering $\leq_{3-icv}$ was introduced in Whitmore (1970) as *third degree stochastic dominance*. The other orderings have been investigated by Rolski (1976), Fishburn (1976) and Denuit et al. (1998). Fishburn (1980) also considered a generalization to non-integer s, based on Theorem 1.6.3 below. Moreover, it should be noted that there are subtle differences between s-convexity on the real line and s-convexity on a finite grid; see Fishburn and Lavalle (1995) for details.

Theorem 1.6.3. $X \leq_{s-cx} Y$ *if and only if*

$$EX^k = EY^k \quad for \ k = 1, \ldots, s-1$$

and

$$E(X - a)_+^{s-1} \leq E(Y - a)_+^{s-1} \quad \textit{for all real } a.$$

Denuit et al. (1998) have shown many properties of these orderings. Most of them can be derived from the general theory of integral stochastic orders presented in Chapter 2; see especially Theorem 2.4.2.

The orderings $\leq_{s-icx}$ and $\leq_{s-icv}$ become weaker and weaker with increasing s. In the case of $\leq_{s-icv}$ we get the so-called Laplace transform order when we take the limit for s to infinity.

Definition 1.6.4. Let X and Y be real random variables. X is said to be less than Y in *Laplace transform order* (written $X \leq_{Lt} Y$), if the Laplace transforms $\hat{f}_X(t) = Ee^{-tX}$ and $\hat{f}_Y(t) = Ee^{-tY}$ exist and satisfy

$$\hat{f}_X(t) \geq \hat{f}_Y(t) \quad for \ all \ t > 0.$$

The existence of the Laplace transforms is trivial if X and Y are non-negative. However, if we want to consider arbitrary distributions on the real line, then the existence of the Laplace transform requires that the distribution has only a thin tail at $-\infty$.

This order relation has been considered for non-negative random variables, see Reuter and Riedrich (1981), Stoyan (1983), Alzaid, Kim, and Proschan (1991), and Caballé and Pomansky (1996).

Note that $X \leq_{Lt} Y$ holds if and only if $Ef(X) \leq Ef(Y)$ for all functions f of the form $f(x) = -e^{-tx}$ for some $t > 0$. However, there is a much larger class of functions that also characterizes $\leq_{Lt}$.

Definition 1.6.5. A real function f is called *completely monotone* if all its derivatives $f^{(n)}$ exist and satisfy

$$(-1)^n f^{(n)}(x) \geq 0 \quad \text{for all } x \text{ and for all } n = 0, 1, 2, \ldots$$

Theorem 1.6.6. $X \leq_{Lt} Y$ *holds if and only if*

$$Ef(X) \leq Ef(Y)$$

for all functions f with a completely monotone derivative, for which the integrals exist.

Theorem 1.6.6 was proved by Reuter and Riedrich (1981). We will give a proof in Example 2.5.4. There we will also derive other interesting properties of Laplace transform order.

A look at the definition of s-convexity reveals the following relation between Laplace transform order and $\leq_{s-icv}$.

Corollary 1.6.7. *If $X \leq_{s-icv} Y$ for some $s \in \mathbb{N}$ then $X \leq_{Lt} Y$.*

For distributions with finite support an even stronger result holds as Fishburn (1980) showed.

Theorem 1.6.8. *If X and Y have distributions with finite support then $X \leq_{Lt} Y$ holds if and only if $X \leq_{s-icv} Y$ for some $s \in \mathbb{N}$.*

Theorem 1.6.8 is not true for continuous distributions. In that case the distributions must fulfill some additional technical conditions. Details can be found in Fishburn (1980).

1.7 Dispersive Order and Relative Inverse Function Orderings

In Theorem 1.2.6 it has been shown that the usual stochastic order holds if and only if the graph of the relative inverse distribution function lies above the $45°$ line. If it crosses this line once from below then convex order holds. This is a consequence of Theorem 1.5.17. In this section further stochastic orders with such characterizations will be considered. First we treat a location free variability order which is called dispersive order.

Definition 1.7.1. A distribution function F is said to be smaller than the distribution function G in *dispersive order* (written $F \leq_{disp} G$) if

$$F^{-1}(t) - F^{-1}(s) \leq G^{-1}(t) - G^{-1}(s) \quad \text{for all } 0 < s < t < 1.$$

It is clear that this relation is reflexive and transitive. However, it is not antisymmetric.

Theorem 1.7.2. *If $X \leq_{disp} Y$ and $Y \leq_{disp} X$ then $Y =_{st} X + a$ for some real a.*

Proof. If $X \leq_{disp} Y$ and $Y \leq_{disp} X$ then

$$F^{-1}(t) - F^{-1}(s) = G^{-1}(t) - G^{-1}(s) \quad \text{for all } 0 < s < t < 1,$$

where F and G are the distribution functions of X and Y. Hence $F^{-1}(t) = G^{-1}(t) - a$ for $a = G^{-1}(1/2) - F^{-1}(1/2)$. Since $X =_{st} F^{-1}(U)$ and $Y =_{st} G^{-1}(U)$ the assertion follows. $\qquad\square$

Dispersive order has been introduced by Doksum (1969) as an ordering of scale. A detailed treatment can be found in Oja (1981) and Shaked (1982). Most of the subsequent results can be found there. Some of their proofs, however, are valid only in the continuous case.

If G is continuous then we can write $s = G(x)$ and $t = G(y)$ for some real $x < y$, and hence for continuous distribution functions $F \leq_{disp} G$ holds if and only if

$$F^{-1}(G(y)) - F^{-1}(G(x)) \leq y - x \quad \text{for all real } x < y.$$

Hence for random variables with continuous distributions $X \leq_{disp} Y$ holds if and only if X has the same distribution as $\phi(Y)$ for some increasing function ϕ with slope at most 1. In order to show that this is true in general, we need the following preliminary result.

Theorem 1.7.3. *$F \leq_{disp} G$ implies $\text{range}(F) \subseteq \text{range}(G)$.*

Proof. The result is trivial if G is continuous. Therefore assume that G has some point mass so that there is some interval $(x, x + p) \subseteq (0, 1)\backslash\text{range}(G)$. Hence

$$G^{-1}(t) - G^{-1}(s) = 0 \quad \text{for all } x \leq s < t < x + p.$$

Since $F \leq_{disp} G$ this implies $F^{-1}(t) - F^{-1}(s) = 0$ and thus $(x, x + p) \subseteq (0, 1)\backslash\text{range}(F)$, too. $\qquad\square$

Theorem 1.7.4. *$X \leq_{disp} Y$ if and only if $X =_{st} \phi(Y)$ for some increasing function ϕ with $\phi(y) - \phi(x) \leq y - x$ for all $x < y$.*

Proof. Assume $X \leq_{disp} Y$. According to Theorem 1.7.3 it is $\text{range}(F_X) \subseteq \text{range}(F_Y)$. Therefore $X =_{st} \phi(Y)$ with $\phi(t) = F_X^{-1}(F_Y(t))$. Moreover, $X \leq_{disp} Y$ implies

$$\phi(y) - \phi(x) \leq y - x \quad \text{for all real } x < y$$

with $x, y \in \text{supp}(Y)$. This proves the only-if-part. The if-part is similar. $\qquad\square$

Remark 1.7.5. When X and Y are both continuous then Theorem 1.7.4 is equivalent to the statement that $X \leq_{disp} Y$ if and only if $Y =_{st} \phi(X)$ for some increasing function ϕ with $\phi(y) - \phi(x) \geq y - x$ for all $x < y$. In this form the result can be found in Oja (1981) or Shaked and Shanthikumar (1994), p. 70. It is not true, however, if X and Y are discrete. As a simple counterexample consider the case when X has a one point distribution, and Y is non-degenerate. Then clearly $X \leq_{disp} Y$, but there cannot be a mapping ϕ such that $Y =_{st} \phi(X)$.

Now it is easy to derive the relationship of dispersive order to usual stochastic order and convex order. Let $\ell_X = \inf\{t : P(X \leq t) > 0\}$ denote the essential infimum of X.

Theorem 1.7.6. a) If $X \leq_{disp} Y$ and $\ell_X \leq \ell_Y$ then $X \leq_{st} Y$.
   b) If $X \leq_{disp} Y$ then $X - EX \leq_{cx} Y - EY$.

Proof. a) As in the proof of Theorem 1.7.4 let $\phi(t) = F_X^{-1}(F_Y(t))$. Then $\phi(\ell_Y) = \ell_X$. Moreover, it has been shown in Theorem 1.7.4 that $X \leq_{disp} Y$ holds if and only if the slope of ϕ is at most one. Hence $\phi(t) \leq t$ for all t and due to Theorem 1.2.6 this implies $X \leq_{st} Y$.
   b) Obviously $X - EX$ and $Y - EY$ have the same mean and since dispersive order is location free, $X \leq_{disp} Y$ implies that the slope of the relative inverse of $X - EX$ with respect to $Y - EY$ is at most one. Hence it crosses the $45°$ line at most once and from above. Therefore Theorem 1.5.17 yields $X - EX \leq_{cx} Y - EY$. $\qquad\square$

As another application of Theorem 1.7.4 it can be shown that dispersive order implies usual stochastic order of the spacings of order statistics. A further result about dispersive ordering of order statistics will be given in Theorem 1.7.11 in a more general context.

Theorem 1.7.7. Let X_i and Y_i, $i = 1, \ldots, n$, be i.i.d. random variables with $X_i \leq_{disp} Y_i$, and let $X_{(k:n)}$ and $Y_{(k:n)}$, $k = 1, \ldots, n$, be the corresponding order statistics. Then

$$X_{(k+1:n)} - X_{(k:n)} \leq_{st} Y_{(k+1:n)} - Y_{(k:n)} \text{ for } k = 1, \ldots, n-1.$$

Proof. According to Theorem 1.7.4 the vector $(X_1, \ldots, X_n)$ has the same distribution as $(\phi(Y_1), \ldots, \phi(Y_n))$, where ϕ is an increasing function with $\phi(y) - \phi(x) \leq y - x$ for all $x < y$. Therefore $(X_{(1:n)}, \ldots, X_{(n:n)})$ also has the same distribution as $(\phi(Y_{(1:n)}), \ldots, \phi(Y_{(n:n)}))$. Hence

$$X_{(k+1:n)} - X_{(k:n)} =_{st} \phi(Y_{(k+1:n)}) - \phi(Y_{(k:n)}) \leq Y_{(k+1:n)} - Y_{(k:n)}$$

almost surely. Thus the assertion follows from Theorem 1.2.4. $\qquad\square$

Now it will be shown that dispersive order is stronger than the randomness order introduced in Definition 1.5.38. This has been shown in Hickey (1986) for the case of distributions with unimodal densities. This restriction is not necessary, however.

Theorem 1.7.8. *Assume that the random variables X and Y are either both discrete or both have densities. Then $X \leq_{disp} Y$ implies $X \leq_{rand} Y$.*

Proof. First assume that X and Y both have densities. Then Theorem 1.7.4 implies $X =_{st} \phi(Y)$ for some function ϕ with $0 \leq \phi' \leq 1$. Therefore

$$f_Y(t) = f_X(\phi(t)) \cdot \phi'(t) \quad \text{for all real } t.$$

Hence for all $a > 0$

$$\int (f_X(t) - a)_+ \mathrm{d}t = \int (f_X(\phi(z)) - a)_+ \cdot \phi'(z)\mathrm{d}z$$

$$\geq \int (f_X(\phi(z)) \cdot \phi'(z) - a)_+ \mathrm{d}z$$

$$= \int (f_Y(t) - a)_+ \mathrm{d}t.$$

Thus the result follows from Theorem 1.5.40.

If X and Y are both discrete then the result follows immediately by combining Theorem 1.7.4 with Theorem 1.5.39 a). □

The previous two theorems yield that dispersive order implies larger variability as well as more randomness. Especially it implies higher variance as well as larger entropy. This is the reason why for many parametric families of distributions ordering by variance is equivalent to ordering by entropy though these two orders are conceptually very different. Variance is related to the location of the mass relative to the mean, whereas entropy only depends on the size of the mass points and is irrespective of their location. This connection between variance and entropy via dispersive ordering has been observed by Ebrahimi, Maasoumi, and Soofi (1999). They also give a series of examples.

There are further orderings that can be defined by considering properties of the P–P and Q–Q plots. We will call all of them *relative inverse function orderings*. To avoid technical difficulties we will assume from now on that all distributions are continuous. Let $\mathcal{F}$ be an arbitrary class of increasing real functions. Then define two relations $\leq_{\mathcal{F}}^{pp}$ and $\leq_{\mathcal{F}}^{qq}$ as

$$F \leq_{\mathcal{F}}^{pp} G \quad \text{if } G(F^{-1}(\cdot)) \in \mathcal{F}$$

and

$$F \leq_{\mathcal{F}}^{qq} G \quad \text{if } G^{-1}(F(\cdot)) \in \mathcal{F}.$$

It is clear that such a relation is reflexive if $\mathcal{F}$ contains the identity function, and it is transitive if the composition of two functions f and g in $\mathcal{F}$ is again in $\mathcal{F}$. It is antisymmetric if $f \in \mathcal{F}$ and $f^{-1} \in \mathcal{F}$ only holds if f is the identity function.

We already know the following examples of stochastic orders of the type $\leq_{\mathcal{F}}^{qq}$:

1. $F \leq_{st} G$ if and only if $G^{-1}(F(x)) \geq x$ for all real x.
2. $F \leq_{disp} G$ if and only if $G^{-1}(F(y)) - G^{-1}(F(x)) \geq y - x$ for all $y \geq x$.

For lifetime distributions also the following relations are of interest. Recall that a function $f : [0, \infty) \to [0, \infty)$ is *star-shaped* (with respect to $(0,0)$), if $t \mapsto f(t)/t$ is increasing for $t > 0$. It is called *superadditive*, if $f(x + y) \geq f(x) + f(y)$ for all $x, y \geq 0$. Clearly, any convex function is star-shaped and any star-shaped function is superadditive.

Definition 1.7.9. Let F and G be distributions with support $\mathbb{R}_+$. Then

(i) F is said to be smaller than G in *convex transformation order* (written $F \leq_{cx}^{qq} G$), if $G^{-1}(F(x))$ is convex,

(ii) F is said to be smaller than G in *star-shaped order* (written $F \leq_*^{qq} G$), if $G^{-1}(F(x))$ is star-shaped (with respect to $(0,0)$),

(iii) F is said to be smaller than G in *superadditive order* (written $F \leq_{su}^{qq} G$), if $G^{-1}(F(x))$ is superadditive.

Orders of this type first appeared in van Zwet (1964). They are *scale free* in the following sense.

Theorem 1.7.10. *Let $\preceq$ be any of the relations $\leq_{cx}^{qq}, \leq_*^{qq}$ and $\leq_{su}^{qq}$. Then $\preceq$ is reflexive and transitive, but not antisymmetric. In fact, we have $X \preceq Y$ and $Y \preceq X$ if and only if $Y =_{st} aX$ for some positive constant a.*

Proof. Any one of these three classes of functions has the following properties: It contains the identity function, it is closed under composition, and $f \in \mathcal{F}$ and $f^{-1} \in \mathcal{F}$ hold simultaneously if and only if $f(x) = ax$ for some $a > 0$. $\quad\square$

It is interesting to observe that any ordering $\leq_{\mathcal{F}}^{qq}$ is preserved under the formation of order statistics.

Theorem 1.7.11. *Let (X_i, Y_i), $i = 1, \ldots, n$, be i.i.d. pairs of random variables with $X_i \leq_{\mathcal{F}}^{qq} Y_i$, and let $X_{(k:n)}$ and $Y_{(k:n)}$, $k = 1, \ldots, n$, be the corresponding order statistics. Then*

$$X_{(k:n)} \leq_{\mathcal{F}}^{qq} Y_{(k:n)}.$$

Proof. It is well known that if X_i has distribution function F then $X_{(k:n)}$ has the distribution function

$$F_{(k:n)}(t) = \sum_{i=0}^{n-k} \binom{n}{i} (1 - F(t))^i F(t)^{n-i}.$$

Hence $F_{(k:n)}(t) = h_k(F(t))$ for some function h_k, and similarly $G_{(k:n)}(t) = h_k(G(t))$. Indeed, it is $h_k(p) = \bar{F}_{n,p}(n-k)$, where $\bar{F}_{n,p}$ is the survival function of a binomial distribution with parameters n and p. Since the binomial distribution is $\leq_{st}$-increasing in p the function h_k is increasing and thus it has an inverse h_k^{-1}. Therefore

$$G_{(k:n)}^{-1} \circ F_{(k:n)} = G^{-1} \circ h_k^{-1} \circ h_k \circ F = G^{-1} \circ F.$$

$\square$

There are also useful orders that have a characterization via P–P plots. We have already seen the following examples.

1. $F \leq_{st} G$ if and only if $G(F^{-1}(x)) \leq x$ for all real $0 < x < 1$.
2. $F \leq_{lr} G$ if and only if $G \circ F^{-1}$ is convex.
3. $F \leq_{hr} G$ if and only if $G \circ F^{-1}$ is star-shaped with respect to $(1,1)$.
4. $F \leq_{rh} G$ if and only if $G \circ F^{-1}$ is star-shaped with respect to $(0,0)$.

All stochastic orders with a characterization in terms of P–P plots have the nice common feature that they are invariant under monotone scale transformations.

Theorem 1.7.12. *If* $X \leq_{\mathcal{F}}^{pp} Y$ *and* ϕ *is an increasing function then* $\phi(X) \leq_{\mathcal{F}}^{pp} \phi(Y)$.

Proof. $F_{\phi(X)}(t) = F_X(\phi^{-1}(t))$ implies

$$F_{\phi(Y)} \circ F_{\phi(X)}^{-1} = F_Y \circ \phi^{-1} \circ \phi \circ F_X^{-1} = F_Y \circ F_X^{-1}.$$

Thus the assertion follows from the definition of $\leq_{\mathcal{F}}^{pp}$. $\square$

1.8   Lifetime Distributions and Notions of Aging

The theory of lifetime distributions plays an important role in fields like reliability, survival analysis and insurance mathematics. Throughout this section we will assume that a lifetime can be described by a non-negative random variable X with an absolutely continuous distribution function F and density f. Recall that $\bar{F}(t) = P(X > t)$ is called the survival function and

$$r(t) = \frac{f(t)}{\bar{F}(t)}$$

is the hazard rate or failure rate. We denote by F_t the distribution function of the residual lifetime after time t, i.e.

$$F_t(x) = P(X \leq t + x | X > t) = \frac{F(t+x) - F(t)}{\bar{F}(t)}.$$

The corresponding densities and hazard rates are called f_t and r_t respectively, and X_t shall be a random variable with distribution function F_t. The function $m(t) = EX_t$ is called the *mean residual life function*. Note that

$$\bar{F}_t(x) = \frac{\bar{F}(t+x)}{\bar{F}(t)}$$

and therefore

$$f_t(x) = \frac{f(t+x)}{\bar{F}(t)} \quad \text{and} \quad r_t(x) = r(t+x).$$

Under the lifetime distributions the exponential distribution with parameter $\alpha > 0$ (denoted as $\text{Exp}(\alpha)$) plays a special role. Its distribution function is given by

$$F(t) = 1 - e^{-\alpha t} \text{ for } t \geq 0.$$

It is easy to see that in this case $r(t) = \alpha$ for all t and therefore $F_t = F$ for all t. This is called *memorylessness property* since it says that the remaining lifetime is independent of age. It can be shown that the exponential distributions are the only lifetime distributions with this property.

Very often it will be more natural to assume that the remaining lifetime decreases (with respect to some stochastic order) with time. This leads to notions of aging. Many classes of lifetime distributions based on notions of aging can be found in the literature, see Barlow and Proschan (1975) or Johnson, Kotz, and Balakrishnan (1994). There are four main ideas as to how to define such notions of aging. A distribution function is said to satisfy some notion of aging,

- if $F_t \preceq F_s$ for all $s < t$, where $\preceq$ is an appropriate stochastic order;
- if $F_t \preceq F$ for all t, where $\preceq$ is an appropriate stochastic order;
- if it is smaller or larger (with respect to some stochastic ordering) than an exponential distribution with the same mean;
- if its hazard rate function or the mean residual life function has some special property (like monotonicity, unimodality, starshapedness etc.).

The most important class of lifetime distributions is that with increasing failure rate.

Definition 1.8.1. A lifetime distribution is said to be an *increasing failure rate* distribution (IFR distribution), if its failure rate $r(t)$ is increasing.

Some authors speak of IHR distributions, standing for *increasing hazard rate*. This notion has several characterizations of the type mentioned above. They are collected in the following theorem.

Theorem 1.8.2. *The following statements are equivalent:*

(i) X has an IFR distribution;

(ii) $X_t \leq_{hr} X_s$ for all $s < t$;

(iii) $X_t \leq_{st} X_s$ for all $s < t$;

(iv) $X_t \leq_{disp} X_s$ for all $s < t$;

(v) $X_t \leq_{hr} X$ for all t;

(vi) $X_t \leq_{disp} X$ for all t;

(vii) $X \leq_{cx}^{qq} Y$ for any exponentially distributed Y;

(viii) $\bar{F}_X$ is log-concave.

Proof. The implications (i) $\Rightarrow$ (ii) $\Rightarrow$ (v) $\Rightarrow$ (i) and (i) $\Leftrightarrow$ (viii) are obvious. The implication (ii) $\Rightarrow$ (iii) is a consequence of Theorem 1.3.8. To show that (iii) implies (i) note that $X_t \leq_{st} X_s$ holds if and only if $F_t(\varepsilon) \geq F_s(\varepsilon)$ for all ε, which implies the assertion because of

$$r(t) = r_t(0) = f_t(0) = \lim_{\varepsilon \to 0} \frac{F_t(\varepsilon)}{\varepsilon}.$$

For $Y \sim \text{Exp}(\alpha)$ the quantile function is $F_Y^{-1}(t) = -\ln(1-t)/\alpha$ and thus

$$F_Y^{-1}(F_X(t)) = -\frac{1}{\alpha} \ln(\bar{F}_X(t)).$$

This implies the equivalence of (vii) and (viii). For the equivalence of (i), (iv) and (vi) we refer to Pellerey and Shaked (1997). □

A stronger notion of aging is obtained if we replace the requirement $X_t \leq_{st} X_s$ for all $s < t$ by the stronger requirement $X_t \leq_{lr} X_s$ for all $s < t$. Distributions with this property are said to have the ILR property, standing for *increasing likelihood ratio*. This notion of aging has interesting applications in scheduling problems; see Righter and Shanthikumar (1992). It has been introduced in Shaked and Shanthikumar (1987a), where the following characterization can be found. The proof is an easy consequence of Theorem 1.4.6.

Theorem 1.8.3. *The following conditions for the ILR property of a random variable X are equivalent:*

(i) $X_t \leq_{lr} X_s$ for all $s < t$;

(ii) $X_t \leq_{lr} X$ for all t;

(iii) f_X is log-concave.

Distributions with the ILR property are also called PF_2 distributions, since log-concave functions are also called *Pólya frequency functions of order 2*, or PF_2 for short.

A weaker notion than IFR is obtained, if only $X_t \leq_{icx} X_s$ for all $s < t$ is required. This concept is called *decreasing mean residual life* (DMRL). The name stems from the equivalence in Theorem 1.8.4 below, which states that DMRL holds if and only if the mean residual life function

$$m(t) = EX_t = \frac{\int_t^\infty \bar{F}_X(x)\mathrm{d}x}{\bar{F}_X(t)} = \frac{\pi_X(t)}{\bar{F}_X(t)} = \frac{-1}{\frac{\mathrm{d}}{\mathrm{d}t}\ln \pi_X(t)} \qquad (1.8.1)$$

is decreasing.

Theorem 1.8.4. *The following conditions for the DMRL property of a random variable X are equivalent:*

(i) $X_t \leq_{icx} X_s$ *for all $s < t$;*

(ii) *the mean residual life function $m(t)$ is decreasing in t;*

(iii) π_X *is log-concave.*

Proof. The equivalence of (ii) and (iii) is obvious from (1.8.1). Moreover it is trivial that (i) implies (ii). Thus it remains to be shown that (ii) implies (i). If $m(t)$ is decreasing and hence π_X is log-concave then it follows from (1.8.1) that

$$\frac{\bar{F}_X(t)}{\bar{F}_X(s)} \geq \frac{\pi_X(t)}{\pi_X(s)} \geq \frac{\pi_X(t+a)}{\pi_X(s+a)}$$

for all $s < t$ and all $a \geq 0$. This implies

$$E(X_t - a)_+ = \frac{\pi_X(t+a)}{\bar{F}_X(t)} \leq \frac{\pi_X(s+a)}{\bar{F}_X(s)} = E(X_s - a)_+. \qquad (1.8.2)$$

According to Theorem 1.5.7 this is equivalent to $X_t \leq_{icx} X_s$. $\square$

Another important concept of aging is the NBU property, an abbreviation for *new better than used*. This holds if $X_t \leq_{st} X$ for all $t > 0$. From characterization (v) in Theorem 1.8.2 it immediately follows that the NBU property is weaker than IFR. Here again different characterizations are available. The proof of the next result is similar to that of Theorem 1.8.2.

Theorem 1.8.5. *The following conditions for the NBU property of a random variable X are equivalent:*

(i) $X_t \leq_{st} X$ *for all t;*

(ii) $X \leq_{su}^{qq} Y$ *for any exponentially distributed Y;*

(iii) $\ln \bar{F}_X$ *is subadditive.*

If only the expectation of the residual lifetime of a device of age t is smaller than the expected lifetime of a new device, i.e. if $m(t) = EX_t \leq EX$ for all $t > 0$, then the distribution is said to be *new better than used in expectation* (NBUE). Clearly, NBUE is weaker than DMRL.

The NBUE property is related to properties of the so called *equilibrium distribution*, which plays an important role in renewal theory; see for example Feller (1971) or Section 7.1. For a non-negative random variable X with finite mean $\mu = EX$ the function

$$F_e(x) = \int_0^x \frac{\bar{F}_X(t)}{\mu} dt \qquad (1.8.3)$$

is a distribution function, which is called the equilibrium distribution. Its hazard rate r_e is given by

$$r_e(t) = \frac{\bar{F}_X(t)}{\int_t^\infty \bar{F}_X(x) dx} = \frac{1}{EX_t} = \frac{1}{m(t)},$$

so that the hazard rate r_e of the equilibrium distribution is the reciprocal of the mean residual life function. The next result shows an interesting connection between the equilibrium distribution and the NBUE property.

Theorem 1.8.6. *A distribution F_X has the NBUE property if and only if $F_e \leq_{st} F_X$.*

Proof. The condition $EX_t \leq EX$ for all $t > 0$ can be written equivalently as

$$\frac{\int_t^\infty \bar{F}_X(s) ds}{\bar{F}_X(t)} \leq EX$$

Rearranging the terms and replacing EX by μ yields

$$\int_t^\infty \frac{\bar{F}_X(s)}{\mu} ds \leq \bar{F}_X(t),$$

which means that $\bar{F}_e(t) \leq \bar{F}_X(t)$ for all t. Thus the assertion follows. $\qquad\square$

Rolski (1975, 1977) introduced the class of distributions which are *harmonic new better than used in expectation* (HNBUE). Remember that a lifetime distribution X with mean μ is NBUE if $m(t) \leq \mu$ for all t, or equivalently, if $r_e(t) \geq 1/\mu$ for all t. It is said to be HNBUE if the harmonic mean of the mean residual life function is smaller than μ for all t, i.e. if

$$\frac{1}{t} \int_0^t r_e(s) ds = \frac{\int_0^t (m(s))^{-1} ds}{t} \geq \frac{1}{\mu} \qquad \text{for all } t > 0. \qquad (1.8.4)$$

It is clear that the HNBUE class is larger than the NBUE class. Moreover, notice that $\int_0^t r_e(s)ds = -\ln \bar{F}_e(t)$, so that (1.8.4) can be rewritten as

$$\bar{F}_e(t) \geq e^{-t/\mu}, \tag{1.8.5}$$

which means that $F_e \leq_{st} \text{Exp}(1/\mu)$. Since (1.8.5) states that

$$\int_x^\infty \frac{\bar{F}(t)}{\mu}dt \leq e^{-x/\mu},$$

or equivalently,

$$\int_x^\infty \bar{F}(t)dt \leq \int_x^\infty e^{-t/\mu}dt,$$

it follows from Theorem 1.5.3 and 1.5.7 that the HNBUE property is also equivalent to the statement $F \leq_{cx} \text{Exp}(1/\mu)$. Hence we have shown the following result.

Theorem 1.8.7. *For a distribution function F with mean μ the following statements are equivalent:*

(i) *F is HNBUE;*

(ii) *$F_e \leq_{st} \text{Exp}(1/\mu)$;*

(iii) *$F \leq_{cx} \text{Exp}(1/\mu)$.*

For further properties and applications of the HNBUE class we refer to Klefsjö (1982). Klefsjö (1983a) and Klar (2000) have developed tests of HNBUE against exponentiality which are based on Theorem 1.8.7. Klefsjö (1983b), Belzunce, Ortega, and Ruiz (1999) and Franco, Ruiz, and Ruiz (2001) discuss generalizations of these concepts, which are obtained by using other stochastic orders like $\leq_{icv}$ or $\leq_{Lt}$.

In a final remark of this subsection we mention that we have considered here only classes of lifetime distributions for devices which exhibit some notion of aging in the sense that they deteriorate with time. All of these notions have counterparts for devices that improve with time. These counterparts are called *decreasing failure rate* (DFR), *increasing mean residual life* (IMRL), *new worse than used* (NWU), *new worse than used in expectation* (NWUE) and *harmonic new worse than used in expectation* (HNWUE). It is obvious how these classes are defined and it is clear that similar results as above can be shown for them. We omit the details and refer the reader to Barlow and Proschan (1975), Szekli (1995) and Kijima (1997).

The exponential distribution has constant failure rate; it is both IFR and DFR. The gamma distribution is IFR for $\alpha \geq 1$ and DFR for $\alpha \leq 1$. The Weibull distribution is IFR for $\alpha \geq 1$ and DFR for $\alpha \leq 1$. Lognormal distributions are DFR, uniform distributions on intervals are IFR. Exponential PERT type distributions (i.e. maxima of sums of independent exponential random variables, see Baccelli and Liu (1992)) are NBU.

1.9   Bivariate Characterizations

A stochastic order relation $\preceq$ is said to admit a bivariate characterization if there is a class $\mathcal{G}$ of bivariate functions $g : \mathbb{R}^2 \to \mathbb{R}$ such that $X \preceq Y$ holds if and only if

$$Eg(X^*, Y^*) \le Eg(Y^*, X^*) \quad \text{for all } g \in \mathcal{G}, \tag{1.9.1}$$

where X^* and Y^* are independent and have the same distribution as X and Y.

Using the abbreviation $\Delta g(x, y) = g(x, y) - g(y, x)$ the condition (1.9.1) can be rewritten in the more concise form

$$E\Delta g(X^*, Y^*) \le 0 \quad \text{for all } g \in \mathcal{G}.$$

Note that the relation $X \preceq Y$ only depends on the marginal distributions of X and Y; it is irrelevant whether or not X and Y are stochastically independent. For the inequality in (1.9.1), however, the assumption of independence of X^* and Y^* is crucial.

Such bivariate characterizations are very important especially for scheduling problems (see Section 7.3) and in portfolio optimization problems (see Theorem 8.2.5). As a simple example suppose that X and Y are the (stochastically independent) processing times of two jobs, and there is a single processor. Processing X first and then Y yields a total flow time (sum of completion times) $2X + Y$; whereas processing Y first and then X yields a total flow time of $X + 2Y$. Hence the second schedule dominates the first one with respect to $\le_{st}$ ($\le_{icx}$) if it can be shown that

$$Ef(2X + Y) \le Ef(X + 2Y) \quad \text{for all increasing (convex) functions } f.$$

The solution of such questions was the motivation for the investigation of bivariate characterizations of stochastic orders; see Shanthikumar and Yao (1991) and Chang and Yao (1993). Other applications can be found in portfolio optimization and mathematical finance; see Landsberger and Meilijson (1990a) and Kijima and Ohnishi (1996, 1999).

Most stochastic orders considered so far have a characterization as *integral stochastic order* with a generator $\mathcal{F}$, i.e. there is a class $\mathcal{F}$ of real functions f such that $X \preceq Y$ holds if and only if $Ef(X) \le Ef(Y)$ for all $f \in \mathcal{F}$. A systematic study of this class of stochastic orders will be given in Chapter 2. Kijima and Ohnishi (1996) have shown the following result saying that every integral stochastic order has a bivariate characterization.

Theorem 1.9.1. *Let $\le_{\mathcal{F}}$ be an integral stochastic order with generator $\mathcal{F}$. Then $X \le_{\mathcal{F}} Y$ holds if and only if*

$$Eg(X^*, Y^*) \le Eg(Y^*, X^*) \quad \text{for all } g \in \mathcal{G}_{\mathcal{F}},$$

where

$$\mathcal{G}_{\mathcal{F}} = \{g : \Delta g(\cdot, y) \in \mathcal{F} \text{ for all } y\}$$

and X^* and Y^* are independent with $X^* =_{st} X$ and $Y^* =_{st} Y$.

Proof. As $Ef(X) \leq Ef(Y)$ holds for any f if and only if $Ef(X) + c \leq Ef(Y) + c$ for any real c we can assume without loss of generality that $f \in \mathcal{F}$ implies $f(\cdot) + c \in \mathcal{F}$ for any real c. Now fix some $f \in \mathcal{F}$ and define $g_f(x, y) = f(x)$ for all y. Then $\Delta g(x, y) = f(x) - f(y)$ and hence $g_f \in \mathcal{G}_{\mathcal{F}}$. Thus

$$Eg(X^*, Y^*) \leq Eg(Y^*, X^*) \quad \text{for all } g \in \mathcal{G}_{\mathcal{F}}$$

implies $Ef(X) \leq Ef(Y)$ for all $f \in \mathcal{F}$, which shows the if-part. For the only-if-part let $\tilde{Y}$ be a random variable with the same distribution as Y, independent of X^* and Y^*. Then $X^* \leq_{\mathcal{F}} \tilde{Y}$ and hence for any $g \in \mathcal{G}_{\mathcal{F}}$

$$E\Delta g(X^*, y) \leq E\Delta g(\tilde{Y}, y) \quad \text{for all } y.$$

This implies

$$
\begin{aligned}
E\Delta g(X^*, Y^*) &= E(E[\Delta g(X^*, Y^*)|Y^*]) \\
&\leq E(E[\Delta g(\tilde{Y}, Y^*)|Y^*]) \\
&= E\Delta g(\tilde{Y}, Y^*) \\
&= 0.
\end{aligned}
$$

$\square$

Stochastic orders which are stronger than $\leq_{st}$ such as likelihood ratio order and the hazard rate order do not have a representation as an integral order, and hence Theorem 1.9.1 is not applicable for them. Nevertheless they also admit bivariate characterizations. The following theorem can be found as Theorem 2.3 in Shanthikumar and Yao (1991) for the case of existing Lebesgue densities. The same idea of proof, however, can be carried over to arbitrary distributions.

Theorem 1.9.2. *$X \leq_{lr} Y$ if and only if*

$$Eg(X^*, Y^*) \leq Eg(Y^*, X^*) \quad \text{for all } g \in \mathcal{G}_{lr},$$

where

$$\mathcal{G}_{lr} = \{g : \Delta g(x, y) \geq 0 \text{ for all } x \geq y\}$$

and X^* and Y^* are independent with $X^* =_{st} X$ and $Y^* =_{st} Y$.

Proof. Assume that $X \leq_{lr} Y$, i.e., X and Y have densities with respect to some dominating measure μ such that

$$f_X(t)f_Y(s) \leq f_X(s)f_Y(t) \quad \text{for all } s \leq t.$$

From $\Delta g(x, y) = -\Delta g(y, x)$ we deduce for any $g \in \mathcal{G}_{lr}$

$$E\Delta g(X^*, Y^*) = \int_y \int_x \Delta g(x, y) f_X(x) f_Y(y) \mu(\mathrm{d}x) \mu(\mathrm{d}y)$$
$$= \int_y \int_{x \geq y} \Delta g(x, y)(f_X(x) f_Y(y) - f_X(y) f_Y(x)) \mu(\mathrm{d}x) \mu(\mathrm{d}y)$$
$$\leq 0.$$

To show the opposite, choose

$$g(x, y) = \begin{cases} 1 & \text{for } c \leq x \leq d, a \leq y \leq b \\ 0 & \text{otherwise} \end{cases}$$

for $a < b < c < d$. Then $g(x, y) = 0$ for $x \leq y$ and hence $g \in \mathcal{G}_{lr}$. As

$$E\Delta g(X^*, Y^*) = P(c \leq X \leq d)P(a \leq Y \leq b) - P(a \leq X \leq b)P(c \leq Y \leq d),$$

the assertion follows from characterization (ii) in Theorem 1.4.6. $\square$

A similar result holds for the hazard rate orders. Part a) of the following theorem is due to Shanthikumar and Yao (1991) and part b) can be found in Kijima and Ohnishi (1996), theorem 4.12.

Theorem 1.9.3. *a)* $X \leq_{hr} Y$ *if and only if*

$$Eg(X^*, Y^*) \leq Eg(Y^*, X^*) \quad \text{for all } g \in \mathcal{G}_{hr},$$

where

$$\mathcal{G}_{hr} = \{g : \Delta g(x, y) \text{ is increasing in } x \text{ for all } x \geq y\}$$

and X^* and Y^* are independent with $X^* =_{st} X$ and $Y^* =_{st} Y$.
b) $X \leq_{rh} Y$ *if and only if*

$$Eg(X^*, Y^*) \leq Eg(Y^*, X^*) \quad \text{for all } g \in \mathcal{G}_{rh},$$

where

$$\mathcal{G}_{rh} = \{g : \Delta g(x, y) \text{ is increasing in } x \text{ for all } x \leq y\}$$

and X^* and Y^* are independent with $X^* =_{st} X$ and $Y^* =_{st} Y$.

Proof. a) Let $X \leq_{hr} Y$ and assume $g \in \mathcal{G}_{lr}$. According to Theorem 1.3.5 we can assume without loss of generality that X and Y have densities f_X and f_Y. As in the proof of Theorem 1.9.2

$$E\Delta g(X^*, Y^*) = \int_y \int_{x \geq y} \Delta g(x, y)(f_X(x) f_Y(y) - f_X(y) f_Y(x)) \mathrm{d}x \mathrm{d}y$$
$$= \int_y \int_{x \geq y} (\bar{F}_X(x) f_Y(y) - f_X(y) \bar{F}_Y(x)) \Delta g(\mathrm{d}x, y) \mathrm{d}y$$
$$\leq 0.$$

Here the second equality follows from integration by parts (with respect to x), taking into account that $\Delta g(x,x) = 0$ for all x. To show the if-part, choose

$$g(x,y) = \begin{cases} 1 & \text{for } x \geq t, \ y \geq s \\ 0 & \text{otherwise} \end{cases}$$

with $s < t$. Then $g \in \mathcal{G}_{hr}$ and

$$E\Delta g(X^*,Y^*) = \bar{F}_X(t)\bar{F}_Y(s) - \bar{F}_X(s)\bar{F}_Y(t).$$

Therefore $\bar{F}_Y(t)/\bar{F}_X(t)$ is increasing and hence $X \leq_{hr} Y$.

The proof of b) is similar and therefore omitted. $\qquad\square$

Note that $\Delta g(x,x) = 0$ for all x and therefore $\mathcal{G}_{lr} \subset \mathcal{G}_{hr} \subset \mathcal{G}_{st}$, and similarly $\mathcal{G}_{lr} \subset \mathcal{G}_{rh} \subset \mathcal{G}_{st}$. Thus we have another possibility to prove Theorem 1.3.8 and Theorem 1.4.5.

In the following theorem we use the notation

$$\mathcal{G}^1_{hr} = \{g : g(x,y) \text{ and } \Delta g(x,y) \text{ are increasing in } x \text{ for all } x \leq y\}.$$

Theorem 1.9.4. *Let X and Y be independent random variables.*
a) If $X \leq_{lr} Y$ then $g(X,Y) \leq_{st} g(Y,X)$ for all $g \in \mathcal{G}_{lr}$.
b) If $X \leq_{hr} Y$ then $g(X,Y) \leq_{icx} g(Y,X)$ for all $g \in \mathcal{G}^1_{hr}$.

Proof. a) Assume that $g \in \mathcal{G}_{lr}$ and let $f : \mathbb{R} \to \mathbb{R}$ be increasing. It has to be shown that the composition $f \circ g$ is in $\mathcal{G}_{lr}$, too. Now $g \in \mathcal{G}_{lr}$ holds if and only if $g(x,y) \geq g(y,x)$ for all $x \geq y$. Hence $f(g(x,y)) \geq f(g(y,x))$ for all $x \geq y$, and thus $f \circ g \in \mathcal{G}_{lr}$.
b) It has to be shown that $g \in \mathcal{G}^1_{hr}$ and f increasing convex implies $f \circ g \in \mathcal{G}_{hr}$. To do so, fix $x' \geq x \geq y$, and let $\delta = g(x,y) - g(y,x)$. For $g \in \mathcal{G}^1_{hr}$ we have $g(x',y) \geq g(x,y)$ and $\Delta g(x',y) \geq \Delta g(x,y)$, or equivalently

$$g(x',y) - \delta \geq g(y,x'). \qquad (1.9.2)$$

For f increasing convex this implies

$$\begin{aligned}
\Delta f(g(x,y)) &= f(g(x,y)) - f(g(x,y) - \delta) \\
&\leq f(g(x',y)) - f(g(x',y) - \delta) \\
&\leq f(g(x',y)) - f(g(y,x')) \\
&= \Delta f(g(x',y)),
\end{aligned}$$

and hence $f \circ g \in \mathcal{G}_{hr}$. $\qquad\square$

Let us return to the simple scheduling problem mentioned at the beginning of this section, i.e. under which conditions can we show $2X + Y \preceq X + 2Y$ for an appropriate stochastic order relation $\preceq$? From Theorem 1.9.4 we can immediately derive the following corollary.

Corollary 1.9.5. *Let X and Y be independent random variables.*
a) If $X \leq_{lr} Y$ then $2X + Y \leq_{st} X + 2Y$.
b) If $X \leq_{hr} Y$ then $2X + Y \leq_{icx} X + 2Y$.

In Ross (1983), p. 268, there is a counterexample which shows that there are independent X and Y with $X \leq_{st} Y$ but $2X + Y \not\leq_{st} X + 2Y$.

1.10 Extremal Elements

There are many applications where it is interesting to have upper and lower bounds with respect to some stochastic order within a specific class of distributions. One important example is the discretization of continuous distributions for computational purposes. If a continuous random variable X is replaced by $[X]$ or $[X] + 1$, where $[x] = \max\{z \in \mathbb{Z} : z \leq x\}$, then it follows immediately from Theorem 1.2.4 that $[X] \leq_{st} X \leq_{st} [X] + 1$. Thus $[X]$ and $[X] + 1$ are always lower and upper bounds with respect to $\leq_{st}$. In fact their distribution can be seen as the solutions of some extremal problem. For given F_X let $\mathbb{M}$ be the set of all distribution functions F with $F(z + 1) - F(z) = F_X(z + 1) - F_X(z)$, $z \in \mathbb{Z}$, endowed with the partial order $\leq_{st}$. Then the distributions of $[X]$ and $[X] + 1$ are the minimum and the maximum of $\mathbb{M}$.

In general, if $\mathbb{M}$ is a set of distributions and $\preceq$ is a partial order on a class $\mathbb{D}$ of distributions with $\mathbb{M} \subseteq \mathbb{D}$, then an element $F_{\max}$ ($F_{\min}$) $\in \mathbb{M}$ is called the maximum (minimum), if

$$F \preceq F_{\max} \ (F_{\min} \preceq F) \quad \text{for all } F \in \mathbb{M}.$$

Similarly, the supremum (infimum) $F_{\sup}$ ($F_{\inf}$) is the smallest (greatest) distribution in $\mathbb{D}$ (with respect to $\preceq$), not necessarily in $\mathbb{M}$, for which

$$F \preceq F_{\sup} \ (F_{\inf} \preceq F) \quad \text{for all } F \in \mathbb{M}.$$

Notice that in general neither a maximum nor a supremum has to exist. However, if there is a maximum, then it is also a supremum, and a supremum $F_{\sup}$ is a maximum if and only if $F_{\sup} \in \mathbb{M}$. Of course, the same holds for minimum and infimum.

If a supremum and an infimum exist for all sets $\mathbb{M} \subset \mathbb{D}$ consisting only of two elements, then $\mathbb{D}$ is called a lattice, and the supremum and the infimum of $\{F, G\}$ are denoted by $F \vee G$ and $F \wedge G$. The following result is obvious.

Theorem 1.10.1. *The set $\mathbb{F}$ of all distribution functions endowed with the partial order $\leq_{st}$ is a lattice with $F \vee G = \min\{F, G\}$ and $F \wedge G = \max\{F, G\}$.*

A similar result holds for convex order and increasing convex order. The following result was first shown by Kellerer (1972), see also Kertz and Rösler (1992).

Theorem 1.10.2. *a) The set* $\mathbb{D}$ *of all distribution functions with finite mean, endowed with the partial order* $\leq_{icx}$, *is a lattice.*

b) The set $\mathbb{D}_\mu$ *of all distribution functions with fixed mean* $\mu \in \mathbb{R}$, *endowed with the partial order* $\leq_{cx}$, *is a lattice.*

Proof. a) The maximum of two integrated survival functions fulfills the properties (i) and (ii) of Theorem 1.5.10, and hence is again an integrated survival function. Thus $F \vee G$ exists and has the integrated survival function $\max\{\pi_F, \pi_G\}$.

Now let $\mathcal{M}_{F,G}$ be the set of all integrated survival functions π with $\pi \leq \pi_F$ and $\pi \leq \pi_G$. This set is non-empty, as it contains, for example, the integrated survival function of the one-point distribution in $\min\{EX, EY\}$. It follows again from Theorem 1.5.10 that the supremum of all functions in $\mathcal{M}_{F,G}$ is an integrated survival function, and it is clear that the corresponding distribution is that of $F \wedge G$.

The proof of b) is similar. □

A set $\mathbb{M}$ is called *bounded* if there exist $F^+, F^- \in \mathbb{D}$ such that

$$F^- \preceq F \preceq F^+ \quad \text{for all } F \in \mathbb{M}.$$

A lattice is called *complete* if all bounded sets have an infimum and a supremum. The Theorems 1.10.1 and 1.10.2 can immediately be generalized to the following statements.

Theorem 1.10.3. *The sets* $\mathbb{F}$ $[\mathbb{D}, \mathbb{D}_\mu]$ *of distribution functions mentioned in Theorems 1.10.1 and 1.10.2 endowed with the partial order* $\leq_{st}$ $[\leq_{icx}, \leq_{cx}]$ *are complete lattices.*

If $\mathbb{M}$ is the set of all distributions on the real line, then extremal elements do not exist for any of the orderings considered so far. Therefore some additional restrictions have to be made. Let $\mathbb{M}^{[a,b]}$ be the set of all distributions on the finite interval $[a, b]$. For this class the following result holds.

Example 1.10.4. (Extremal elements of $\mathbb{M}^{[a,b]}$.) For the class $\mathbb{M}^{[a,b]}$ of distributions on $[a, b]$,

(i) δ_a is the minimum with respect to $\leq_{lr}, \leq_{hr}, \leq_{st}$ and $\leq_{icx}$;

(ii) δ_b is the maximum with respect to $\leq_{lr}, \leq_{hr}, \leq_{st}$ and $\leq_{icx}$.

Distributions can be compared with respect to convex order only if they have the same mean. Therefore with respect to $\leq_{cx}$ neither an infimum nor a supremum can exist, as long as we do not consider a class of distributions with fixed mean μ. In this case, however, comparison with respect to $\leq_{st}$ or any stronger order doesn't make sense, since for distributions with the same mean $X \leq_{st} Y$ only holds if X and Y have the same distribution, see Theorem 1.2.9 b).

Example 1.10.5. (Extremal elements of $\mathbb{M}_\mu$) For the class $\mathbb{M}_\mu$ of all distributions with mean μ,

(i) δ_μ is the minimum with respect to $\leq_{cx}$ and $\leq_{icx}$;

(ii) there is no supremum with respect to $\leq_{cx}$ or $\leq_{icx}$. In $\mathbb{M}_\mu \cap \mathbb{M}^{[a,b]}$, however, the maximum with respect to $\leq_{cx}$ and $\leq_{icx}$ exists and is given by the two-point distribution

$$F_{\max} = \frac{b - \mu}{b - a}\delta_a + \frac{\mu - a}{b - a}\delta_b.$$

The situation becomes more difficult if one considers the class $\mathbb{M}_{\mu,\sigma}$ of all distributions with fixed mean μ and fixed variance σ^2. First it will be shown that distributions within that class cannot be compared with respect to $\leq_{cx}$.

Theorem 1.10.6. *Let X and X be random variables with $X \leq_{cx} Y$ and $Var(X) = Var(Y)$. Then X and Y have the same distribution.*

Proof. Using integration by parts yields the formula

$$EX^2 = \int_{-\infty}^0 2t(-F_X(t))\mathrm{d}t + \int_0^\infty 2t(1 - F_X(t))\mathrm{d}t$$

$$= \int_{-\infty}^0 2(\pi_X(t) + t - EX)\mathrm{d}t + \int_0^\infty 2\pi_X(t)\mathrm{d}t.$$

In the case $EX = EY$ this yields

$$EY^2 - EX^2 = 2\int_{-\infty}^\infty (\pi_Y(t) - \pi_X(t))\mathrm{d}t.$$

If $X \leq_{cx} Y$, then $EX = EY$ holds, and thus $Var(X) = Var(Y)$ implies

$$2\int_{-\infty}^\infty (\pi_Y(t) - \pi_X(t))\mathrm{d}t = EY^2 - EX^2 = Var(Y) - Var(X) = 0.$$

But $X \leq_{cx} Y$ implies $\pi_X \leq \pi_Y$ and therefore necessarily $\pi_X = \pi_Y$ and hence $X =_{st} Y$. $\qquad\square$

A generalization of Theorem 1.10.6 can be found in Li and Zhu (1994). They show that the condition $Var(X) = Var(Y)$ can be replaced by $EX^r = EY^r$ for some $r \geq 2$. Moreover, they prove similar results for the weaker s-convex orders. Theorem 1.10.6 shows that $\mathbb{M}_{\mu,\sigma}$ does not have a maximum with respect to $\leq_{cx}$. A supremum, however, exists.

Theorem 1.10.7. *The class $\mathbb{M}_{\mu,\sigma}$ of all distributions with mean μ and variance σ^2 has a supremum with respect to $\leq_{cx}$. It has the distribution function*

$$F(x) = F_0\left(\frac{x - \mu}{\sigma}\right) \quad \text{where} \quad F_0(x) = \frac{1}{2} + \frac{1}{2}\frac{x}{\sqrt{x^2 + 1}}.$$

Proof. a) Let $F \in \mathbb{M}_{0,1}$ and fix $a \in \mathbb{R}$. If $\alpha \geq 0$ and $\beta \leq a$ are given such that

$$(x - a)_+ \leq \alpha(x - \beta)^2 \qquad \text{for all real } x, \qquad (1.10.1)$$

then

$$\pi_F(a) = \int (x - a)_+ F(\mathrm{d}x) \leq \int \alpha(x - \beta)^2 F(\mathrm{d}x) = \alpha(1 + \beta^2).$$

Now minimize the function $h(\alpha, \beta) = \alpha(1 + \beta^2)$ subject to the restriction (1.10.1). This can be done sequentially. For fixed β one gets $\alpha^* = (4(a - \beta))^{-1}$ and then $\beta^* = a - \sqrt{a^2 + 1}$. With these values equality holds in (1.10.1) for $x_1 = \beta^* = a - \sqrt{a^2 + 1}$ and $x_2 = a + \sqrt{a^2 + 1}$.
Now choose $\gamma = \frac{1}{2}(1 + a/\sqrt{a^2 + 1})$. Then $G = \gamma \delta_{x_1} + (1 - \gamma)\delta_{x_2} \in \mathbb{M}_{0,1}$, and hence

$$\pi_F(a) \leq h(\alpha^*, \beta^*) = -\frac{1}{2}a + \frac{1}{2}\sqrt{a^2 + 1}$$

for all $F \in \mathbb{M}_{0,1}$, und equality holds for G. Thus

$$\rho(a) = -\frac{1}{2}a + \frac{1}{2}\sqrt{a^2 + 1} = \max_{F \in \mathbb{M}_{0,1}} \pi_F(a).$$

Since

$$D^+\rho(a) + 1 = \frac{1}{2} + \frac{1}{2}\frac{a}{\sqrt{a^2 + 1}} = F_0(a),$$

the assertion follows for $\mathbb{M}_{0,1}$ from Theorem 1.5.10 and Theorem 1.10.2.

b) The result for general $\mathbb{M}_{\mu,\sigma}$ can be deduced from a), since $X \leq_{cx} Y$ holds, if and only if $(X - \mu)/\sigma \leq_{cx} (Y - \mu)/\sigma$. $\qquad\square$

Remark 1.10.8. The first proof of Theorem 1.10.7 was probably given in Stoyan (1973). There the theory of Chebychev systems was used to show that it is sufficient to consider distributions with finite support. The proof given here is more elementary. For non-negative random variables this result as well as various extensions can also be found in the actuarial literature; see De Vylder and Goovaerts (1982), Jansen, Haezendonck, and Goovaerts (1986) and Hürlimann (1997a, 1997b). In an application to some scheduling problems, Kamburowski (1989) derived the minimal and maximal elements in the class $\mathbb{M}_{\mu,\sigma^2}^{[a,b]}$ with respect to $\leq_{cx}$. They are three- and five-point distributions, respectively.

The distribution of maximal entropy in the set of all distributions on the non-negative reals with fixed mean μ is the exponential distribution with mean μ, while in the class $\mathbb{M}_{\mu,\sigma^2}$ the maximal element is the normal distribution with mean μ and variance σ^2; see Cover and Thomas (1991), chapter 11.

1.11 Monotone Approximations

The results of Examples 1.10.4 and 1.10.5 can be used to find monotone discrete approximations with respect to $\leq_{st}$ and $\leq_{cx}$. To do so, divide the real line into intervals of length $h > 0$. Then there are four reasonable ways of discretizing a distribution F such that the mass in each of the intervals $(kh, (k+1)h]$ is concentrated in one or two points. They can be described as follows:

1. Denote by F_h^l the distribution that is obtained by moving all probability to the *left* endpoints of the intervals.
2. Denote by F_h^r the distribution that is obtained by moving all probability to the *right* endpoints of the intervals.
3. Moving the mass in each interval to its centre yields a distribution that will be denoted by F_h^e.
4. If the mass in each interval is moved to its two endpoints in such a way that the mean is preserved, then the resulting discretization is called F_h^d.

The formal expressions for these four procedures are as follows. Let X be a random variable with distribution function F. For given $h > 0$ and $k \in \mathbb{Z}$ define

$$p_{hk} = F((k+1)h) - F(kh)$$

and

$$m_{hk} = \begin{cases} E[X|kh < X \leq (k+1)h] & \text{for } p_{hk} > 0 \\ (k+1)h & \text{otherwise} \end{cases}.$$

For $* \in \{l, r, e, d\}$ define then $F_h^* = \sum_{k \in \mathbb{Z}} a_{hk}^* \cdot \delta_{x_{hk}^*}$ where

1. $x_{hk}^l = kh$; $\quad a_{hk}^l = p_{hk}$.
2. $x_{hk}^r = (k+1)h$; $\quad a_{hk}^r = p_{hk}$.
3. $x_{hk}^e = m_{hk}$; $\quad a_{hk}^e = p_{hk}$.
4. $x_{hk}^d = kh$; $\quad a_{hk}^d = \alpha_{hk} p_{hk} + (1 - \alpha_{h,k-1}) p_{h,k-1}$,
   where $\alpha_{hk} = 1 - (m_{hk} - kh)/h$.

These discrete approximations are upper and lower bounds with respect to $\leq_{st}$ and $\leq_{cx}$ respectively. Moreover, the next result reveals that finer discretizations lead to better approximations.

Theorem 1.11.1. *Let F be an arbitrary distribution, $h > 0$ and $i \in \mathbb{N}$. Then*
a) $F_{ih}^l \leq_{st} F_h^l \leq_{st} F \leq_{st} F_h^r \leq_{st} F_{ih}^r$.
b) If F has a finite mean, then

$$F_{ih}^e \leq_{cx} F_h^e \leq_{cx} F \leq_{cx} F_h^d \leq_{cx} F_{ih}^d.$$

Proof. a) If $F \in \mathbb{M}^{[kh,(k+1)h]}$ then it follows from Example 1.10.4 that $F_h^l \leq_{st} F \leq_{st} F_h^r$. Since an arbitrary F can be written as a mixture

$$F = \sum_{k \in \mathbb{Z}} p_{hk} \cdot F_{hk}$$

of such distributions, the general case then follows from Theorem 1.2.15. As $F_{ih}^l = (F_h^l)_{ih}^l$ this also implies $F_{ih}^l \leq_{st} F_h^l \leq_{st} F$. Similarly, $F \leq_{st} F_h^r \leq_{st} F_{ih}^r$.

b) Since it has been shown in Theorem 1.5.6 that $\leq_{cx}$ is closed under mixtures, too, the proof is similar to a) taking into account Example 1.10.5.   □

Remark 1.11.2. a) Theorem 1.11.1 a) does not remain true if $\leq_{st}$ is replaced by one of the stronger orderings $\leq_{hr}$ or $\leq_{lr}$, since they are not closed with respect to mixtures.

b) Part b) of Theorem 1.11.1 is very often used in stochastic programming for computing upper and lower bounds, see Kall and Wallace (1994), section 3.4. In this context the lower bound is called the Jensen bound, since it follows from Jensen's inequality. The upper bound is named Edmundson–Madansky bound, since it is based on work of Edmundson (1956) and Madansky (1959). Kamburowski (1992) gave an upper bound with respect to $\leq_{cx}$ which has equal weights in irregularly spaced atoms.

1.12 Relationships and Comparison Criteria for Univariate Stochastic Orders

This section presents Figure 1.1 which describes the relationships between the more important stochastic orders and Tables 1.1 and 1.2 with criteria for $F \preceq G$ for univariate discrete and continuous distributions.

Tables 1.1 and 1.2 give criteria for the comparison of some important univariate distributions. Only distributions of the same type are compared and only the strongest orders are given. More criteria can be found in Lisek (1978), Taylor (1983), Stoyan (1983), and Denuit and Lefèvre (2001). These papers consider also other distributions and compare also pairs of different distributions. We note in passing that it is usually easier to show relations for the stronger order $\leq_{lr}$ rather than for the weaker relation $\leq_{st}$.

In Table 1.1 m_F and m_G denote the means of the first and second distribution, respectively.

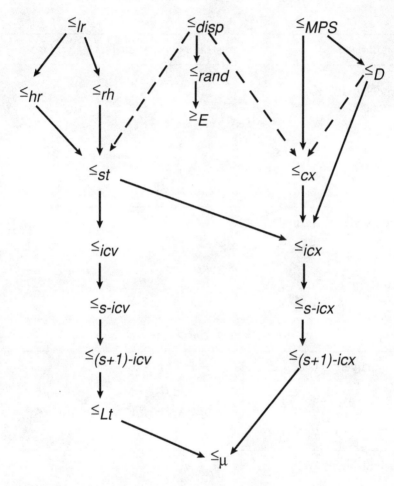

Figure 1.1 Relations between various univariate stochastic orders. $A \longrightarrow B$ means that A implies B and $A \dashrightarrow B$ that A implies B under certain conditions. Observe the central roles of $\leq_{st}$ and $\leq_{cx}$.

Table 1.1 Comparison criteria for continuous distributions

Comparison	Constraints on parameters	Order
Exponential versus Exponential		
	$\lambda \geq \mu$	$\leq_{disp}$
$\lambda \exp(-\lambda(x - a))$	$a \leq b,\ \lambda \geq \mu$	$\leq_{lr}$
$\mu \exp(-\mu(x - b))$	$\lambda \geq \mu,\ a + \lambda^{-1} \leq b + \mu^{-1}$	$\leq_{icx}$
	$a \leq b,\ a + \lambda^{-1} \leq b + \mu^{-1}$	$\leq_{icv}$
Normal versus Normal		
	$\sigma \leq \tau$	$\leq_{disp}$
$\frac{1}{\sigma\sqrt{2\pi}} \exp\left(-\frac{(x-\mu)^2}{2\sigma^2}\right)$	$\mu \leq \nu,\ \sigma = \tau$	$\leq_{lr}$
	$\mu = \nu,\ \sigma \leq \tau$	$\leq_{cx}$
$\frac{1}{\tau\sqrt{2\pi}} \exp\left(-\frac{(x-\nu)^2}{2\tau^2}\right)$	$\mu \leq \nu,\ \sigma \leq \tau$	$\leq_{icx}$
	$\mu \leq \nu,\ \sigma \geq \tau$	$\leq_{icv}$
Gamma versus Gamma		
$\lambda^\alpha x^{\alpha-1} \exp(-\lambda x)/\Gamma(\alpha)$	$\alpha \leq \beta,\ \lambda \geq \mu$	$\leq_{lr}$
	$\alpha \geq \beta,\ \alpha\lambda^{-1} \leq \beta\mu^{-1}$	$\leq_{icx}$
$\mu^\beta x^{\beta-1} \exp(-\mu x)/\Gamma(\beta)$	$\alpha \leq \beta,\ \lambda \leq \mu,\ \alpha\lambda^{-1} \leq \beta\mu^{-1}$	$\leq_{icv}$
Weibull versus Weibull		
$\lambda\alpha x^{\alpha-1} \exp(-\lambda x^\alpha)$	$\lambda \geq \mu,\ \alpha = \beta$	$\leq_{lr}$
	$\alpha \geq \beta,\ m_F \leq m_G$	$\leq_{icx}$
$\mu\beta x^{\beta-1} \exp(-\mu x^\beta)$	$\alpha \leq \beta,\ m_F \leq m_G$	$\leq_{icv}$

(**Table 1.1** continued)

Lognormal versus Lognormal		
$\frac{1}{x\sigma\sqrt{2\pi}} \exp\left(-\frac{(\ln x - \mu)^2}{2\sigma^2}\right)$	$\mu \le \nu,\ \sigma = \tau$	$\le_{lr}$
	$\sigma \le \tau,\ m_F \le m_G$	$\le_{icx}$
$\frac{1}{x\tau\sqrt{2\pi}} \exp\left(-\frac{(\ln x - \nu)^2}{2\tau^2}\right)$	$\sigma \ge \tau,\ m_F \le m_G$	$\le_{icv}$
Beta versus Beta		
$x^p(1-x)^r C_{pr}$	$p \le q,\ r \ge s$	$\le_{lr}$
	$p \ge q,\ r \ge s,\ m_F \le m_G$	$\le_{icx}$
$x^q(1-x)^s C_{qs}$	$p \le q,\ r \le s,\ m_F \le m_G$	$\le_{icv}$
$0 \le x \le 1$		
Uniform versus Uniform		
	$a \le c,\ b \le d$	$\le_{st}$
$\mathbf{1}_{(a,b)}$	$a + b \le c + d,\ b \le d,$	$\le_{icx}$
$\mathbf{1}_{(c,d)}$	$b - a \le d - c$	$\le_{disp}$

Table 1.2 Comparison criteria for discrete distributions

Comparison	Constraints on parameters	Order
Poisson versus Poisson		
$\frac{\lambda^k}{k!}\exp(-\lambda)$	$\lambda \le \mu$	$\le_{lr}, \le_{rand}$
$\frac{\mu^k}{k!}\exp(-\mu)$		
Binomial versus Binomial		
$\binom{m}{k}p^k(1-p)^{m-k}$	$m \le n,\ p = q$	$\le_{rand}$
$\binom{n}{k}q^k(1-q)^{n-k}$	$m \le n,\ p \le q$	$\le_{lr}$
Geometric versus Geometric		
$(1-p)p^k$	$p \le q$	$\le_{lr},\ \le_{disp}$
$(1-q)q^k$		

2

Theory of Integral Stochastic Orders

2.1 Introduction

Many stochastic orders can be characterized by a class $\mathcal{F}$ of functions as follows. For random variables X and Y (which may have values in more general spaces than the real line)

$$X \preceq Y \text{ if } Ef(X) \leq Ef(Y) \text{ for all } f \in \mathcal{F}. \tag{2.1.1}$$

For the stochastic orders treated in Chapter 1 this applies to $\leq_{st}$, $\leq_{cx}$, $\leq_{icx}$ and $\leq_{Lt}$, among others. Whitt (1986) introduced the term *integral stochastic orders* for this class of orders, since equation (2.1.1) can be rewritten in terms of probability measures as

$$P \preceq Q \text{ if } \int f \mathrm{d}P \leq \int f \mathrm{d}Q \text{ for all } f \in \mathcal{F}. \tag{2.1.2}$$

The set $\mathcal{F}$ is called a *generator* of the integral stochastic order, and from now on we will use $\leq_{\mathcal{F}}$ as a generic symbol for the corresponding integral stochastic order. Notice, however, that there may be different generators for the same integral stochastic order.

There are two main fields where this class of stochastic orders is of particular interest, since one is directly confronted with the comparison of integrals as in (2.1.2).

The first one is applied probability where characteristics of a stochastic model can be identified with an integral of a function f with respect to some probability measure P. In this context the importance of integral stochastic orders will be demonstrated in Section 4.3.1. A typical application is given in Theorem 4.3.1 where we describe the so-called *functional method*.

The other field where they play a dominant role is the theory of decision under risk. This branch of economics deals with the problem how an agent decides if she has to choose between different risky alternatives. A decision for one of the alternatives leads to a random amount of some goods. This means that the agent has to compare random variables. The most studied approach to this problem is the so called subjective expected utility theory. This concept was already used by Bernoulli in the 18th century, and has been axiomatically derived in the famous work of von Neumann and Morgenstern (1947). This theory states that the decision maker has a utility function u such that she prefers the alternative with a random yield Y to one that yields X, if and only if $Eu(X) \leq Eu(Y)$. In practice, however, it is nearly impossible to determine somebody's utility function exactly. Therefore the following important question naturally arises. Is it possible to predict which alternative will be chosen by the agent, if we have only partial knowledge of her utility function, say that we only know that it belongs to some class $\mathcal{F}$ of utility functions? According to (2.1.1) it is obvious that she will prefer Y to X if $X \leq_{\mathcal{F}} Y$.

There is another interpretation, which ends up with the same solution. Assume that there is a group of decision makers, each with his own utility function u. If $\mathcal{F}$ is the set of all utility functions of the members of the group, then all members will agree that Y should be preferred to X, if and only if $X \leq_{\mathcal{F}} Y$. In the theory of decision making under risk, integral stochastic orders can be found under the key word *stochastic dominance rules* (see Section 8.1).

In this chapter we will develop a general theory of integral stochastic orders. Especially the following problems will be treated. First we will be interested in finding 'small' and 'large' generators. Consider as an example the usual stochastic order $\leq_{st}$ on the real line. We know that $X \leq_{st} Y$ holds if $P(X > t) \leq P(Y > t)$ for all real t. Hence there is a 'small' generator consisting of the indicator functions $\mathbf{1}_{(t,\infty)}$ for all real t. On the other hand, we know that there are much larger generators. As Theorem 1.2.8 shows, a 'large' generator is given by the set of all increasing functions. We will see below that this generator is in some sense the largest one possible.

Small generators are of interest for checking whether or not two random variables are ordered. For example, in the context of $\leq_{st}$ it is difficult to check whether $Ef(X) \leq Ef(Y)$ holds for all increasing functions, but very often it is easy to show this inequality for the small generator mentioned above, which means nothing else than checking whether or not the distribution functions are comparable pointwise.

The second question to be considered here is that of showing interesting properties of stochastic orders by verifying some properties of the generators which are easy to check. Indeed, such properties of stochastic orders like closure under weak convergence, closure under convolution, closure under mixtures etc. can easily be deduced from properties of the generators.

Finally it will be shown that many of the considered integral stochastic orders have generators consisting of infinitely differentiable functions. This property is helpful in many proofs.

The natural way to derive these results is by using some tools from functional analysis. The reason can briefly be described as follows. Condition

(2.1.2) can be rewritten as

$$P \preceq Q \quad \text{if and only if} \quad \int f \mathrm{d}(Q - P) \geq 0 \text{ for all } f \in \mathcal{F}, \qquad (2.1.3)$$

where now $Q - P$ is a signed measure. For signed measures μ, however, the expression $\int f \mathrm{d}\mu$ is linear in f as well as in μ so that we have a duality of a function space and a (signed) measure space equipped with a bilinear form, and an integral stochastic order can be considered as an order generated by the dual cone of the set $\mathcal{F}$ of functions. Thus we can use the well developed theory of duality, as it can be found, for example, in Choquet (1969).

This elegant approach seems not to be widely known. As far as we know, it was introduced into the literature on mathematical economics by Border (1991) and into the applied probability literature by Müller (1997b). Border (1991) considers only probability measures on a compact interval $[a, b] \subset \mathbb{R}$ and functions of bounded variation. However, we want to include the case of unbounded functions and of outcomes in an arbitrary measure space. This is useful since distributions with unbounded support arise naturally in many situations. For example, any distribution derived from a continuous time stochastic process with independent and stationary increments is infinitely divisible and thus has an unbounded support (see Chow and Teicher (1978), p. 413). Therefore we will consider general measure spaces and we will allow unbounded functions.

In this chapter sets of functions are typically denoted by capital *script* letters as $\mathcal{F}, \mathcal{G}, \mathcal{R}, \mathcal{B}, \dots$, while we use *calligraphic* letters as $\mathcal{A}, \mathcal{B}, \dots$ for σ-algebras. Sets of (signed) measures are denoted by *open-face* letters like $\mathbb{M}, \mathbb{P}, \dots$.

2.2   Tools from Functional Analysis

Let $(S, \mathcal{S})$ be an arbitrary measure space. We will only assume that the σ-algebra $\mathcal{S}$ contains all singletons so that probability measures with a finite support are always well defined. Moreover, let $b : S \to [1, \infty)$ be a measurable function, called *weight function*. We consider the set $\mathcal{B}_b$ of measurable functions $f : S \to \mathbb{R}$ with finite weighted supremum norm (called *b-norm*)

$$\|f\|_b = \sup_{s \in S} \frac{|f(s)|}{b(s)} < \infty. \qquad (2.2.1)$$

Notice that $\mathcal{B}_b$ contains unbounded functions if b is unbounded. For a signed measure μ on $\mathcal{S}$ we denote the positive and negative part by μ^+ and μ^-, respectively. As usual $|\mu| = \mu^+ + \mu^-$ is the total variation. Integrals will often be written as a bilinear form $\langle f, \mu \rangle = \int f \, \mathrm{d}\mu = \int f \, \mathrm{d}\mu^+ - \int f \, \mathrm{d}\mu^-$. Notice that $\langle f, \mu \rangle$ exists and is finite if and only if $\langle |f|, |\mu| \rangle = \langle |f|, \mu^+ \rangle + \langle |f|, \mu^- \rangle < \infty$.

The set of all signed measures μ on $\mathcal{S}$ with $\langle b, |\mu| \rangle < \infty$ is denoted by $\mathbb{M}_b$. We write $\mathbb{P}$ for the set of all probability measures on $\mathcal{S}$, and $\mathbb{P}_b = \mathbb{P} \cap \mathbb{M}_b$ is the restriction of M_b to $\mathbb{P}$; thus $\mathbb{P}_b$ is the set of all probability measures for which $\int b \, dP$ is finite. $\mathbb{P}_b$ is nonvoid as it contains all probability measures with finite support. M_b^N is the set of all signed measures in M_b with $\mu(S) = 0$. Notice that the difference of two probability measures in $\mathbb{P}_b$ is contained in M_b^N and that every signed measure in M_b^N is a multiple of such a difference, i.e. M_b^N is the linear span of $\mathbb{P}_b - \mathbb{P}_b$.

Next we present the necessary facts from functional analysis, which can be found, for example, in Choquet (1969), section 22.

A pair (E, F) of vector spaces is said to be in *duality*, if there is a bilinear mapping $\langle \cdot, \cdot \rangle : E \times F \to \mathbb{R}$. The duality is said to be *strict*, if for each $0 \neq x \in E$ there is a $y \in F$ with $\langle x, y \rangle \neq 0$ and if for each $0 \neq y \in F$ there is an $x \in E$ with $\langle x, y \rangle \neq 0$.

Using the notation introduced above the following Lemma can be stated.

Lemma 2.2.1. M_b *and* $\mathcal{B}_b$ *are in strict duality under the bilinear mapping*

$$
\begin{aligned}
\langle \cdot, \cdot \rangle &: M_b \times \mathcal{B}_b \to \mathbb{R} \\
\langle \mu, f \rangle &= \int f \, d\mu
\end{aligned}
\qquad (2.2.2)
$$

Proof. Obviously $\mathcal{B}_b$ and M_b are vector spaces. For $f \in \mathcal{B}_b$ we have $|f| \leq \|f\|_b \cdot b$, and hence

$$
|\langle f, \mu \rangle| \leq \langle |f|, \mu^+ \rangle + \langle |f|, \mu^- \rangle \leq \|f\|_b \cdot (\langle b, \mu^+ \rangle + \langle b, \mu^- \rangle) < \infty
$$

for $\mu \in M_b$. Thus the mapping $\langle \cdot, \cdot \rangle$ is well defined. Obviously, it is bilinear. It remains to show the strictness of the duality.

(i) $\mathcal{B}_b$ contains the indicator functions of all sets $A \in \mathcal{S}$, as $b \geq 1$. Therefore $\langle f, \mu \rangle = 0$ for all $f \in \mathcal{B}_b$ implies $\mu(A) = 0$ for all $A \in \mathcal{S}$, and thus $\mu \equiv 0$.

(ii) M_b contains the one-point distributions δ_s for all $s \in S$. Hence $\langle f, \mu \rangle = 0$ for all $\mu \in M_b$ implies $\langle f, \delta_s \rangle = f(s) = 0$ for all $s \in S$ and consequently $f \equiv 0$. $\qquad \square$

Note that in part (i) of the proof we needed the requirement $b \geq 1$ for the weight function. Sometimes there is a naturally given weight function b', which only fulfills $b' \geq 0$. Then we can use $b = b' + 1$, leading to $M_b = M_{b'}$ and $\mathcal{B}_{b'} \subset \mathcal{B}_b$, i.e. the measure space remains the same and even more functions can be handled.

Unfortunately the duality $(M_b^N, \mathcal{B}_b)$ is not strict, as $\langle f, \mu \rangle = 0$ for all $\mu \in M_b^N$ only implies f to be constant. But strict duality can be obtained by identifying functions which differ only by a constant. Formally, we define an equivalence relation $f \sim g$ if $f - g$ is constant. Denoting the corresponding quotient space by $\mathcal{B}_{b/\sim}$ we get the following lemma.

Lemma 2.2.2. M_b^N *and* $\mathcal{B}_{b/\sim}$ *are vector spaces in strict duality under the bilinear mapping* (2.2.2).

The bipolar theorem for convex cones plays a crucial role in our further investigations. Therefore we introduce now the notion of polars. We follow the notation of Choquet (1969).

The *polar* M° of a set $M \subset E$ (in the duality (E, F)) is defined as

$$M^\circ = \{y \in F : \langle x, y \rangle \geq -1 \text{ for all } x \in M\}. \qquad (2.2.3)$$

The polar of a set $N \subset F$ is defined analogously.

As usual a subset K of a vector space is called a *cone* if $x \in K$ implies $\alpha x \in K$ for all $\alpha \geq 0$. The *dual cone* of an arbitrary set $M \subset E$ is the set

$$M^+ = \{y \in F : \langle x, y \rangle \geq 0 \text{ for all } x \in M\}.$$

It is easy to see that M^+ is a convex cone. Moreover, notice that for a convex cone K it holds $K^\circ = K^+$.

Finally, we need the concept of *weak topology*. For any duality (E, F) we can define the weak topology $\sigma(E, F)$ on E as the weakest topology on E such that the mappings $x \mapsto \langle x, y \rangle$ are continuous for all $y \in F$. Now the bipolar theorem for convex cones can be stated as follows.

Theorem 2.2.3. *(Corollary 22.10 in Choquet (1969)).*
Suppose E and F are in strict duality and $X \subset E$ is a convex cone. Then $X^{\circ\circ}$ is the weak closure of X and $X^\circ = X^+$.

2.3   Maximal Generators of Integral Stochastic Orders

Many authors define integral stochastic orders as follows:
Let $\mathcal{F}$ be a class of real-valued functions. Then

$$P \preceq Q \quad \text{if} \quad \int f \, \mathrm{d}P \leq \int f \, \mathrm{d}Q \qquad (2.3.1)$$

for all $f \in \mathcal{F}$ for which the integrals exist. This definition can be found, for example, in Stoyan (1983), Definition 1.1.2, or Marshall (1991), p. 231.

A relation defined like this is not necessarily transitive, as the following example on the real line shows. If $\mathcal{F}$ contains only the identity $f(x) = x$ and Q denotes the Cauchy distribution (for which no mean exists) then $\delta_1 \preceq Q$ and $Q \preceq \delta_0$ (since the integral $\int f \mathrm{d}Q$ doesn't exist!), but $\delta_1 \npreceq \delta_0$. However, if all occurring integrals would be finite then this flaw could not happen.

In order to ensure the finiteness of all integrals we assume the existence of a weight function b as introduced in Subsection 2.2. Then by Lemma 2.2.1 $\int f \, \mathrm{d}P$ exists and is finite for all $f \in \mathcal{B}_b$ and $P \in \mathbb{P}_b$. Therefore we can define an order relation $\leq_{\mathcal{F}}$ on $\mathbb{P}_b$ via (2.3.1) for $\mathcal{F} \subset \mathcal{B}_b$.

Definition 2.3.1. For $\mathcal{F} \subset \mathcal{B}_b$ the relation $\leq_{\mathcal{F}}$ is defined on $\mathbb{P}_b$ by

$$P \leq_{\mathcal{F}} Q \quad \text{if} \quad \int f \, \mathrm{d}P \leq \int f \, \mathrm{d}Q \text{ for all } f \in \mathcal{F}. \qquad (2.3.2)$$

Using the notation introduced above, the right-hand side of (2.3.2) can be rewritten as $\langle f, Q - P \rangle \geq 0$ for all $f \in \mathcal{F}$, or equivalently $Q - P \in \mathcal{F}^+$.

The following properties can be derived easily from the definition.

Lemma 2.3.2. *For arbitrary $\mathcal{F} \subset \mathcal{B}_b$*
a) $\leq_{\mathcal{F}}$ is transitive and reflexive, i.e. a pre-order;
b) $\leq_{\mathcal{F}}$ is a (partial) order if and only if $\mathcal{F}$ is separating in $\mathbb{P}_b$, i.e. if the following holds: if for some $P, Q \in \mathbb{P}_b$

$$\int f \, \mathrm{d}P = \int f \, \mathrm{d}Q \quad \text{for all } f \in \mathcal{F}, \tag{2.3.3}$$

then $P = Q$.

We already mentioned that there may be different classes of functions which are generators of the same stochastic order. For checking $P \leq_{\mathcal{F}} Q$ it is desirable to have 'small' generators, whereas 'large' generators are interesting for applications. Among others, Stoyan (1983) and Marshall (1991) posed the question of characterizing the largest generator, which Marshall (1991) called *stochastic completion*.

Definition 2.3.3. Let b be a weight function and let $\mathcal{F} \subset \mathcal{B}_b$ be an arbitrary generator of an order $\leq_{\mathcal{F}}$ on $\mathbb{P}_b$.
a) The *maximal generator* $\mathcal{R}_{\mathcal{F}}$ of $\leq_{\mathcal{F}}$ (in $\mathcal{B}_b$) is the set of all functions $f \in \mathcal{B}_b$ with the property that

$$P \leq_{\mathcal{F}} Q \quad \text{implies} \quad \int f \, \mathrm{d}P \leq \int f \, \mathrm{d}Q.$$

b) Denote by $\tilde{\mathcal{R}}_{\mathcal{F}}$ the set of all measurable functions $f : S \to \mathbb{R}$ (not necessarily in $\mathcal{B}_b$) with the property that $P, Q \in \mathbb{P}_b$ and $P \leq_{\mathcal{F}} Q$ imply $\int f \, \mathrm{d}P \leq \int f \, \mathrm{d}Q$, if the integrals exist. $\tilde{\mathcal{R}}_{\mathcal{F}}$ is called the *extended maximal generator*.

This definition can be interpreted in the context of decisions under risks as follows: If Q is preferred to P by all decision makers with a utility function $u \in \mathcal{F}$ then any decision maker with a utility function in $\mathcal{R}_{\mathcal{F}}$ will also prefer Q to P.

A characterization of $\tilde{\mathcal{R}}_{\mathcal{F}}$ would immediately give a characterization of $\mathcal{R}_{\mathcal{F}}$, too, since $\mathcal{R}_{\mathcal{F}}$ is just the restriction of $\tilde{\mathcal{R}}_{\mathcal{F}}$ to the set $\mathcal{B}_b$ of b-bounded functions. Notice that this means that $\mathcal{R}_{\mathcal{F}}$ implicitly depends on the weight function b. However, it will turn out that it is much easier to characterize $\mathcal{R}_{\mathcal{F}}$.

It is also clear that any generator of $\leq_{\mathcal{F}}$ is a subset of the maximal generator $\mathcal{R}_{\mathcal{F}}$. Hence $\mathcal{R}_{\mathcal{F}}$ is the unique maximal element of all generators of a given integral stochastic order.

Lemma 2.3.4. *Let $\mathcal{F} \subset \mathcal{G} \subset \mathcal{B}_b$ and $P, Q \in \mathbb{P}_b$.*
a) If $P \leq_{\mathcal{G}} Q$ then $P \leq_{\mathcal{F}} Q$.
b) $\mathcal{R}_{\mathcal{F}} \subset \mathcal{R}_{\mathcal{G}}$ and $\tilde{\mathcal{R}}_{\mathcal{F}} \subset \tilde{\mathcal{R}}_{\mathcal{G}}$.
c) If $\mathcal{F} \subset \mathcal{G} \subset \mathcal{R}_{\mathcal{F}}$ then $\leq_{\mathcal{G}}$ and $\leq_{\mathcal{F}}$ are identical.

The next two results are counterparts to proposition 3.3 and 3.5 in Marshall (1991).

Theorem 2.3.5. *Let $\mathcal{F}$ be an arbitrary generator of the order $\leq_{\mathcal{F}}$. Then:*
a) $\mathcal{R}_{\mathcal{F}}$ *contains the convex cone spanned by $\mathcal{F}$;*
b) $\mathcal{R}_{\mathcal{F}}$ *contains all constant functions;*
c) *if the sequence $(f_n) \subset \mathcal{R}_{\mathcal{F}}$ converges uniformly to f, then $f \in \mathcal{R}_{\mathcal{F}}$.*

Proof. a) Let $f_1, \ldots, f_n \in \mathcal{F}$ and $a_1, \ldots, a_n \geq 0$. As $\mathcal{B}_b$ is a vector space, $f = \sum a_i f_i \in \mathcal{B}_b$. If $P, Q \in \mathbb{P}_b$ with $P \leq_{\mathcal{F}} Q$ then by definition $\langle f_i, P \rangle \leq \langle f_i, Q \rangle$ and thus

$$\langle f, P \rangle = \sum_{i=1}^{n} a_i \langle f_i, P \rangle \leq \sum_{i=1}^{n} a_i \langle f_i, Q \rangle = \langle f, Q \rangle.$$

Hence $f \in \mathcal{R}_{\mathcal{F}}$.
b) is obvious.
c) Let $(f_n) \subset \mathcal{R}_{\mathcal{F}}$ be a sequence with $\|f_n - f\|_\infty \to 0$ for some $f : S \to \mathbb{R}$. Then for every $\varepsilon > 0$ there is some $m \in \mathbb{N}$ with $\|f_m - f\|_\infty \leq \varepsilon$. Thus, as $b \geq 1$, $|f(x)| \leq |f_m(x)| + \varepsilon \leq (\|f_m\|_b + \varepsilon) b(x)$. Consequently $f \in \mathcal{B}_b$. Therefore, if $P \leq_{\mathcal{F}} Q$ then

$$\langle f, P \rangle = \langle \lim f_n, P \rangle = \lim \langle f_n, P \rangle \leq \lim \langle f_n, Q \rangle = \langle f, Q \rangle,$$

establishing $f \in \mathcal{R}_{\mathcal{F}}$.   □

The next result is a generalization of Theorem 2.3.5 a).

Theorem 2.3.6. *Let $(\Omega, \mathcal{A}, \rho)$ be a σ-finite measure space and let $f : \Omega \times S \to \mathbb{R}$ be an $\mathcal{A} \otimes \mathcal{S}$-measurable function which fulfills the following assumptions:*

(i) $f(\omega, \cdot) \in \mathcal{F}$ *for all $\omega \in \Omega$;*

(ii) *there exists a ρ-integrable function $c : \Omega \to [0, \infty)$ with*
$|f(\omega, s)| \leq c(\omega) b(s)$ *for all $\omega \in \Omega$ and $s \in S$.*

Then $g(\cdot) = \int f(\omega, \cdot) \rho(d\omega)$ is finite and belongs to $\mathcal{R}_{\mathcal{F}}$.

Proof. Since $|f(\omega, x)| \leq c(\omega) b(x)$ it is for all $\mu \in \mathbb{M}_b$

$$\iint |f(\omega, x)| \, \rho(d\omega) |\mu|(dx) \leq \int c(\omega) \rho(d\omega) \int b(x) |\mu|(dx) < \infty. \tag{2.3.4}$$

Specializing $\mu = \delta_s$ for $s \in S$ yields the existence and finiteness of

$$g(s) = \int f(\omega, s) \rho(d\omega).$$

Now (2.3.4) and (ii) imply $\|g\|_b \leq \int c \, d\rho < \infty$. Hence $g \in \mathcal{B}_b$ and we can apply Fubini's theorem. Thus for $P, Q \in \mathbb{P}_b$ with $P \leq_{\mathcal{F}} Q$

$$
\begin{aligned}
\int g(x) P(dx) &= \iint f(\omega, x) \rho(d\omega) P(dx) = \iint f(\omega, x) P(dx) \rho(d\omega) \\
&\leq \iint f(\omega, x) Q(dx) \rho(d\omega) = \int g(x) Q(dx).
\end{aligned}
$$

This yields $g \in \mathcal{R}_\mathcal{F}$.   □

Now we are ready for one of the main results of this section, which gives a characterization of the maximal generator. This solves an open problem stated by Marshall (1991). Recall that $\sigma(\mathcal{B}_b, \mathbb{M}_b)$ defines the weak topology as introduced on p. 69.

Theorem 2.3.7. $\mathcal{R}_\mathcal{F}$ *is the* $\sigma(\mathcal{B}_b, \mathbb{M}_b)$-*closure of the convex cone which is spanned by* $\mathcal{F}$ *and the constant functions.*

Proof. First we consider the strict duality $(\mathbb{M}_b^N, \mathcal{B}_{b/\sim})$ described in Lemma 2.2.2. As $\mathbb{M}_b^N$ is the linear span of $\mathbb{P}_b - \mathbb{P}_b$, the definition of $\mathcal{R}_\mathcal{F}$ implies that $f \in \mathcal{R}_{\mathcal{F}/\sim}$ holds if and only if

$$\langle f, \mu \rangle \geq 0 \quad \text{for all } \mu \in (\mathcal{F}_{/\sim})^+, \tag{2.3.5}$$

which is equivalent to $f \in (\mathcal{F}_{/\sim})^{++}$. Theorem 2.3.5 yields that $\mathcal{R}_{\mathcal{F}/\sim}$ contains the convex cone spanned by $\mathcal{F}_{/\sim}$. Now a look at Lemma 2.3.4 c) shows that without loss of generality $\mathcal{F}_{/\sim}$ can be assumed to be a convex cone. But by Lemma 2.2.2 $\mathbb{M}_b^N$ and $\mathcal{B}_{b/\sim}$ are in strict duality, and thus Theorem 2.2.3 and (2.3.5) imply that $\mathcal{R}_{\mathcal{F}/\sim}$ is the $\sigma(\mathcal{B}_{b/\sim}, \mathbb{M}_b^N)$-closure of $\mathcal{F}_{/\sim}$. The result finally follows from the definition of the equivalence relation $\sim$.   □

Theorem 2.3.7 is of great theoretical value. To use it practically, it is necessary to find easy characterizations for closed sets in the $\sigma(\mathcal{B}_b, \mathbb{M}_b)$-topology. Here the following generalization of theorem 4.1 in Border (1991) is available.

Theorem 2.3.8. *On any* $\| \cdot \|_b$-*norm bounded subset of* $\mathcal{B}_b$ *the* $\sigma(\mathcal{B}_b, \mathbb{M}_b)$-*topology and the topology of pointwise convergence coincide.*

Proof. Since $\mathbb{M}_b$ contains all one-point probability measures, convergence in the $\sigma(\mathcal{B}_b, \mathbb{M}_b)$-topology implies pointwise convergence. Conversely, since sets bounded in $\| \cdot \|_b$-norm are also pointwise bounded, we can deduce from Lebesgue's dominated convergence theorem that pointwise convergence implies $\sigma(\mathcal{B}_b, \mathbb{M}_b)$-convergence.   □

Combining this result with Theorem 2.3.7 we get the following sufficient condition for $\mathcal{F} = \mathcal{R}_\mathcal{F}$, which is due to Müller (1997b). It is sometimes very easy to check.

Corollary 2.3.9. *If* $\mathcal{F} \subset \mathcal{G} \subset \mathcal{R}_\mathcal{F}$, *and* $\mathcal{G}$ *is a convex cone containing the constant functions and being closed under pointwise convergence, then* $\mathcal{G} = \mathcal{R}_\mathcal{F}$.

In general the extended maximal generator $\tilde{\mathcal{R}}_\mathcal{F}$ is larger than the maximal generator $\mathcal{R}_\mathcal{F}$ for a given weight function b. As an example consider $\leq_{st}$ for distributions on the real line. If we choose $b \equiv 1$ then we get for $\mathcal{R}_{\mathcal{F}}$ the set of all bounded increasing functions. It follows immediately from the next theorem that $\tilde{\mathcal{R}}_{\mathcal{F}}$ is the set of all increasing functions. Notice that this implies that it is not possible to find a weight function b such that $\tilde{\mathcal{R}}_{\mathcal{F}} \subset \mathcal{B}_b$.

Theorem 2.3.10. *Let (f_n) be a monotone sequence in $\mathcal{R}_{\mathcal{F}}$, converging pointwise to a real-valued function f (not necessarily in $\mathcal{B}_b$). Then $f \in \tilde{\mathcal{R}}_{\mathcal{F}}$.*

Proof. For probability measures $P, Q \in \mathbb{P}_b$ with $P \leq_{\mathcal{F}} Q$ and $\langle |f|, P \rangle, \langle |f|, Q \rangle$ finite we have to show that $\langle f, P \rangle \leq \langle f, Q \rangle$ holds. If $f_n \in \mathcal{R}_{\mathcal{F}}$ then $\langle f_n, P \rangle$ and $\langle f_n, Q \rangle$ are finite and $\langle f_n, P \rangle \leq \langle f_n, Q \rangle$. Hence the monotone convergence theorem implies $\langle f, P \rangle = \lim \langle f_n, P \rangle \leq \lim \langle f_n, Q \rangle = \langle f, Q \rangle$.   □

2.4   Properties of Stochastic Orders

Now we define some interesting properties of (integral) stochastic orders and show how they arise directly as consequences of corresponding properties of the generator $\mathcal{F}$. Most of these properties were originally defined by Stoyan (1983).

Sometimes it is more convenient to formulate the results for random variables instead of probability measures. Therefore we write $X \leq_{\mathcal{F}} Y$ if $P_X \leq_{\mathcal{F}} P_Y$ holds for the corresponding distributions. For some of the properties defined below we need an order structure on S, for some others we need a topology or a vector space structure. Therefore we will assume from now on that S is a Polish vector space with a closed order. Recall that a topological space is called Polish if it admits a separable and complete metric. An order $\preceq$ on S is said to be closed, if the set $\{(x, y) \in S \times S : x \preceq y\}$ is a closed subset of the product space $S \times S$, or in other words, if $x_n \preceq y_n$ for all n and $x_n \to x$, $y_n \to y$ implies $x \preceq y$. This assumption includes all practically relevant cases, including the function spaces C and D considered in Billingsley (1999).

Moreover, from now on we tacitly assume that all occurring distributions belong to $\mathbb{P}_b$.

Definition 2.4.1. Let S be some Polish vector space with a closed order and let $b : S \to [1, \infty)$ be any weight function. Let $\leq_{\mathcal{F}}$ be some stochastic order on $\mathbb{P}_b$. Then $\leq_{\mathcal{F}}$ has

- Property **(R)** if $a \leq b$ implies $\delta_a \leq_{\mathcal{F}} \delta_b$;
- Property **(E)** if $X \leq_{\mathcal{F}} Y$ implies $EX \leq EY$ (assuming the expectations exist);
- Property **(M)** if $X \leq_{\mathcal{F}} Y$ implies $\alpha X \leq_{\mathcal{F}} \alpha Y$ for all scalars $\alpha \geq 0$;
- Property **(T)** if $X \leq_{\mathcal{F}} X + a$ holds for all r.v. X and all positive a;
- Property **(C)** if $P_1 \leq_{\mathcal{F}} P_2$ implies $P_1 * Q \leq_{\mathcal{F}} P_2 * Q$ for all probability measures Q;

- Property **(MI)** if $P_\theta \leq_{\mathcal{F}} Q_\theta$ for all $\theta \in \Theta$ implies $P \leq_{\mathcal{F}} Q$, where $P = \int P_\theta\ \mu(d\theta)$ and $Q = \int Q_\theta\ \mu(d\theta)$ for some measure μ on Θ, i.e., P and Q are the μ-mixtures of the families (P_θ) and (Q_θ), respectively.
- Property **(W)** if $\leq_{\mathcal{F}}$ is closed with respect to weak convergence, i.e., if $P_n \leq_{\mathcal{F}} Q_n$ holds for all $n \in \mathbb{N}$ and the sequences (P_n) and (Q_n) converge weakly to P and Q, respectively, then $P \leq_{\mathcal{F}} Q$.

Property (R) means that the stochastic order is consistent with the order structure on S, i.e. if we restrict the stochastic order to the one-point measures then there should be an order isomorphism between this space and the ordered space S.

Most of the other properties go also under the name closure properties. For example, a stochastic order with property (E) could be called closed with respect to the formation of expectations. If (C) holds then we will sometimes say that the order is *closed under convolution*, (MI) is also called *closed under mixture* and for (W) the expression *closed under weak convergence* is also common.

It turns out that for integral stochastic orders all these properties can be traced back to properties of the generator $\mathcal{F}$ or the maximal generator $\mathcal{R}_{\mathcal{F}}$, respectively. Part e) of the following theorem can be found in similar form in Stoyan (1983), prop. 1.1.2, and the if-part of g) is contained in proposition 3.13 of Marshall (1991). As usual the set of all bounded continuous functions will be denoted by $\mathcal{C}_b$.

Theorem 2.4.2. *a) Property (R) holds if and only if all functions in $\mathcal{F}$ are increasing.*
b) Property (E) holds if and only if $\tilde{\mathcal{R}}_{\mathcal{F}}$ contains all increasing linear functions.
c) Property (M) holds if and only if $\mathcal{R}_{\mathcal{F}}$ is scale invariant, i.e. $f \in \mathcal{R}_{\mathcal{F}}$ and
  $\alpha > 0$ implies $f_\alpha \in \mathcal{R}_{\mathcal{F}}$, where $f_\alpha(x) = f(\alpha x)$.
d) Property (T) holds if and only if all functions in $\mathcal{F}$ are increasing.
e) Property (C) holds if and only if $\mathcal{R}_{\mathcal{F}}$ is invariant under translations.
f) Property (MI) always holds!
g) Property (W) holds if and only if there is a generator $\mathcal{G} \subset \mathcal{C}_b$ for $\leq_{\mathcal{F}}$.

Proof. Parts a) - f) and the if-part of g) are easy to check. Therefore we only prove the only-if-part of g). Suppose $\leq_{\mathcal{F}}$ has property (W) and there is no generator $\mathcal{G} \subset \mathcal{C}_b$. Then define $\mathcal{F}_1 = \mathcal{R}_{\mathcal{F}} \cap \mathcal{C}_b$,

$$\mathbb{F} = \{\mu \in \mathbb{M}_b^N : \mu \in \mathcal{F}^+\} \quad \text{and} \quad \mathbb{F}_1 = \{\mu \in \mathbb{M}_b^N : \mu \in \mathcal{F}_1^+\}.$$

We endow $\mathbb{M}_b^N$ with the weak topology $\sigma(\mathbb{M}_b^N, \mathcal{C}_b)$. Then, by assumption, $\mathbb{F}$ is a closed convex cone and a proper subset of $\mathbb{F}_1$. Hence, by a separation theorem for closed convex sets (Choquet (1969), Theorem 21.12), for $\mu \in \mathbb{F}_1 \setminus \mathbb{F}$ there is a continuous linear functional L on $\mathbb{M}_b^N$ with

$$L(\mu) < \inf_{\nu \in \mathcal{F}} L(\nu) = \alpha \tag{2.4.1}$$

for some real α. Since $\mathbb{F}$ is a convex cone, $\alpha = 0$. According to proposition 22.4 in Choquet (1969) there is a representation of L as a function $f \in \mathcal{C}_b$ with $L(\mu) = \langle f, \mu \rangle$. Thus (2.4.1) implies $\inf_{\nu \in \mathcal{F}} \langle f, \nu \rangle = 0$ and $\langle f, \mu \rangle < 0$. Hence $f \in \mathbb{F}^+ = \mathcal{R}_{\mathcal{F}}$ and $f \notin \mathcal{F}_1$, a contradiction. $\qquad \square$

As a consequence of Theorem 2.4.2 it turns out that for integral stochastic orders property (R) holds if and only if property (T) holds. This is not true for other stochastic orders which do not have a characterization by integrals. For the likelihood ratio order $\leq_{lr}$, for example, property (R) holds, but (T) doesn't. Indeed, it is well known that $X \leq_{lr} X + a$ holds for all $a > 0$ only if X has a PF$_2$ density (see Shaked and Shanthikumar (1994), theorem 1.C.22).

The situation is similar in the case of property (MI). This property does not hold for the likelihood ratio order either, since e.g. $\delta_0 \leq_{lr} \delta_1$ and $\delta_2 \leq_{lr} \delta_3$, but

$$\frac{1}{2}(\delta_0 + \delta_2) \not\leq_{lr} \frac{1}{2}(\delta_1 + \delta_3).$$

2.5 Small Generators

In contrast to the maximal generator there is in general no unique minimal element in the set of all generators of a given order $\leq_{\mathcal{F}}$ (see Example 2.5.2). Nevertheless there often exist 'small' generators, which give rise to the same order (and are better suited for checking $P \leq_{\mathcal{F}} Q$). In the next theorem we use the following notation:

A subset B of a cone K in an arbitrary vector space is called a *base*, if for every $x \in K$ with $x \neq 0$ there is a unique $y \in B$ and a unique $\alpha > 0$ with $x = \alpha y$. Moreover, a point $x \in B$ is called an *extreme point* of B, if $x = \alpha y + (1 - \alpha)z$ for some $\alpha \in (0, 1)$ and $y, z \in B$ is only possible for $x = y = z$.

Theorem 2.5.1.
a) If $\mathcal{F}$ is a convex cone, then the following subsets of $\mathcal{F}$ also generate $\leq_{\mathcal{F}}$:

(i) *each base of $\mathcal{F}$;*

(ii) *the set of extreme points of a convex base, if this is a compact subset of the normed space $(\mathcal{B}_b, \|\cdot\|_b)$.*

b) Each set which is dense in $\mathcal{F}$ with respect to uniform convergence is a generator of $\leq_{\mathcal{F}}$.

Proof. a) (i) If $\mathcal{G}$ is a base of $\mathcal{F}$, then Theorem 2.3.5 implies that $\mathcal{R}_{\mathcal{G}}$ contains the convex cone which is spanned by $\mathcal{G}$. Thus $\mathcal{G} \subset \mathcal{F} \subset \mathcal{R}_{\mathcal{G}}$. Lemma 2.3.4 c) yields that $\leq_{\mathcal{F}}$ and $\leq_{\mathcal{G}}$ are identical.
(ii) Let $\mathcal{G}$ be the convex compact base of the convex cone $\mathcal{F}$, and let $\mathcal{W}$ be the set of extreme points of $\mathcal{G}$. Then by Lemma 2.3.4 c) it is sufficient to show that $\mathcal{F} \subset \mathcal{R}_{\mathcal{W}}$. But Theorem 2.3.5 implies that $\mathcal{R}_{\mathcal{W}}$ is convex and $\|\cdot\|_b$-closed.

Hence from the Krein–Milman theorem we can deduce $\mathcal{G} \subset \mathcal{R}_W$. As $\mathcal{F}$ is the convex cone spanned by $\mathcal{G}$, the assertion follows from Theorem 2.3.7.
b) follows like a) (i) from 2.3.4 c) and 2.3.5 b).   □

The following examples demonstrate what the results of this section mean for the important examples of the stochastic orders $\leq_{st}$, $\leq_{icx}$ and $\leq_{Lt}$ on the real line. Multivariate examples will be considered in detail in Chapter 3.

Example 2.5.2. (Generators and properties of stochastic order $\leq_{st}$.) As a first example consider the usual stochastic order $\leq_{st}$ for distributions on the real line. Here $S = \mathbb{R}$, of course endowed with the Borel σ-algebra. As a suitable weight function the trivial function $b \equiv 1$ can be chosen so that $\mathbb{P}_b$ contains all probability measures, and $\mathcal{B}_b$ is the set of all bounded functions. It follows from Theorem 1.2.8 that the set of all increasing functions is the extended maximal generator $\tilde{\mathcal{R}}_{\mathcal{F}}$, and hence the set of all bounded increasing functions is a generator of $\leq_{st}$. Since this set is a convex cone containing the constant functions and closed under pointwise convergence, we deduce from Corollary 2.3.9 that it is indeed the maximal generator. Hence it follows from Theorem 2.4.2 that $\leq_{st}$ has all the properties (R), (E), (M), (T), (C), (MI), and (W).

The results of this section can also be used to obtain small generators of $\leq_{st}$. A base of the maximal generator is obtained by fixing the limits

$$\lim_{x \to -\infty} f(x) \quad \text{and} \quad \lim_{x \to \infty} f(x).$$

We will fix these limits to 0 and 1, respectively. Thus it follows from Theorem 2.5.1 a) (i) that a generator of $\leq_{st}$ is already given by the set of all increasing functions f with

$$\lim_{x \to -\infty} f(x) = 0 \quad \text{and} \quad \lim_{x \to \infty} f(x) = 1. \tag{2.5.1}$$

Any such function can be approximated uniformly by right-continuous step functions with finitely many jumps fulfilling (2.5.1). Hence it follows from Theorem 2.5.1 b) that this set again is a smaller generator. Since these functions are convex combinations of indicator functions, it finally follows that a really small generator is given by the set of all indicator functions $\mathbf{1}_{[t,\infty)}$ with $t \in \mathbb{R}$. Indeed, these functions are the extreme points in the set of right-continuous increasing functions fulfilling (2.5.1). Alternatively, the same result could be also derived from Theorem 2.3.6. This is due to the fact that any right-continuous increasing function g with the property (2.5.1) is a distribution function and hence the identity

$$g(s) = \int \mathbf{1}_{[\omega,\infty)}(s) g(d\omega)$$

holds. Consequently, we can apply Theorem 2.3.6 with $f(\omega, s) = \mathbf{1}_{[\omega,\infty)}(s)$, $\rho = g$ and $c \equiv 1$ to show that the set of all increasing right-continuous functions generates the same stochastic order as the indicator functions $\mathbf{1}_{[t,\infty)}$ with $t \in \mathbb{R}$. Notice, however, that in contrast to the maximal generator there is no *minimal* generator. Take any dense subset $A \subset \mathbb{R}$. Then the indicators $\mathbf{1}_{[t,\infty)}$ with $t \in A$ are still generators of $\leq_{st}$ since an indicator $\mathbf{1}_{[t,\infty)}$ with $t \notin A$ is a monotone limit of a sequence of functions $\mathbf{1}_{[t_n,\infty)}$ with $t_n \in A$. It is well known, however, that there is no minimal dense subset of $\mathbb{R}$.

Example 2.5.3. (Generators and properties of convex order $\leq_{icx}$.) Now we consider the increasing convex order $\leq_{icx}$ for distributions on the real line. In Definition 1.5.1 this stochastic order was introduced to be the one which is generated by the set $\mathcal{F}_{icx}$ of all increasing convex functions. Indeed, it is easy to see that $\mathcal{F}_{icx}$ is the extended maximal generator of $\leq_{icx}$. However, here we need a non-trivial weight function because there are no bounded convex functions on the real line except the constant ones. Therefore we have to introduce an unbounded weight function and thus there will be a restriction in the set of probability measures that can be considered. This, however, is not surprising. In Definition 1.5.1 we have already restricted the consideration to random variables with finite mean. This will now be reflected by using the weight function $b(s) = 1 + |s|$. Then $\mathbb{P}_b$ consists exactly of all probability measures with a finite mean. The maximal generator $\mathcal{R}_{\mathcal{F}}$ then consists of all increasing convex functions which are Lipschitz continuous. Thus it follows immediately from Theorem 2.4.2 that $\leq_{icx}$ has the properties (R), (E), (M), (T), (C), and (MI), but (W) does not hold, as we have already seen in Example 1.5.8 and Theorem 1.5.9.

The maximal generator in particular contains all functions of the form $f(x) = (x - t)_+$ for some real t. It has been shown in Theorem 1.5.7 that the set of these functions is a generator of $\leq_{icx}$, too. Already in the proof given there we used the facts that convex combinations of these functions are piecewise linear convex functions and that any convex function can be approximated monotonically by such functions. Here again an alternative derivation is available from Theorem 2.3.6 since for any increasing convex function g the identity

$$g(s) = \int (\omega - s)_+ h(\mathrm{d}\omega)$$

holds with h the right derivative of g. Notice that the right derivative of a convex function is increasing and right-continuous and hence defines a measure. Moreover, it is easy to see that these so called *wedge functions* $f(x) = (x - t)_+$ are the extreme points of the set of all increasing convex functions.

There is some sort of a duality between the generators of $\leq_{st}$ and $\leq_{icx}$ mentioned in Examples 2.5.2 and 2.5.3. If $\mathcal{F}$ is any generator of $\leq_{st}$ then we obtain a generator of $\leq_{icx}$ by considering the functions

$$s \mapsto \int_{-\infty}^{s} f(x)\mathrm{d}x$$

for all $f \in \mathcal{F}$. This is the reason why the arguments in the two examples are so similar.

Example 2.5.4. (Generators and properties of Laplace transform order $\leq_{Lt}$.) In Definition 1.6.4 the Laplace transform order $\leq_{Lt}$ has been introduced. It is generated by the functions $\phi_t(x) = -e^{-tx}$ with $t > 0$. In Theorem 1.6.6 it has been stated that its extended maximal generator is given by the set $\mathcal{F}_{\infty}$ of all real functions with a completely monotone derivative, i.e.

$$\mathcal{F}_{\infty} = \{f : (-1)^{n+1} f^{(n)} \geq 0 \text{ for all } n = 1, 2, \dots\}. \qquad (2.5.2)$$

Reuter and Riedrich (1981) have shown this for distributions on $(0, \infty)$, but the result remains true for distributions on the whole real line. To show this, we use a famous theorem of Bernstein which says that a function g is completely monotone if and only if it is the Laplace transform of some measure μ on $[0, \infty)$, i.e. if and only if

$$g(x) = \int_0^{\infty} e^{-sx} \mu(\mathrm{d}s).$$

A proof of this general version of Bernstein's theorem for arbitrary functions g on the real line (and not only for functions on $(0, \infty)$!) can be found in Choquet (1969), theorem 32.6.

Now let us show that $\mathcal{F}_{\infty}$ is the extended maximal generator of Laplace transform order. First observe that $\mathcal{F}_{\infty}$ is a convex cone containing the constant functions and closed with respect to pointwise convergence. The latter property follows from the fact that the derivatives in (2.5.2) can be replaced by the corresponding differences. It is also easy to see that $\phi_t \in \mathcal{F}_{\infty}$ for all $t > 0$. Therefore consider now an arbitrary function $f \in \mathcal{F}_{\infty}$. We have to show that $X \leq_{Lt} Y$ implies $Ef(X) \leq Ef(Y)$. According to Bernstein's theorem

$$f'(x) = \int_0^{\infty} e^{-sx} \mu(\mathrm{d}s) = \mu(\{0\}) + \int 1_{(0,\infty)}(s) e^{-sx} \mu(\mathrm{d}s)$$

for some measure μ on $[0, \infty)$. Hence

$$f(x) = \mu(\{0\}) \cdot x + \int 1_{(0,\infty)}(s) \frac{-e^{-sx}}{s} \mu(\mathrm{d}s) + c$$

for some real constant c. Thus the assertion follows from Theorem 2.3.6 and Corollary 2.3.9, if we can show that the maximal generator contains the identity function. But the function $f(x) = x$ is the monotone limit of the sequence (f_n) given by $f_n(x) = n(1 - e^{-x/n})$, and thus the proof is finished.

It follows immediately that $\leq_{Lt}$ has the properties (R), (E), (M), (T), (C), and (MI). Notice, however, that property (W) is lost by the extension to arbitrary real random variables! Consider the following counterexample. Assume that F_n is a two-point distribution with mass $(n - 1)/n$ in 0 and mass $1/n$ in the point $-n$. Then $F_n \leq_{Lt} \delta_{-1}$, but for $n \to \infty$ the sequence (F_n) converges weakly to δ_0 and $\delta_0 \not\leq_{Lt} \delta_{-1}$! On the other hand, it follows immediately from Theorem 2.4.2 that $\leq_{Lt}$ is closed with respect to weak convergence, if it is restricted to non-negative random variables, since the functions $x \mapsto e^{-tx}$ are obviously bounded and continuous on $[0, \infty)$ for all $t > 0$.

The small generators obtained from Theorem 2.5.1 often have the disadvantage that they contain functions which are not differentiable. There are many applications, however, where the verification of a result about stochastic order can be simplified considerably by assuming that there is a generator consisting of functions that are sufficiently often differentiable. In particular, partial integration is often helpful, as we will see in Section 3.13.

It will be shown below that most of the integral stochastic orders considered in the literature have a generator consisting of infinitely differentiable functions. In the following S will always be some Euclidean space $S = \mathbb{R}^d$ endowed with the Borel σ-algebra, and as usual $\mathcal{C}^\infty$ denotes the set of all functions on S that are infinitely differentiable. We want to show that under weak assumptions $\mathcal{R}_{\mathcal{F}} \cap \mathcal{C}^\infty$ is a generator of $\mathcal{F}$. Typically this can be achieved by a two-step procedure as follows: First show that the generator can be restricted to continuous functions. Then find for an arbitrary continuous $f \in \mathcal{R}_{\mathcal{F}}$ a sequence (f_n) of infinitely differentiable functions $f_n \in \mathcal{R}_{\mathcal{F}}$ such that

$$\lim_{n \to \infty} \int f_n \mathrm{d}P = \int f \mathrm{d}P \qquad (2.5.3)$$

for all $P \in \mathbb{P}_b$. In order to obtain the f_n's, we resort on the following well-known construction. Let g be an infinitely differentiable probability density with compact support, and let $g_n(\mathbf{x}) = 1/n^d \cdot g(1/d \cdot \mathbf{x})$. Then it is well known that the sequence

$$f_n(\mathbf{x}) = f * g_n(\mathbf{x}) = \int f(\mathbf{x} - \mathbf{y}) g_n(\mathbf{y}) \mathrm{d}\mathbf{y} \qquad (2.5.4)$$

satisfies (2.5.3); see, for example, Rudin (1987), theorem 9.10. Thus the only problem is to show that $f_n \in \mathcal{R}_{\mathcal{F}}$. This question, however, is related to the question whether or not $\leq_{\mathcal{F}}$ is closed under convolution and weak convergence. In fact, we have the following general result.

Theorem 2.5.5. *Let $\leq_{\mathcal{F}}$ be an integral stochastic order for distributions on some Euclidean space with the property (C) and having a generator consisting of continuous functions. Then there is a generator $\mathcal{G}$ of $\leq_{\mathcal{F}}$ consisting only of infinitely differentiable functions. Especially, $\mathcal{G} = \mathcal{R}_{\mathcal{F}} \cap \mathcal{C}^{\infty}$ is such a generator.*

Proof. As a consequence of Lemma 2.3.4 we can assume without loss of generality that $\mathcal{F}$ consists of all continuous functions in $\mathcal{R}_{\mathcal{F}}$. It follows from Theorem 2.4.2 e) that (C) holds if and only if $\mathcal{F}$ is invariant under translations, and therefore $f(\cdot - \mathbf{y}) \in \mathcal{F}$ for all $\mathbf{y} \in \mathbb{R}^d$. Thus Theorem 2.3.6 ensures that the function f_n defined in (2.5.4) is in $\mathcal{R}_{\mathcal{F}}$. Now set $\mathcal{G} = \mathcal{R}_{\mathcal{F}} \cap \mathcal{C}^{\infty}$. Since $\mathcal{G} \subset \mathcal{R}_{\mathcal{F}}$ it is obvious that $P \leq_{\mathcal{F}} Q$ implies $P \leq_{\mathcal{G}} Q$. On the other hand, we have just shown that $f_n \in \mathcal{G}$ for all $f \in \mathcal{F}$. Hence $P \leq_{\mathcal{G}} Q$ implies $\int f_n \mathrm{d}P \leq \int f_n \mathrm{d}Q$ and by equation (2.5.3) $\int f \mathrm{d}P \leq \int f \mathrm{d}Q$ for all $f \in \mathcal{F}$. Therefore we also have that $P \leq_{\mathcal{G}} Q$ implies $P \leq_{\mathcal{F}} Q$, i.e., $\leq_{\mathcal{F}}$ and $\leq_{\mathcal{G}}$ are identical. $\qquad\square$

It has been shown in Theorem 2.4.2 g) that an integral stochastic order is closed under weak convergence if and only if it has a generator consisting of bounded continuous functions. Hence we can derive the following corollary from Theorem 2.5.5.

Corollary 2.5.6. *If $\leq_{\mathcal{F}}$ has the properties (C) and (W) then it has a generator $\mathcal{G} \subset \mathcal{C}^{\infty}$.*

Corollary 2.5.6 is particularly appealing since many stochastic orders possess these two closure properties.

2.6 Strassen Type Theorems

This section is devoted to a famous result called *Strassen's theorem*, since a very general version of it has been shown in Strassen (1965). It deals with the existence of special transition kernels that characterize integral stochastic order relations. The result as given below in Theorem 2.6.1 is equivalent to theorem T53 in Meyer (1966) formulated in our notation for the setting of integral stochastic orders. For a transition kernel Q from S to S and two probability measures P' and P'' on S we will write $P'' = P' \times Q$, if $P''(A) = \int Q(x, A) P'(\mathrm{d}x)$ for all $A \in \mathcal{S}$.

Theorem 2.6.1. *Assume that S is a Polish space and that the maximal generator $\mathcal{R}_{\mathcal{F}}$ is closed with respect to maximization, i.e. $f, g \in \mathcal{R}_{\mathcal{F}}$ implies $\max\{f, g\} \in \mathcal{R}_{\mathcal{F}}$. Then the following statements are equivalent:*

(i) $P' \leq_{\mathcal{F}} P''$,

(ii) *there is a transition kernel Q from S to S such that $P'' = P' \times Q$ and for all $x \in S$ the probability measure $Q(x, \cdot)$ fulfills*

$$\int f(s) Q(x, \mathrm{d}s) \geq f(x) \quad \text{for all } f \in \mathcal{R}_{\mathcal{F}}.$$

We will not give a proof of this result here. The interested reader is referred to Meyer (1966) and Strassen (1965). We mention, however, that the proof relies on the Hahn–Banach theorem and on the Riesz representation theorem, and thus relies on the axiom of choice. Therefore in this general version there is no hope of finding a direct construction of the transition kernel. Constructive proofs exist only in some special cases. For distributions on the real line we have given constructive proofs for the case of $\leq_{st}$ in Theorem 1.2.4 and for $\leq_{cx}$ in Theorem 1.5.20. Already in the case of finite-dimensional Euclidean spaces, however, constructive proofs are not known in general. Only in the case of distributions with finite support such proofs are known. For $\leq_{st}$ this is due to Hansel and Troallic (1978), who show that the Ford–Fulkerson algorithm can be used to find the transition kernel. In the case of $\leq_{cx}$ Elton and Hill (1998) have given a proof for distributions in $\mathbb{R}^d$ with finite support, using geometric ideas.

Theorem 2.6.1 can also be applied if the generator is closed with respect to minimization. One can consider the order relation $\leq'_{\mathcal{F}}$ defined by $P' \leq'_{\mathcal{F}} P''$, if and only if $P'' \leq_{\mathcal{F}} P'$. Then $\leq'_{\mathcal{F}}$ is generated by the functions $-f$ with $f \in \mathcal{R}_{\mathcal{F}}$, and this generator is closed with respect to maximization if $\mathcal{R}_{\mathcal{F}}$ is closed with respect to minimization.

The two most important cases where Theorem 2.6.1 can be applied is when $\mathcal{R}_{\mathcal{F}}$ consists of all increasing or all convex functions, respectively.

Suppose that S and S' are Polish spaces with closed partial orders $\preceq$ and $\preceq'$. A function $f : S \to S'$ is said to be *increasing* if $x \preceq y$ implies $f(x) \preceq' f(y)$. Typically S is (a subset of) a product space $\mathbb{R}^A$ for some index set A endowed with the product order, i.e. for $\mathbf{x} = (x_a)_{a \in A}$ and $\mathbf{y} = (y_a)_{a \in A}$ it holds $\mathbf{x} \preceq \mathbf{y}$ if and only if $x_a \leq y_a$ for all $a \in A$. In this case the product order is closed when A is countable as, for example, in the case of Euclidean spaces $\mathbb{R}^n$ or the space $\mathbb{R}^{\mathbb{N}}$ of all real sequences. Examples with uncountable A are subspaces of functions $f : A \to \mathbb{R}$ for $A = [0,1]$ or $A = [0, \infty)$. Here the product order is closed only if a restriction is made to functions which are determined by the values on a countable subset of A. Important examples are the spaces $C(A)$ and $D(A)$ of continuous function respectively cadlag functions on A; see Billingsley (1999).

Definition 2.6.2. Let $(S, \mathcal{S})$ be a Polish space with a closed partial order, endowed with the Borel σ-algebra. For probability measures on $(S, \mathcal{S})$ denote by $\leq_{st}$ the integral stochastic order generated by the set $\mathcal{F}_{st}$ of all bounded and measurable increasing functions $f : S \to \mathbb{R}$.

For $S = \mathbb{R}$ this is the usual stochastic order considered in Chapter 1. In full generality this order has been considered in Kamae, Krengel, and O'Brien (1977) and Kamae and Krengel (1978). Notice that $P' \leq_{st} P''$ if and only if

$$\int f' \mathrm{d}P' \leq \int f'' \mathrm{d}P'' \qquad (2.6.1)$$

holds for all measurable functions $f', f'' : S \to \mathbb{R}$ such that $f'(x) \leq f''(y)$ whenever $x \preceq y$. This can be seen as follows. Define $f(x) = \inf_{y \succeq x} f''(y)$. Then $f' \leq f \leq f''$ and f is increasing. Hence

$$\int f' \mathrm{d}P' \leq \int f \mathrm{d}P' \leq \int f \mathrm{d}P'' \leq \int f'' \mathrm{d}P''.$$

On the other hand, it is clear that (2.6.1) implies $P' \leq_{st} P''$, since for any increasing f the functions $f' = f'' = f$ fulfill $f'(x) \leq f''(y)$ whenever $x \preceq y$. If (2.6.1) is true some authors speak of P'' being a *strong dilatation* of P' (see Kemperman (1977)).

The following theorem is taken from Kamae et al. (1977). For distributions on the real line a simple proof has been given in Theorem 1.2.4.

Theorem 2.6.3. *Assume that S is a Polish space with a closed partial order. Then the following statements are equivalent:*

(i) $P' \leq_{st} P''$,

(ii) *there are random variables $X' \sim P'$ and $X'' \sim P''$ such that $X' \leq X''$ holds almost surely.*

Proof. Corollary 2.3.9 yields that $\mathcal{F}_{st}$ is the maximal generator of $\leq_{st}$, and it is clear that it fulfills the assumptions of Theorem 2.6.1. Next observe that the condition $\int f(s) Q(x, \mathrm{d}s) \geq f(x)$ holds for all $f \in \mathcal{R}_{st}$ if and only if $Q(x, \{y : y \geq x\}) = 1$. To see the necessity consider the indicator function $y \mapsto \mathbf{1}_{[y \geq x]}$, while the sufficiency is obvious. Hence it follows from Theorem 2.6.1 that $P'' = P' \times Q$ for some transition kernel Q with $Q(x, \{y : y \geq x\}) = 1$. In terms of random variables this means that there are random variables $X' \sim P'$ and $X'' \sim P''$ such that $X' \leq X''$ holds almost surely. The implication from (ii) to (i) is trivial. $\qquad\square$

Other important characterizations of this order are collected in the next result. For a proof we refer to Kamae et al. (1977). Extensions to more general settings can be found in Kellerer (1984), Skala (1993) and Ramachandran and Rüschendorf (1995). In Theorem 3.3.4 we will give a proof for the case $S = \mathbb{R}^n$.

Recall that a subset $U \subset S$ is called an *upper set* (or *increasing* set), if its indicator function $\mathbf{1}_U$ is increasing. Thus a set U is upper if and only if $\mathbf{x} \in U$ and $\mathbf{x} \leq \mathbf{y}$ imply $\mathbf{y} \in U$.

Theorem 2.6.4. *For probability measures P' and P'' on a partially ordered Polish space the following statements are equivalent:*

(i) $P' \leq_{st} P''$;

(ii) $\int f \mathrm{d}P' \leq \int f \mathrm{d}P''$ *for all bounded continuous increasing functions f;*

(iii) $P'(U) \leq P''(U)$ *for all upper sets U;*

(iv) $P(U) \leq P(U)$ *for all closed upper sets U.*

In the following the convex order on separable Banach spaces S will be considered. Recall that a function $f : S \to \mathbb{R}$ is said to be *convex* if

$$f(\alpha x + (1 - \alpha)y) \leq \alpha f(x) + (1 - \alpha)f(y)$$

for all $x, y \in S$ and all $0 < \alpha < 1$. These functions define an integral stochastic order as follows.

Definition 2.6.5. Let S be a separable Banach space and define the bounding function $b(s) = 1 + \|s\|$. The *convex order* $\leq_{cx}$ is given on $\mathbb{P}_b$ as the integral stochastic order generated by the set $\mathcal{F}_{cx}$ of all convex functions in $\mathcal{B}_b$.

Note that $\mathbb{P}_b$ is the set of all probability measures with a finite mean.

For the convex order on the real line a Strassen type theorem has already been given in Theorem 1.5.20. The next result is a more general version for separable Banach spaces, thus including the important examples of Euclidean spaces $\mathbb{R}^n$ as well as some function spaces like $C(0, 1)$. The assumption of a Banach space is needed for the conditional expectation to make sense, see, for example, Araujo and Giné (1980).

Theorem 2.6.6. *Assume that S is a separable Banach space with a closed partial order. Then the following statements are equivalent:*

(i) $P' \leq_{cx} P''$,

(ii) *there are random variables $X' \sim P'$ and $X'' \sim P''$ such that $X' = E[X''|X']$ almost surely.*

Proof. Again it is clear that $\mathcal{F}_{cx}$ is the maximal generator and that it fulfills the assumptions of Theorem 2.6.1. Hence there is a transition kernel Q such that $P'' = P' \times Q$ and

$$\int f(y)Q(x, \mathrm{d}y) \geq f(x) \quad \text{for all convex } f. \tag{2.6.2}$$

Therefore,

$$\int \ell(y)Q(x, \mathrm{d}y) = \ell(x) \quad \text{for all linear } \ell,$$

and hence there is a pair (X', X'') of random variables with $X' \sim P'$ and $X'' \sim P''$ such that $E[X''|X'] = X'$ almost surely; see Araujo and Giné (1980). The converse immediately follows from the general version of Jensen's inequality as it can be found, for example, in Perlman (1974). $\square$

Similarly the following result can be shown for increasing convex and increasing concave order.

Theorem 2.6.7. *Assume that S is a separable Banach space with a closed partial order. Then the following statements are equivalent:*

(i) $P' \leq_{icx} P''$,

(ii) *there are random variables $X' \sim P'$ and $X'' \sim P''$ such that*
$$X' \leq E[X''|X'] \text{ almost surely.}$$

Theorem 2.6.8. *Assume that S is a separable Banach space with a closed partial order. Then the following statements are equivalent:*

(i) $P' \leq_{icv} P''$,

(ii) *there are random variables $X' \sim P'$ and $X'' \sim P''$ such that*
$$X'' \geq E[X'|X''] \text{ almost surely.}$$

All these results can immediately be generalized to sequences (P_n) of probability measures. Composing the corresponding transition kernels yields (non-stationary) Markov chains with appropriate properties. Thus we get the following results. If (P_n) is a $\leq_{st}$-increasing sequence of probability measures, then there is a sequence (X_n) of random variables with marginal distributions P_n, such that $X_n \leq X_{n+1}$ a.s. for all n. Similarly, if (P_n) is a $\leq_{cx}$-increasing sequence, then there is a martingale (X_n) with marginals P_n, and for $\leq_{icx}$-increasing ($\leq_{icv}$-decreasing) sequences one finds a corresponding submartingale (supermartingale). In all cases the sequences (X_n) are Markovian.

Kellerer (1972) proved the following generalization to the case of a family of distributions (P_t) with real-valued parameter t.

Theorem 2.6.9. *Let T be a subset of the real line, and let $(P_t : t \in T)$ be a family of probability measures on the real line with finite mean such that*

$$s < t \quad \text{implies } P_s \leq_{icx} P_t.$$

Then there is a stochastic process $(X_t : t \in T)$ such that X_t has the distribution P_t for all t, and (X_t) is a Markovian submartingale.

3

Multivariate Stochastic Orders

3.1 Preliminaries

This chapter deals with orders of multivariate distributions and random vectors. A random vector $\mathbf{X} = (X_1, \ldots, X_n)$ is a collection of n univariate random variables on an arbitrary probability space $(\Omega, \mathcal{A}, P)$, which usually will not be mentioned explicitly. Its distribution thus is a probability measure on $(\mathbb{R}^n, \mathcal{B}^n)$, where $\mathbb{R}^n$ denotes the n-dimensional Euclidean space and $\mathcal{B}^n$ is its Borel σ-algebra. The usual componentwise partial order on $\mathbb{R}^n$ will be denoted by $\leq$, i.e. for $\mathbf{x} = (x_1, \ldots, x_n)$ and $\mathbf{y} = (y_1, \ldots, y_n)$ it holds $\mathbf{x} \leq \mathbf{y}$ if $x_i \leq y_i$ for all $i = 1, \ldots, n$. We will write $\mathbf{x} < \mathbf{y}$ if $x_i < y_i$ for all $i = 1, \ldots, n$. A function $f : \mathbb{R}^n \to \mathbb{R}^k$ is said to be *increasing*, if $\mathbf{x} \leq \mathbf{y}$ implies $f(\mathbf{x}) \leq f(\mathbf{y})$, meaning that we use the notion 'increasing' in the weak sense.

The distribution of a random vector $\mathbf{X} = (X_1, \ldots, X_n)$ can be characterized by its *distribution function*

$$F_{\mathbf{X}}(\mathbf{x}) = P(\mathbf{X} \leq \mathbf{x}) = P(X_1 \leq x_1, \ldots, X_n \leq x_n)$$

as well as its (multivariate) *survival function*

$$\bar{F}_{\mathbf{X}}(\mathbf{x}) = P(\mathbf{X} > \mathbf{x}) = P(X_1 > x_1, \ldots, X_n > x_n).$$

The distribution functions $F_i(x) = P(X_i \leq x)$ for $i = 1, \ldots, n$ are called the *marginals*. Notice that in contrast to the univariate case the identity $\bar{F}_{\mathbf{X}}(\mathbf{x}) = 1 - F_{\mathbf{X}}(\mathbf{x})$ is not true in general. In the bivariate case the relation

$$\bar{F}_{\mathbf{X}}(x_1, x_2) = 1 - F_1(x_1) - F_2(x_2) + F_{\mathbf{X}}(x_1, x_2) \tag{3.1.1}$$

holds for all x_1, x_2. For $n > 2$, however, there is no simple relation between the distribution function and the survival function.

In general, there are many multivariate distributions with the same marginals, which exhibit different kinds of dependence between the components. The dependence structure of a random vector can be described by a so-called *copula*. A copula is the distribution function of a random vector

$U = (U_1, \dots, U_n)$ whose components are all uniformly distributed on $(0,1)$. Given a copula C and marginals $F_1, \dots, F_n$, it is easy to see that the function

$$F(\mathbf{x}) = C(F_1(x_1), \dots, F_n(x_n)) \qquad (3.1.2)$$

is a distribution function with marginals $F_1, \dots, F_n$. Moreover, if $\mathbf{U}$ is a random vector with distribution function C then

$$\mathbf{X} = (F_1^{-1}(U_1), \dots, F_n^{-1}(U_n))$$

has the distribution function F given in (3.1.2).

On the other hand, to every multivariate distribution function F with marginals $F_1, \dots, F_n$ there exists a copula such that (3.1.2) holds. This result is known as Sklar's theorem; see Sklar (1959). If F is continuous then this copula is unique; otherwise it is determined uniquely only on the range of F. An introduction to the theory of copulas can be found in Nelsen (1999), and a comprehensive treatment at a more advanced level is given in Joe (1997).

Typical examples of copulas are the *independent copula*

$$C(\mathbf{u}) = u_1 \cdots u_n$$

of the random vector $\mathbf{U} = (U_1, \dots, U_n)$ consisting of independent uniformly distributed components, and the *upper Fréchet bound copula*

$$C^+(\mathbf{u}) = \min\{u_1, \dots, u_n\}$$

of the random vector $\mathbf{U} = (U_1, \dots, U_1)$ consisting of identical uniformly distributed components, and thus exhibiting a strong form of dependence.

Denote by $\Gamma(F_1, \dots, F_n)$ the set of all distribution functions with given marginals $F_1, \dots, F_n$. This set is called the *Fréchet class* of $F_1, \dots, F_n$. For any distribution in a given Fréchet class there are sharp upper and lower bounds for the distribution function, which are called *Fréchet bounds*. They are given in the next result. A proof can be found in Joe (1997), theorem 3.1.

Theorem 3.1.1. *Let F be a distribution function with marginals $F_1, \dots, F_n$. Then*

$$F^-(\mathbf{x}) \leq F(\mathbf{x}) \leq F^+(\mathbf{x})$$

for all $\mathbf{x} \in \mathbb{R}^n$, where

$$F^-(\mathbf{x}) = \max\{0, F_1(x_1) + \dots + F_n(x_n) - (n-1)\}$$

and

$$F^+(\mathbf{x}) = \min\{F_1(x_1), \dots, F_n(x_n)\}.$$

F^+ is called the *upper Fréchet bound*. It is a distribution function with marginals $F_1, \ldots, F_n$. A random vector $\mathbf{X}$ with this distribution can be constructed as follows. Let U be a random variable uniformly distributed on $(0, 1)$, and define

$$\mathbf{X} = (F_1^{-1}(U), \ldots, F_n^{-1}(U)).$$

Then $F_{\mathbf{X}} = F^+$, since the random vector $\mathbf{U}$ related to the copula C^+ of F^+ has the form $\mathbf{U} = (U, \ldots, U)$. A random vector with distribution F^+ is sometimes called *comonotone*, since it has the property that if $X_i(\omega) < X_i(\omega')$, then $X_j(\omega) \le X_j(\omega')$, too, for all $j \ne i$. For the distribution this means that the support of $\mathbf{X}$ is contained in a completely ordered subset of $\mathbb{R}^n$. Thus the upper Fréchet bound exhibits very strong positive dependence.

F^- is called the *lower Fréchet bound*. In the bivariate case it is a distribution function, too. A random vector with distribution F^- is given by

$$(X_1, X_2) = (F_1^{-1}(U), F_2^{-1}(1 - U))$$

with U uniformly distributed on $(0, 1)$. This random vector is sometimes called *countermonotonic*. For dimension $n \ge 3$ the function F^- is not necessarily a distribution function; additional conditions on $F_1, \ldots, F_n$ are needed to guarantee this. They are stated in the following theorem which is due to Dall'Aglio (1972); for a proof see Joe (1997), theorem 3.7.

Theorem 3.1.2. *For dimension $n \ge 3$ the lower Fréchet bound F^- is a distribution function if and only if one of the following conditions holds:*

(i) $\sum_{j=1}^n F_j(x_j) \le 1$ *for all $\mathbf{x}$ with $F_j(x_j) < 1$ for all j;*

(ii) $\sum_{j=1}^n F_j(x_j) \ge n - 1$ *for all $\mathbf{x}$ with $F_j(x_j) > 0$ for all j.*

This result has the following interpretation: Either there are (case (i)) finite upper bounds ξ_j^+ for the supports of F_j with point masses in ξ_j^+ that add up to at least $n-1$; or there are (case (ii)) finite lower bounds ξ_j^- for the supports of F_j with this property. An important special case of the latter condition is that of non-negative random variables, where the lower bounds are $\xi_j^- = 0$ for all j. Then the condition is fulfilled if and only if with probability 1 at most one of the random variables assumes a strictly positive value. Such random vectors appear in the context of insurance risks, where random variables with this property are sometimes called *mutually exclusive risks*; see Dhaene and Denuit (1999).

Next we will describe a transformation that transforms a random vector $\mathbf{X} = (X_1, \ldots, X_n)$ into independent uniformly distributed random variables $U_1, \ldots, U_n$. It is sometimes called *Rosenblatt's transformation*, since it was first published in Rosenblatt (1952). Recall that F_1 is the marginal distribution of X_1. Denote by $F_{i+1|1,2,\ldots,i}(\cdot | x_1, x_2, \ldots, x_i)$ the conditional distribution function of X_{i+1} given $X_1 = x_1, X_2 = x_2, \ldots, X_i = x_i$. As before, the generalized inverse of F_1 will be called $F_1^{-1}(\cdot)$ and the generalized inverse of $F_{i+1|1,2,\ldots,i}(\cdot|x_1, x_2, \ldots, x_i)$ will be denoted by $F_{i+1|1,2,\ldots,i}^{-1}(\cdot|x_1, x_2, \ldots, x_i)$, $i = 1, 2, \ldots, n-1$. Consider the transformation $\Psi_F : \mathbb{R}^n \to (0,1)^n$ (which depends on F) defined by

$$\Psi_F(x_1, x_2, \ldots, x_n) = (F_1(x_1), F_{2|1}(x_2|x_1), \ldots, F_{n|1,2,\ldots,n-1}(x_n|x_1, \ldots, x_{n-1})),$$
$$(3.1.3)$$

for all $(x_1, x_2, \ldots, x_n)$ in the support of $(X_1, X_2, \ldots, X_n)$.

Lemma 3.1.3. *Let the distribution function F of the random vector $(X_1, X_2, \ldots, X_n)$ be absolutely continuous. Define*

$$(U_1, U_2, \ldots, U_n) = \Psi_F(X_1, X_2, \ldots, X_n). \qquad (3.1.4)$$

Then $U_1, U_2, \ldots, U_n$ are independent random variables, uniformly distributed on $(0,1)$.

Note that the transformation defined in (3.1.4) is only one of many transformations which transform the random variables $X_1, X_2, \ldots, X_n$ into n independent uniformly on $(0,1)$ distributed random variables. For example, we can permute the indices $1, 2, \ldots, n$ to get other transformations. The assumption of absolute continuity can be relaxed; see Rüschendorf (1981b).

By 'inverting' Ψ_F we can express the X_i's as functions of the independent uniform random variables $U_1, U_2, \ldots, U_n$ (see, for example, Rüschendorf and Valk (1993)). Denote

$$x_1 = F^{-1}(u_1), \qquad (3.1.5)$$

and, by induction,

$$x_i = F_{i|1,2,\ldots,i-1}^{-1}(u_i|x_1, x_2, \ldots, x_{i-1}) \quad \text{for } i = 2, 3, \ldots, n. \qquad (3.1.6)$$

Consider the transformation $\Psi_F^* : (0,1)^n \to \mathbb{R}^n$ defined by

$$\Psi_F^*(u_1, u_2, \ldots, u_n) = (x_1, x_2, \ldots, x_n), \qquad (u_1, u_2, \ldots, u_n) \in (0,1)^n, \quad (3.1.7)$$

where the x_is are functions of the u_is as given in (3.1.5) and (3.1.6).

If

$$(\hat{X}_1, \hat{X}_2, \ldots, \hat{X}_n) \equiv \Psi_F^*(U_1, U_2, \ldots, U_n) \qquad (3.1.8)$$

then

$$(\hat{X}_1, \hat{X}_2, \ldots, \hat{X}_n) =_{st} (X_1, X_2, \ldots, X_n). \qquad (3.1.9)$$

Note that no continuity assumptions are needed for the validity of (3.1.9).

The construction described in (3.1.8) is called the *standard construction*; it is a well-known method of multivariate simulation, see Rubinstein and Melamed (1998).

3.2   Properties of Multivariate Stochastic Orders

In addition to the properties of stochastic orders introduced in Definition 2.4.1 there are further properties which only make sense for multivariate stochastic orders that are defined for arbitrary dimensions n. The most important ones will be introduced in the next definition, where the following notation is used. For a set $K = \{k_1, \dots, k_m\} \subset \{1, \dots, n\}$ with $k_1 < \dots < k_m$ and an arbitrary vector $\mathbf{x} = (x_1, \dots, x_n)$ let $\mathbf{x}_K = (x_{k_1}, \dots, x_{k_m})$ be the subvector of $\mathbf{x}$ containing the components in K.

Definition 3.2.1. A stochastic order $\preceq$ for random vectors on $\mathbb{R}^n$ is said to have

- property (MA) if $\mathbf{X} \preceq \mathbf{Y}$ implies $\mathbf{X}_K \preceq \mathbf{Y}_K$ for all $K \subset \{1, \dots, n\}$;
- property (ID) if $\mathbf{X} \preceq \mathbf{Y}$ implies $(\mathbf{X}_K, \mathbf{X}_L) \preceq (\mathbf{Y}_K, \mathbf{Y}_L)$ for all $K, L \subset \{1, \dots, n\}$;
- property (IN) if $\mathbf{X}_i \preceq \mathbf{Y}_i$ for $i = 1, 2$ implies $(\mathbf{X}_1, \mathbf{X}_2) \preceq (\mathbf{Y}_1, \mathbf{Y}_2)$ whenever $\mathbf{X}_1, \mathbf{X}_2$ and $\mathbf{Y}_1, \mathbf{Y}_2$ are independent random vectors, respectively.

A stochastic order with property (MA) is said to be *closed with respect to* **ma**rginalization. The symbol (ID) is an abbreviation for *closed with respect to* **id**entical concatenation and (IN) for *closed with respect to* **in**dependent concatenation. Notice that in the definition of (ID) it is allowed that $K = L = \{1, \dots, n\}$. By induction it follows that if (ID) holds then $\mathbf{X} \preceq \mathbf{Y}$ implies $(\mathbf{X}, \dots, \mathbf{X}) \preceq (\mathbf{Y}, \dots, \mathbf{Y})$. The proof of the following result is obvious.

Lemma 3.2.2. *For a stochastic order $\preceq$ with property (MA) the property (ID) holds if and only if $\mathbf{X} \preceq \mathbf{Y}$ implies $(\mathbf{X}, \mathbf{X}) \preceq (\mathbf{Y}, \mathbf{Y})$.*

In the next result sufficient conditions are given for these properties in the case of integral stochastic orders. Let $\preceq$ be an integral stochastic order, defined for random vectors of arbitrary dimension. Denote by $\mathcal{R}_{\mathcal{F}}^n$ the maximal generator for the order when considered as an order of random vectors of dimension n.

Theorem 3.2.3. *a) If for all $K \subset \{1, \dots, n\}$ and every $h \in \mathcal{R}_{\mathcal{F}}^{|K|}$ the function f given by $\mathbf{x} \mapsto f(\mathbf{x}) = h(\mathbf{x}_K)$ is in $\mathcal{R}_{\mathcal{F}}^n$, then (MA) holds.*
b) If for every $h \in \mathcal{R}_{\mathcal{F}}^{|K|+|L|}$ with $K, L \subset \{1, \dots, n\}$ the function $\mathbf{x} \mapsto f(\mathbf{x}) = h(\mathbf{x}_K, \mathbf{x}_L)$ is in $\mathcal{R}_{\mathcal{F}}^n$, then (ID) holds.
c) If $f(\mathbf{x}_1, \cdot) \in \mathcal{R}_{\mathcal{F}}^n$ and $f(\cdot, \mathbf{x}_2) \in \mathcal{R}_{\mathcal{F}}^m$ for all $f \in \mathcal{R}_{\mathcal{F}}^{m+n}$, all $\mathbf{x}_1 \in \mathbb{R}^m$ and all $\mathbf{x}_2 \in \mathbb{R}^n$, then (IN) holds.

Proof. Part a) and b) are obvious. For c) observe that if $f(\mathbf{x}_1, \cdot) \in \mathcal{R}_{\mathcal{F}}^n$ for all $\mathbf{x}_1$ then Theorem 2.3.6 yields

$$h(\cdot) = \int f(\mathbf{x}_1, \cdot) P_{\mathbf{X}_1}(\mathrm{d}\mathbf{x}_1) \in \mathcal{R}_{\mathcal{F}}^n.$$

Hence $\mathbf{X}_2 \preceq \mathbf{Y}_2$ implies

$$Ef(\mathbf{X}_1, \mathbf{X}_2) = Eh(\mathbf{X}_2) \leq Eh(\mathbf{Y}_2) = Ef(\mathbf{X}_1, \mathbf{Y}_2).$$

Similarly it can be shown that $Ef(\mathbf{X}_1, \mathbf{Y}_2) \leq Ef(\mathbf{Y}_1, \mathbf{Y}_2)$, and therefore

$$(\mathbf{X}_1, \mathbf{X}_2) \preceq (\mathbf{Y}_1, \mathbf{Y}_2)$$

holds.   □

3.3   Usual Stochastic Order and Orthant Orders

Recall that in the univariate case the following conditions for the usual stochastic order $\leq_{st}$ are equivalent:

- $Ef(X) \leq Ef(Y)$ for all bounded increasing f;
- $F_X(t) \geq F_Y(t)$ for all real t;
- $\bar{F}_X(t) \leq \bar{F}_Y(t)$ for all real t.

When the univariate random variables X and Y are replaced by random vectors $\mathbf{X}$ and $\mathbf{Y}$, these conditions are no longer equivalent. Therefore there are three different generalizations to the multivariate case.

Definition 3.3.1. Let $\mathbf{X}$ and $\mathbf{Y}$ be random vectors with values in $\mathbb{R}^n$. $\mathbf{X}$ and $\mathbf{Y}$ are said to be comparable with respect to
a) *usual stochastic order* (written $\mathbf{X} \leq_{st} \mathbf{Y}$) if $Ef(\mathbf{X}) \leq Ef(\mathbf{Y})$ for all bounded increasing $f : \mathbb{R}^n \to \mathbb{R}$;
b) *upper orthant order* (written $\mathbf{X} \leq_{uo} \mathbf{Y}$) if $\bar{F}_{\mathbf{X}}(\mathbf{t}) \leq \bar{F}_{\mathbf{Y}}(\mathbf{t})$ for all $\mathbf{t}$;
c) *lower orthant order* (written $\mathbf{X} \leq_{lo} \mathbf{Y}$) if $F_{\mathbf{X}}(\mathbf{t}) \geq F_{\mathbf{Y}}(\mathbf{t})$ for all $\mathbf{t}$.

The first occurrence of $\leq_{st}$ seems to be Lehmann (1955). Other important contributions were given by Kamae et al. (1977), Marshall and Olkin (1979) and Veinott (1965). Early references to the orthant orders are Veinott (1965), Cambanis, Simons, and Stout (1976), Bergmann (1978), Marshall and Olkin (1979), Rüschendorf (1980), Tchen (1980), and others.
    All three orders are integral stochastic orders. The upper orthant order is generated by the indicator functions $f(\mathbf{x}) = \mathbf{1}_{(\mathbf{t}, \infty)}(\mathbf{x})$, $\mathbf{t} \in \mathbb{R}^n$, and the lower orthant order is generated by the functions $f(\mathbf{x}) = -\mathbf{1}_{(-\infty, \mathbf{t}]}(\mathbf{x})$, $\mathbf{t} \in \mathbb{R}^n$. Since these functions are increasing, the following result holds.

Theorem 3.3.2. *If* $\mathbf{X} \leq_{st} \mathbf{Y}$ *then* $\mathbf{X} \leq_{uo} \mathbf{Y}$ *and* $\mathbf{X} \leq_{lo} \mathbf{Y}$.

Example 3.3.3. (Usual stochastic order is stronger than the orthant orders.) Let $n = 2$ and assume $P(\mathbf{X} = (1,0)) = P(\mathbf{X} = (0,1)) = 1/2$ and $P(\mathbf{Y} = (0,0)) = P(\mathbf{Y} = (1,1)) = 1/2$. Then $\mathbf{X} \leq_{uo} \mathbf{Y}$ and $\mathbf{X} \leq_{lo} \mathbf{Y}$, but $\mathbf{X} \not\leq_{st} \mathbf{Y}$.

As Example 3.3.3 shows, it is possible that $\mathbf{X} \leq_{uo} \mathbf{Y}$ and $\mathbf{X} \leq_{lo} \mathbf{Y}$ hold simultaneously in cases where $\mathbf{X} \leq_{st} \mathbf{Y}$ is not true. This could be a reason to define a further stochastic order. The simultaneous validity of $\mathbf{X} \geq_{lo} \mathbf{Y}$ and $\mathbf{X} \leq_{uo} \mathbf{Y}$ (which is only possible in the case of equal marginals of $\mathbf{X}$ and $\mathbf{Y}$) defines the concordance order; see Definition 3.8.5.

The set $\mathcal{F}_{st}$ of bounded increasing functions is a convex cone, closed under pointwise convergence and containing the constant functions. Hence it follows from Theorem 2.3.9 that $\mathcal{F}_{st}$ is the maximal generator of $\leq_{st}$. The extended maximal generator consists of all increasing functions. Some important smaller generators are given in the next theorem. Recall that a subset $U \subset \mathbb{R}^n$ is called an *upper set*, if the indicator function $\mathbf{1}_U$ is increasing. Thus a set is upper if and only if $\mathbf{x} \in U$ and $\mathbf{x} \leq \mathbf{y}$ imply $\mathbf{y} \in U$.

Theorem 3.3.4. *The following statements are equivalent:*

(i) $\mathbf{X} \leq_{st} \mathbf{Y}$*;*

(ii) $Ef(\mathbf{X}) \leq Ef(\mathbf{Y})$ *for all bounded continuous increasing functions* f*;*

(iii) $Ef(\mathbf{X}) \leq Ef(\mathbf{Y})$ *for all bounded differentiable increasing functions* f*;*

(iv) $P(\mathbf{X} \in U) \leq P(\mathbf{Y} \in U)$ *for all upper sets* U*;*

(v) $P(\mathbf{X} \in U) \leq P(\mathbf{Y} \in U)$ *for all closed upper sets* U*.*

Proof. The implication (i) $\Rightarrow$ (ii) is obvious. The equivalence of (ii) and (iii) is a consequence of Theorem 2.5.5. Thus it is sufficient to show (ii) $\Rightarrow$ (v) $\Rightarrow$ (iv) $\Rightarrow$ (i).

(ii) $\Rightarrow$ (v): Let U be a closed upper set, denote by d the usual Euclidean metric on $\mathbb{R}^n$, and as usual define $d(\mathbf{x}, U) = \inf_{\mathbf{u} \in U} d(\mathbf{x}, \mathbf{u})$. Then the function $f_n(\mathbf{x}) = (1 - nd(\mathbf{x}, U))_+$ is bounded, continuous and increasing, and the sequence (f_n) converges monotonically to $\mathbf{1}_U$. Thus (ii) implies $Ef_n(\mathbf{X}) \leq Ef_n(\mathbf{Y})$ for all n, and the monotone convergence theorem yields $E\mathbf{1}_U(\mathbf{X}) \leq E\mathbf{1}_U(\mathbf{Y})$; hence (v) holds.

(v) $\Rightarrow$ (iv): Let U be an arbitrary upper set and fix some $\varepsilon > 0$. Since any probability measure on $\mathbb{R}^n$ is tight, there is some compact $K_\varepsilon \subset U$ such that $P(\mathbf{X} \in U) - P(\mathbf{X} \in K_\varepsilon) < \varepsilon$ and $P(\mathbf{Y} \in U) - P(\mathbf{Y} \in K_\varepsilon) < \varepsilon$. Define $H = \{\mathbf{y} : \mathbf{y} \geq \mathbf{x} \text{ for some } \mathbf{x} \in K_\varepsilon\}$. Then H is a closed upper set (see Nachbin (1965)), and $K_\varepsilon \subset H \subset U$. Hence

$$P(\mathbf{X} \in U) \leq P(\mathbf{X} \in H) + \varepsilon \leq P(\mathbf{Y} \in H) + \varepsilon \leq P(\mathbf{Y} \in U) + \varepsilon.$$

Since ε is arbitrary this implies $P(\mathbf{X} \in U) \leq P(\mathbf{Y} \in U)$.

(iv) $\Rightarrow$ (i): Let f be a bounded increasing function with $\gamma = \inf_{\mathbf{x} \in \mathbb{R}^n} f$. For any real α let $U_\alpha = \{\mathbf{x} \in \mathbb{R}^n : f(\mathbf{x}) \geq \alpha\}$. Clearly U_α is an upper set. Define for all $n \in \mathbb{N}$

$$f_n(\mathbf{x}) = \gamma + \sum_{k=0}^{\infty} \frac{1}{n} \mathbf{1}_{U_{\gamma + k/n}}(\mathbf{x}).$$

Since f is bounded, the series in the definition of f_n is a finite sum, and hence (iv) implies

$$Ef_n(\mathbf{X}) = \gamma + \sum_{k=0}^{\infty} \frac{1}{n} P(\mathbf{X} \in U_{\gamma+k/n}) \leq \gamma + \sum_{k=0}^{\infty} \frac{1}{n} P(\mathbf{Y} \in U_{\gamma+k/n}) = Ef_n(\mathbf{Y}).$$

Since $|f_n(\mathbf{x}) - f(\mathbf{x})| \leq 1/n$ for all $\mathbf{x}$, taking the limit $n \to \infty$ yields $Ef(\mathbf{X}) \leq Ef(\mathbf{Y})$. $\square$

Theorem 2.6.3 yields immediately the following almost sure characterization of $\leq_{st}$.

Theorem 3.3.5. *The following conditions are equivalent:*

(i) $\mathbf{X} \leq_{st} \mathbf{Y}$;

(ii) *there are random vectors* $\hat{\mathbf{X}} =_{st} \mathbf{X}$ *and* $\hat{\mathbf{Y}} =_{st} \mathbf{Y}$ *such that* $P(\hat{\mathbf{X}} \leq \hat{\mathbf{Y}}) = 1$.

In contrast to the univariate case, we are not aware of a general constructive proof of this result for the multivariate case. Only for distributions with finite support constructive proofs are available. In this case the Ford–Fulkerson algorithm can be used to find the coupling construction; see Nawrotzki (1962) and Hansel and Troallic (1978). As a natural generalization of the univariate case one could conjecture that the standard construction described in (3.1.8) yields a constructive proof. Unfortunately this is not true in general.

Example 3.3.6. (The standard construction does not necessarily yield comparable random vectors.) Let

$$P(\mathbf{X} = (1,0)) = P(\mathbf{Y} = (1,0)) = \frac{1}{2}$$

and

$$P(\mathbf{X} = (0,1)) = P(\mathbf{Y} = (2,1)) = \frac{1}{2}.$$

Then clearly $\mathbf{X} \leq_{st} \mathbf{Y}$, and random vectors $\hat{\mathbf{X}} =_{st} \mathbf{X}$ and $\hat{\mathbf{Y}} =_{st} \mathbf{Y}$ with $P(\hat{\mathbf{X}} \leq \hat{\mathbf{Y}}) = 1$ are easily obtained if with probability $1/2$

$$\hat{\mathbf{X}} = \hat{\mathbf{Y}} = (1,0)$$

and with probability $1/2$

$$\hat{\mathbf{X}} = (0,1) \text{ and } \hat{\mathbf{Y}} = (2,1).$$

Under the standard construction, however, we get with probability $1/2$

$$\tilde{\mathbf{X}} = (0,1) \text{ and } \tilde{\mathbf{Y}} = (1,0)$$

and
$$\tilde{\mathbf{X}} = (1,0) \text{ and } \tilde{\mathbf{Y}} = (2,1),$$
respectively, so that $P(\tilde{\mathbf{X}} \leq \tilde{\mathbf{Y}}) = 1/2 < 1$.

Nevertheless the standard construction can be used to derive interesting sufficient conditions for $\leq_{st}$. An important case is given in the following result due to Veinott (1965). Generalizations can be found in Kamae et al. (1977), Rüschendorf (1981b) and Cohen and Sackrowitz (1995).

Theorem 3.3.7. *Let* $\mathbf{X}$ *and* $\mathbf{Y}$ *be* n-*dimensional random vectors. If*

$$X_1 \leq_{st} Y_1$$

and for $i = 2, 3, \dots, n$

$$[X_i | X_1 = x_1, \dots, X_{i-1} = x_{i-1}] \leq_{st} [Y_i | Y_1 = y_1, \dots, Y_{i-1} = y_{i-1}]$$

whenever $x_j \leq y_j$ *for all* $j = 1, \dots, i-1$, *then*

$$\mathbf{X} \leq_{st} \mathbf{Y}.$$

Proof. Let F and G be the distribution functions of $\mathbf{X}$ and $\mathbf{Y}$, respectively. Define $\hat{\mathbf{X}} = \Psi_F^*(\mathbf{U})$ and $\hat{\mathbf{Y}} = \Psi_G^*(\mathbf{U})$ as in (3.1.8). Then clearly $P(\hat{\mathbf{X}} \leq \hat{\mathbf{Y}}) = 1$. Since $\hat{\mathbf{X}} =_{st} \mathbf{X}$ and $\hat{\mathbf{Y}} =_{st} \mathbf{Y}$ the result follows from Theorem 3.3.5. $\square$

Another important case is that of two random vectors with a common copula. The following result has first been shown by Scarsini (1988). The proof given here is taken from Müller and Scarsini (2001). Implicitly this construction appears already in the proof of proposition 7 in Rüschendorf (1981b).

Theorem 3.3.8. *Let* $\mathbf{X} = (X_1, X_2, \dots, X_n)$ *and* $\mathbf{Y} = (Y_1, Y_2, \dots, Y_n)$ *have a common copula. If* $X_i \leq_{st} Y_i$ *for all* $i = 1, 2, \dots, n$, *then* $\mathbf{X} \leq_{st} \mathbf{Y}$.

Proof. Under the given assumptions there is a copula C such that

$$F_{\mathbf{X}}(\mathbf{x}) = C(F_1(x_1), \dots, F_n(x_n)) \text{ and } F_{\mathbf{Y}}(\mathbf{x}) = C(G_1(x_1), \dots, G_n(x_n)).$$

$X_i \leq_{st} Y_i$ yields $F_i^{-1} \leq G_i^{-1}$. Hence, if $\mathbf{U}$ is a random vector with distribution function C, then

$$\hat{\mathbf{X}} = (F_1^{-1}(U_1), \dots, F_n^{-1}(U_n)) \leq (G_1^{-1}(U_1), \dots, G_n^{-1}(U_n)) = \hat{\mathbf{Y}}$$

almost surely, and since $\hat{\mathbf{X}} =_{st} \mathbf{X}$ and $\hat{\mathbf{Y}} =_{st} \mathbf{Y}$, the assertion follows from Theorem 3.3.5. $\square$

On the other hand, distributions with the same univariate marginals cannot be ordered with respect to $\leq_{st}$. The following result can be found as proposition 6 in Rüschendorf (1981b).

Theorem 3.3.9. *If* $\mathbf{X}$ *and* $\mathbf{Y}$ *are random vectors with the same univariate marginals then* $\mathbf{X} \leq_{st} \mathbf{Y}$ *implies* $\mathbf{X} =_{st} \mathbf{Y}$.

Proof. According to Theorem 3.3.5 there are random vectors with $\hat{\mathbf{X}} =_{st} \mathbf{X}$ and $\hat{\mathbf{Y}} =_{st} \mathbf{Y}$ such that $\hat{\mathbf{X}} \leq \hat{\mathbf{Y}}$ almost surely. Hence $\hat{X}_i \leq \hat{Y}_i$ a.s. for all $i = 1, \ldots, n$ and $\hat{X}_i$ and $\hat{Y}_i$ have the same distributions. This is only possible if $\hat{X}_i = \hat{Y}_i$ almost surely. Therefore $\hat{\mathbf{X}} = \hat{\mathbf{Y}}$ a.s. and thus $\mathbf{X} =_{st} \mathbf{Y}$.   □

The next result shows that $\leq_{st}$ has all the properties of a stochastic order stated in the Definitions 2.4.1 and 3.2.1. We omit the proof since it is an immediate consequence of Theorem 2.4.2 and Theorem 3.2.3, taking into account the characterization (ii) of Theorem 3.3.4 for (W).

Theorem 3.3.10. *The multivariate stochastic order* $\leq_{st}$ *has all the properties (R), (E), (M), (T), (C), (MI), (W), (MA), (ID), and (IN).*

Usual stochastic order is preserved under monotone transformations.

Theorem 3.3.11. *If* $\mathbf{X} \leq_{st} \mathbf{Y}$ *and* $g : \mathbb{R}^n \to \mathbb{R}^k$ *is increasing, then* $g(\mathbf{X}) \leq_{st} g(\mathbf{Y})$.

Proof. This is a consequence of the fact that for increasing functions $f : \mathbb{R}^k \to \mathbb{R}$ and $g : \mathbb{R}^n \to \mathbb{R}^k$ the composition $f \circ g$ is again increasing.   □

The next result is known as *Efron's theorem*. It can be found in Efron (1965). Proofs and extensions are also given in Shanthikumar (1987), Daduna and Szekli (1996) and Liggett (2000).

Theorem 3.3.12. *Let* $X_1, \ldots, X_n$ *be non-negative independent random variables with log-concave densities, and let* $S = \sum_{i=1}^{n} X_i$. *Then*

$$[(X_1, \ldots, X_n)|S = s] \leq_{st} [(X_1, \ldots, X_n)|S = t] \quad \text{for all } s < t.$$

Now we will characterize $\leq_{st}$ within the normal distribution family. The result is a folklore theorem, but its only explicit statement in the literature seems to be in Müller (2001b).

Theorem 3.3.13. *Let* $\mathbf{X} \sim \mathcal{N}(\boldsymbol{\mu}, \boldsymbol{\Sigma})$ *and* $\mathbf{X}' \sim \mathcal{N}(\boldsymbol{\mu}', \boldsymbol{\Sigma}')$ *be* n-*dimensional normally distributed random vectors. Then* $\mathbf{X} \leq_{st} \mathbf{X}'$ *if and only if* $\mu_i \leq \mu_i'$ *for all* $i = 1, \ldots, n$ *and* $\boldsymbol{\Sigma} = \boldsymbol{\Sigma}'$.

Proof. The if-part follows immediately from Theorem 3.3.5, since $\mathbf{X}'$ has the same distribution as $\mathbf{X} + \boldsymbol{\mu}' - \boldsymbol{\mu}$. Therefore assume that $\mathbf{X} \leq_{st} \mathbf{X}'$. This implies that $X_i \leq_{st} X_i'$ holds for all marginals. Hence we can deduce from the characterization of stochastic order for univariate normals that $\mu_i \leq \mu_i'$ and $\sigma_{ii} = \sigma_{ii}'$. Moreover, $X_i + X_j \leq_{st} X_i' + X_j'$ holds for all $1 \leq i < j \leq n$, and since $X_i + X_j \sim \mathcal{N}(\mu_i + \mu_j, \sigma_{ii}^2 + \sigma_{jj}^2 + 2\sigma_{ij})$, we must also have $\sigma_{ii}^2 + \sigma_{jj}^2 + 2\sigma_{ij} = \sigma_{ii}'^2 + \sigma_{jj}'^2 + 2\sigma_{ij}'$ and thus necessarily $\sigma_{ij} = \sigma_{ij}'$.   □

Next the maximal generators of the orthant orders are characterized. The generator used in the definition of the upper orthant order is given by the distribution functions of one-point probability measures in $\mathbb{R}^n$. The closed convex cone generated by these functions clearly contains all distribution functions of probability measures with finite support. Since they are dense in the set of all probability measures, the maximal generator should contain all distribution functions. Similarly the maximal generator of lower orthant order must contain all survival functions.

It is well known that multivariate distribution functions and survival functions can be characterized by difference operators.

Definition 3.3.14. a) For a function $f : \mathbb{R}^n \to \mathbb{R}$ define the *difference operators*

$$\Delta_i^\varepsilon f(\mathbf{x}) = f(\mathbf{x} + \varepsilon \mathbf{e}_i) - f(\mathbf{x}),$$

where $\mathbf{e}_i$ is the i-th unit vector and $\varepsilon > 0$.
b) A function $f : \mathbb{R}^n \to \mathbb{R}$ is said to be Δ-*monotone*, if for every subset $J = \{i_1, \dots, i_k\} \subset \{1, \dots, n\}$ and every $\varepsilon_1, \dots, \varepsilon_k > 0$

$$\Delta_{i_1}^{\varepsilon_1} \dots \Delta_{i_k}^{\varepsilon_k} f(\mathbf{x}) \geq 0 \quad \text{for all } \mathbf{x}.$$

c) A function $f : \mathbb{R}^n \to \mathbb{R}$ is said to be Δ-*antitone*, if $g(\mathbf{x}) = f(-\mathbf{x})$ is Δ-monotone, i.e. if for every subset $J = \{i_1, \dots, i_k\} \subset \{1, \dots, n\}$ and every $\varepsilon_1, \dots, \varepsilon_k > 0$

$$(-1)^k \Delta_{i_1}^{\varepsilon_1} \dots \Delta_{i_k}^{\varepsilon_k} f(\mathbf{x}) \geq 0 \quad \text{for all } \mathbf{x}.$$

Remarks: 1. It is well known that distribution functions are Δ-monotone, see, for example, Rüschendorf (1980). From this it follows immediately that multivariate survival functions are Δ-antitone.
2. If f is n times differentiable, then f is Δ-monotone if and only if all mixed partial derivatives up to order n are non-negative.

It follows easily from the definition that the set of all Δ-monotone (Δ-antitone) functions is a convex cone closed under pointwise convergence. Hence Corollary 2.3.9 implies the following result.

Theorem 3.3.15.
a) The maximal generator of the upper orthant order is the set of all bounded Δ-monotone functions.
b) The maximal generator of the lower orthant order is the set of all bounded functions f such that $-f$ is Δ-antitone.

There are some other classes of functions that generate the lower and upper orthant order, respectively. They are described in the next two results, which can be deduced easily from Theorem 3.3.15. They are due to Bergmann (1978).

Theorem 3.3.16. *Let* $\mathbf{X}$ *and* $\mathbf{Y}$ *be two* n*-dimensional random vectors. Then*
a) $\mathbf{X} \leq_{uo} \mathbf{Y}$ *if and only if*

$$E\left(\prod_{i=1}^{n} f_i(X_i)\right) \leq E\left(\prod_{i=1}^{n} f_i(Y_i)\right)$$

for every collection $f_1, \ldots, f_n$ *of univariate non-negative increasing functions.*
b) $\mathbf{X} \leq_{lo} \mathbf{Y}$ *if and only if*

$$E\left(-\prod_{i=1}^{n} f_i(X_i)\right) \leq E\left(-\prod_{i=1}^{n} f_i(Y_i)\right)$$

for every collection $f_1, \ldots, f_n$ *of univariate non-negative decreasing functions.*

Theorem 3.3.17. *Let* $\mathbf{X}$ *and* $\mathbf{Y}$ *be two* n*-dimensional non-negative random vectors. Then*
a) $\mathbf{X} \leq_{uo} \mathbf{Y}$ *if and only if*

$$\min\{a_1 X_1, \ldots, a_n X_n\} \leq_{st} \min\{a_1 Y_1, \ldots, a_n Y_n\}$$

for all $a_1, \ldots, a_n > 0$.
b) $\mathbf{X} \leq_{lo} \mathbf{Y}$ *if and only if*

$$\max\{a_1 X_1, \ldots, a_n X_n\} \leq_{st} \max\{a_1 Y_1, \ldots, a_n Y_n\}$$

for all $a_1, \ldots, a_n > 0$.

Notice that it is not possible for the orthant orders $\leq_{lo}$ and $\leq_{uo}$ to deduce an almost sure characterization by means of Strassen's theorem (Theorem 2.6.1), since the maximal generators are not closed with respect to the formation of maxima or minima.

The next result shows that the orthant orders are preserved under increasing transformations of the components. This can easily be deduced either from the definition or from Theorem 3.3.16.

Theorem 3.3.18. *Suppose that* $g_1, \ldots, g_n : \mathbb{R} \to \mathbb{R}$ *are increasing functions. Then* $(X_1, \ldots, X_n) \leq_{lo} (Y_1, \ldots, Y_n)$ *implies*

$$(g_1(X_1), \ldots, g_n(X_n)) \leq_{lo} (g_1(Y_1), \ldots, g_n(Y_n))$$

and $(X_1, \ldots, X_n) \leq_{uo} (Y_1, \ldots, Y_n)$ *implies*

$$(g_1(X_1), \ldots, g_n(X_n)) \leq_{uo} (g_1(Y_1), \ldots, g_n(Y_n)).$$

The following theorem collects some properties of these orders; some of them are stated in Li and Xu (2000).

Theorem 3.3.19. *The order relations $\leq_{uo}$ and $\leq_{lo}$ have all the properties (R), (E), (M), (T), (C), (MI), (W), (MA), (ID), and (IN).*

Proof. Most of these properties follow immediately from Theorem 2.4.2 and Theorem 3.2.3. Therefore only the proof of (W) is given here. According to Theorem 2.4.2 it is sufficient to show that for all $\mathbf{t} \in \mathbb{R}^n$ the indicator function $f^{(\mathbf{t})}(\mathbf{x}) = \mathbf{1}_{(\mathbf{t},\infty)}(\mathbf{x})$ can be approximated by bounded continuous Δ-monotone functions. But this can be achieved by convolutions with suitable uniform distributions on small enough intervals. In detail, define

$$g_k(\mathbf{x}) = \begin{cases} k^n & \text{for } 0 \leq x_i \leq 1/k, \ 1 \leq i \leq n \\ 0, & \text{otherwise} \end{cases}$$

and let $f_k^{(\mathbf{t})}(\mathbf{x}) = \int f^{(\mathbf{t})}(\mathbf{x}-\mathbf{z})g_k(\mathbf{z})d\mathbf{z}$. Then $0 \leq f_k^{(\mathbf{t})} \leq 1$, i.e. $f_k^{(\mathbf{t})}$ is bounded. Moreover, $f_k^{(\mathbf{t})}$ is Δ-monotone since it is a mixture of Δ-monotone functions, and it is continuous, since it is obtained by convolution with a continuous distribution. For $k \to \infty$ the sequence $f_k^{(\mathbf{t})}$ converges monotonically to $f^{(\mathbf{t})}$. Hence the set $\{f_k^{(\mathbf{t})} : k \in \mathbb{N}, \mathbf{t} \in \mathbb{R}^n\}$ of all such functions forms a generator of $\leq_{uo}$ consisting of bounded and continuous functions. For $\leq_{lo}$ property (W) can be shown analogously. $\qquad\square$

Even for multivariate normal distributions upper and lower orthant order are strictly weaker than usual stochastic order. The following result is known as *Slepian's inequality*; see Tong (1990), theorem 5.1.7. A nice proof can be found in Tong (1980), p. 8. Based on the same idea we will give a proof of a more general result in Theorem 3.13.5.

Theorem 3.3.20. *Let $\mathbf{X} \sim \mathcal{N}(\boldsymbol{\mu},\boldsymbol{\Sigma})$ and $\mathbf{X}' \sim \mathcal{N}(\boldsymbol{\mu}',\boldsymbol{\Sigma}')$ be n-dimensional normally distributed random vectors with $\mu_i = \mu_i'$ and $\sigma_{ii} = \sigma_{ii}'$ for all $i = 1,\ldots,n$. If $\sigma_{ij} \leq \sigma_{ij}'$, $1 \leq i < j \leq n$, then $\mathbf{X} \leq_{uo} \mathbf{X}'$ and $\mathbf{X} \geq_{lo} \mathbf{X}'$.*

Theorem 3.3.20 can immediately be generalized to random vectors $\mathbf{X}$ with elliptically contoured distributions, i.e. distributions for which there are $\boldsymbol{\mu} \in \mathbb{R}^n$, a non-negative definite $n \times n$-matrix $\boldsymbol{\Sigma}$ and a function $\phi : [0,\infty) \to \mathbb{R}$ such that the random vector $\mathbf{X}$ has a characteristic function of the form

$$\Psi(\mathbf{s}) = Ee^{i\mathbf{s}^T\mathbf{X}} = \exp(i\boldsymbol{\mu}^T\mathbf{s}) \cdot \phi(\mathbf{s}^T\boldsymbol{\Sigma}\mathbf{s});$$

see Cambanis and Simons (1982) for details.

Combining Theorem 3.3.20 with Theorem 3.3.13 yields the following result.

Theorem 3.3.21. *Let $\mathbf{X} \sim \mathcal{N}(\boldsymbol{\mu},\boldsymbol{\Sigma})$ and $\mathbf{X}' \sim \mathcal{N}(\boldsymbol{\mu}',\boldsymbol{\Sigma}')$ be n-dimensional normally distributed random vectors.*
a) If $\mu_i \leq \mu_i'$ for all $1 \leq i \leq n$, $\sigma_{ii} = \sigma_{ii}'$, $1 \leq i \leq n$, and $\sigma_{ij} \leq \sigma_{ij}'$, $1 \leq i < j \leq n$, then $\mathbf{X} \leq_{uo} \mathbf{X}'$ and $\mathbf{X} \geq_{lo} \mathbf{X}'$.
b) If $\mathbf{X} \leq_{uo} \mathbf{X}'$ or $\mathbf{X} \geq_{lo} \mathbf{X}'$, then $\mu_i \leq \mu_i'$ for all $1 \leq i \leq n$ and $\sigma_{ii} = \sigma_{ii}'$.

Proof. Part a) is an immediate consequence of Theorem 3.3.13 and Theorem 3.3.20. Part b) follows from the fact that $\mathbf{X} \leq_{uo} \mathbf{X}'$ ($\mathbf{X} \geq_{lo} \mathbf{X}'$) implies $X_i \leq_{st} Y_i$ for all $1 \leq i \leq n$. □

Unfortunately we are not able to give an if-and-only-if characterization of the orthant orders for multinormal distributions.

An extension of the theory of orthant orders to general product spaces is given in Bergmann (1991). It is based on the characterization in Theorem 3.3.16.

3.4 Convex Orders

As in the univariate case the most natural orders to compare variability of random vectors are those generated by convex functions. The univariate definitions of convex, increasing convex and increasing concave order extend immediately to higher dimensions.

Definition 3.4.1. Let $\mathbf{X}$ and $\mathbf{Y}$ be n-dimensional random vectors with finite expectations. Then
a) $\mathbf{X}$ is less than $\mathbf{Y}$ in *increasing convex order* (written $\mathbf{X} \leq_{icx} \mathbf{Y}$), if $Ef(\mathbf{X}) \leq Ef(\mathbf{Y})$ for all increasing convex functions $f : \mathbb{R}^n \to \mathbb{R}$ such that the expectations exist;
b) $\mathbf{X}$ is less than $\mathbf{Y}$ in *increasing concave order* (written $\mathbf{X} \leq_{icv} \mathbf{Y}$), if $Ef(\mathbf{X}) \leq Ef(\mathbf{Y})$ for all increasing concave functions $f : \mathbb{R}^n \to \mathbb{R}$ such that the expectations exist;
c) $\mathbf{X}$ is less than $\mathbf{Y}$ in *convex order* (written $\mathbf{X} \leq_{cx} \mathbf{Y}$), if $Ef(\mathbf{X}) \leq Ef(\mathbf{Y})$ for all convex functions $f : \mathbb{R}^n \to \mathbb{R}$ such that the expectations exist.

In the definition the extended maximal generators are used. To fit the definition to the theory of Chapter 2, we have to introduce a suitable weight function. The most natural one is $b(\mathbf{x}) = 1 + \|\mathbf{x}\|$, where $\| \cdot \|$ is any norm on $\mathbb{R}^n$. Then $\mathbb{P}_b$ consists of all probability measures with a finite mean. This is the least possible restriction. All three generators used in the definition are convex cones containing the constant functions and being closed under pointwise convergence. Hence in all three cases the maximal generator is just the restriction of the extended maximal generator to b-bounded functions.

Unfortunately, in the multivariate case there is no easy characterization of these orders as in Theorem 1.5.7. Johansen (1972, 1974) showed that the extreme rays of the convex cone of convex functions are dense in the cone. In fact, every maximum of a finite set of affine functions is such an extreme point. Therefore there is no hope of finding a 'small' generator. Thus the easiest way to show that these orders hold is through the almost-sure constructions that can be derived from Theorem 2.6.1. Since the generators of $\leq_{cx}$ and $\leq_{icx}$ are closed with respect to the formation of maxima, and the one of $\leq_{icv}$ is closed under minimization, we get the following results.

Theorem 3.4.2.

a) $\mathbf{X} \leq_{cx} \mathbf{Y}$ *if and only if there are random vectors* $\hat{\mathbf{X}} =_{st} \mathbf{X}$ *and* $\hat{\mathbf{Y}} =_{st} \mathbf{Y}$
such that $E[\hat{\mathbf{Y}}|\hat{\mathbf{X}}] = \hat{\mathbf{X}}$ *almost surely.*

b) $\mathbf{X} \leq_{icx} \mathbf{Y}$ *if and only if there are random vectors* $\hat{\mathbf{X}} =_{st} \mathbf{X}$ *and* $\hat{\mathbf{Y}} =_{st} \mathbf{Y}$
such that $E[\hat{\mathbf{Y}}|\hat{\mathbf{X}}] \geq \hat{\mathbf{X}}$ *almost surely.*

c) $\mathbf{X} \leq_{icv} \mathbf{Y}$ *if and only if there are random vectors* $\hat{\mathbf{X}} =_{st} \mathbf{X}$ *and* $\hat{\mathbf{Y}} =_{st} \mathbf{Y}$
such that $E[\hat{\mathbf{X}}|\hat{\mathbf{Y}}] \leq \hat{\mathbf{Y}}$ *almost surely.*

For distributions with finite support Elton and Hill (1998) derived a
constructive proof of Theorem 3.4.2 a), which is based on the concept of
fusions introduced in Elton and Hill (1992). It can be considered as a
multivariate generalization of the local mean preserving spreads introduced
in Definition 1.5.28.

As a consequence of Theorem 3.4.2 a separation theorem similar to Theorem
1.5.14 can be derived.

Theorem 3.4.3. *If* $\mathbf{X} \leq_{icx} \mathbf{Y}$, *then there is a random vector* $\mathbf{Z}$ *such that*

$$\mathbf{X} \leq_{st} \mathbf{Z} \leq_{cx} \mathbf{Y}.$$

Proof. Assume $\mathbf{X} \leq_{icx} \mathbf{Y}$. According to Theorem 3.4.2 b) there are random
vectors $\hat{\mathbf{X}} =_{st} \mathbf{X}$ and $\hat{\mathbf{Y}} =_{st} \mathbf{Y}$ such that $E[\hat{\mathbf{Y}}|\hat{\mathbf{X}}] \geq \hat{\mathbf{X}}$ almost surely. Now
define $\mathbf{Z} = E[\hat{\mathbf{Y}}|\hat{\mathbf{X}}]$. Then obviously $\mathbf{Z} = E[\hat{\mathbf{Y}}|\mathbf{Z}]$ almost surely and $\mathbf{Z} \geq \hat{\mathbf{X}}$
almost surely. Hence the assertion follows immediately from Theorem 3.4.2 a)
and Theorem 3.3.5. $\qquad\square$

From Theorem 2.4.2 and Theorem 3.2.3 the following properties can be
derived.

Theorem 3.4.4. *The order* $\leq_{cx}$ *has the properties (E), (M), (C), (MI),
(MA), (ID), and (IN). The orders* $\leq_{icx}$ *and* $\leq_{icv}$ *in addition have the
properties (R) and (T).*

None of these orders is closed with respect to weak convergence. This cannot
hold, since it is already false in the univariate case, as Example 1.5.8 shows.
However, we will show a multivariate generalization of Theorem 1.5.9 below.

Recall that in Landau's O-notation a function f is said to be $O(g(\mathbf{x}))$ at $\mathbf{x}_0$,
if there is a constant C such that $|f(\mathbf{x})| \leq C|g(\mathbf{x})|$ for all $\mathbf{x}$ in a neighborhood
of $\mathbf{x}_0$. Moreover, let $\|\cdot\|$ be any norm on $\mathbb{R}^n$.

Lemma 3.4.5. *For random vectors* $\mathbf{X}$ *and* $\mathbf{X}^{(k)}$, $k \in \mathbb{N}$, *the following
statements are equivalent:*

(i) $\mathbf{X}^{(k)} \to \mathbf{X}$ *in distribution and* $E\|\mathbf{X}^{(k)}\| \to E\|\mathbf{X}\|$;

(ii) $Ef(\mathbf{X}^{(k)}) \to Ef(\mathbf{X})$ *for all continuous* f *such that* $f(\mathbf{x}) = O(\|\mathbf{x}\|)$ *at
infinity.*

Proof. This is a special case of lemma 8.3 in Bickel and Freedman (1981). □

Note that $E\|\mathbf{X}^{(k)}\| \to E\|\mathbf{X}\|$ holds if and only if $E|X_i^{(k)}| \to E|X_i|$ for all $i = 1, \ldots, n$. Lemma 3.4.5 immediately implies the following result.

Theorem 3.4.6. *Let $(\mathbf{X}^{(k)})$ and $(\mathbf{Y}^{(k)})$ be sequences of random vectors with $\mathbf{X}^{(k)} \leq_{cx} [\leq_{icx}] \mathbf{Y}^{(k)}$ for all k. If $\mathbf{X}^{(k)} \to \mathbf{X}$ and $\mathbf{Y}^{(k)} \to \mathbf{Y}$ in distribution, and if moreover $E\mathbf{X}^{(k)} \to E\mathbf{X}$ and $E\mathbf{Y}^{(k)} \to E\mathbf{Y}$, then $\mathbf{X} \leq_{cx} [\leq_{icx}] \mathbf{Y}$.*

For multivariate normal distributions the following characterization of convex order holds.

Theorem 3.4.7. *Let $\mathbf{X}$ and $\mathbf{X}'$ be n-dimensional random vectors with normal distributions $\mathcal{N}(\boldsymbol{\mu}, \boldsymbol{\Sigma})$ and $\mathcal{N}(\boldsymbol{\mu}', \boldsymbol{\Sigma}')$, respectively. Then the following conditions are equivalent:*

(i) $\mathbf{X} \leq_{cx} \mathbf{X}'$;

(ii) $\boldsymbol{\mu} = \boldsymbol{\mu}'$ *and* $\boldsymbol{\Sigma}' - \boldsymbol{\Sigma}$ *is non-negative definite.*

Proof. a) If $\boldsymbol{\mu} = \boldsymbol{\mu}'$ and $\boldsymbol{\Sigma}' - \boldsymbol{\Sigma}$ is non-negative definite, then $\mathbf{X}'$ has the same distribution as $\mathbf{Y} = \mathbf{X} + \mathbf{Z}$, where $\mathbf{X}$ and $\mathbf{Z}$ are independent and $\mathbf{Z} \sim \mathcal{N}(\mathbf{0}, \boldsymbol{\Sigma}' - \boldsymbol{\Sigma})$. Hence $E[\mathbf{Y}|\mathbf{X}] = \mathbf{X}$ and therefore $\mathbf{X} \leq_{cx} \mathbf{Y}$.
b) Assume that $\mathbf{X} \leq_{cx} \mathbf{X}'$ holds. Since the functions $f(\mathbf{x}) = x_i$ and $f(\mathbf{x}) = -x_i$ are convex for $i = 1, \ldots, n$, it is clear that the condition $\boldsymbol{\mu} = \boldsymbol{\mu}'$ is necessary. Therefore assume that $\boldsymbol{\Sigma}' - \boldsymbol{\Sigma}$ is not non-negative definite, i.e. there is some $\mathbf{a} \in \mathbb{R}^n$ such that $\mathbf{a}^T(\boldsymbol{\Sigma}' - \boldsymbol{\Sigma})\mathbf{a} < 0$. Then

$$E(\mathbf{a}^T(\mathbf{X} - \boldsymbol{\mu}))^2 = \mathbf{a}^T\boldsymbol{\Sigma}\mathbf{a} > \mathbf{a}^T\boldsymbol{\Sigma}'\mathbf{a} = E(\mathbf{a}^T(\mathbf{X}' - \boldsymbol{\mu}))^2.$$

Hence $\mathbf{X} \leq_{cx} \mathbf{X}'$ does not hold, since the function $f(\mathbf{x}) = (\mathbf{a}^T(\mathbf{x} - \boldsymbol{\mu}))^2$ is convex. □

This result can be found in Scarsini (1998), theorem 4, but the proof given there is more complicated.

For the case of $\leq_{icx}$ an if-and-only-if characterization seems to be unknown; there is still a gap between the necessary and the sufficient conditions in this case. We only can show the following result.

Theorem 3.4.8. *Let $\mathbf{X}$ and $\mathbf{X}'$ be n-dimensional random vectors with normal distributions $\mathcal{N}(\boldsymbol{\mu}, \boldsymbol{\Sigma})$ and $\mathcal{N}(\boldsymbol{\mu}', \boldsymbol{\Sigma}')$, respectively. Then the following conditions hold:*
a) If $\boldsymbol{\mu} \leq \boldsymbol{\mu}'$ and $\boldsymbol{\Sigma}' - \boldsymbol{\Sigma}$ is non-negative definite, then $\mathbf{X} \leq_{icx} \mathbf{X}'$.
b) If $\mathbf{X} \leq_{icx} \mathbf{X}'$, then $\boldsymbol{\mu} \leq \boldsymbol{\mu}'$ and $\mathbf{a}^T(\boldsymbol{\Sigma}' - \boldsymbol{\Sigma})\mathbf{a} \geq 0$ for all $\mathbf{a} \geq \mathbf{0}$.

Proof. a) follows immediately from Theorem 3.3.13 and Theorem 3.4.7, taking into account the transitivity of $\leq_{icx}$.

b) If $\mathbf{X} \leq_{icx} \mathbf{X}'$, then $\boldsymbol{\mu} \leq \boldsymbol{\mu}'$ since the functions $f(\mathbf{x}) = x_i$ are increasing convex for all $i = 1, \ldots, n$. Therefore let $\mathbf{a} \geq \mathbf{0}$. Then the function $f_{\mathbf{a}}(\mathbf{x}) = f(\mathbf{a}^T \mathbf{x})$ is increasing convex for all increasing convex functions $f : \mathbb{R} \to \mathbb{R}$. Hence $\mathbf{X} \leq_{icx} \mathbf{X}'$ implies $\mathbf{a}^T \mathbf{X} \leq_{icx} \mathbf{a}^T \mathbf{X}'$ and thus necessarily $Var(\mathbf{a}^T \mathbf{X}) = \mathbf{a}^T \boldsymbol{\Sigma} \mathbf{a} \leq \mathbf{a}^T \boldsymbol{\Sigma}' \mathbf{a} = Var(\mathbf{a}^T \mathbf{X}')$. $\qquad\square$

Remark 3.4.9. A matrix $\mathbf{A}$ with the property that $\mathbf{x}^T \mathbf{A} \mathbf{x} \geq 0$ for all $\mathbf{x} \geq \mathbf{0}$ is usually called *copositive*; see Cottle, Habetler, and Lemke (1970). Theorem 3.4.8 b) thus states that for multivariate normal distributions $\mathbf{X} \leq_{icx} \mathbf{X}'$ implies that $\boldsymbol{\mu} \leq \boldsymbol{\mu}'$ and that $\boldsymbol{\Sigma}' - \boldsymbol{\Sigma}$ is copositive.

3.5 Linear Convex Orders

In economic applications a random vector $\mathbf{X} = (X_1, \ldots, X_n)$ often represents a random bundle of commodities. Assuming that a unit of commodity i has the price a_i, then the value of the whole bundle is $\mathbf{a}^T \mathbf{X}$, where $\mathbf{a} = (a_1, \ldots, a_n)$ is the price vector. Therefore it is natural to compare two bundles $\mathbf{X}$ and $\mathbf{Y}$ by comparing their values $\mathbf{a}^T \mathbf{X}$ and $\mathbf{a}^T \mathbf{Y}$. It will often be the case that we are not able to fix the price vector $\mathbf{a}$; it may only be known that the price vector belongs to some set $A \subset \mathbb{R}^n$. The two following cases are of special interest.

1. $A = \mathbb{R}^n_+ = \{\mathbf{a} \in \mathbb{R}^n : \mathbf{a} \geq \mathbf{0}\}$. This means that it is only known that each commodity has a non-negative value.

2. $A = \mathbb{R}^n$. This means that nothing is known about the price vector. It may even happen that prices are negative, i.e. something has to be paid for getting rid of the commodity.

If univariate (increasing) convex order is used to compare the values of the bundles then we are naturally led to the following definitions.

Definition 3.5.1. Let $\mathbf{X}$ and $\mathbf{Y}$ be n-dimensional random vectors.
a) $\mathbf{X}$ is less than $\mathbf{Y}$ in *linear convex order* (written $\mathbf{X} \leq_{lcx} \mathbf{Y}$), if $\mathbf{a}^T \mathbf{X} \leq_{cx} \mathbf{a}^T \mathbf{Y}$ for all $\mathbf{a} \in \mathbb{R}^n$;
b) $\mathbf{X}$ is less than $\mathbf{Y}$ in *positive linear convex order* (written $\mathbf{X} \leq_{plcx} \mathbf{Y}$), if $\mathbf{a}^T \mathbf{X} \leq_{cx} \mathbf{a}^T \mathbf{Y}$ holds for all $\mathbf{a} \geq \mathbf{0}$;
c) $\mathbf{X}$ is less than $\mathbf{Y}$ in *increasing positive linear convex order* (written $\mathbf{X} \leq_{iplcx} \mathbf{Y}$), if $\mathbf{a}^T \mathbf{X} \leq_{icx} \mathbf{a}^T \mathbf{Y}$ for all $\mathbf{a} \geq \mathbf{0}$.

Remark 3.5.2. It may seem to be reasonable to define in addition the following increasing linear convex order: $\mathbf{X} \leq_{ilcx} \mathbf{Y}$ if $\mathbf{a}^T \mathbf{X} \leq_{icx} \mathbf{a}^T \mathbf{Y}$ for all $\mathbf{a} \in \mathbb{R}^n$. However, as has been shown in Scarsini (1998), lemma 1, this order relation is equivalent to $\leq_{lcx}$.

It is clear that all stochastic orders in the above definition are integral stochastic orders, which are generated by the functions obtained as compositions $f(\ell(\cdot))$ of an (increasing) convex function $f : \mathbb{R} \to \mathbb{R}$ and an (increasing) linear function $\ell : \mathbb{R}^n \to \mathbb{R}$. Using this fact, we can immediately derive the following properties of these orderings from Theorem 2.4.2 and Theorem 3.2.3.

Theorem 3.5.3. *The orderings $\leq_{lcx}$ and $\leq_{plcx}$ have the properties (E), (M), (C), (MI), (MA), (ID), and (IN). In addition, the ordering $\leq_{iplcx}$ has the properties (R) and (T).*

The ordering $\leq_{lcx}$ has recently been considered by Koshevoy and Mosler (1998) under the name *lift zonoid order*. They also showed the properties (C) and (MI) for this order relation in their theorems 5.6 and 5.7 respectively, but they used a different approach; their proof is based on geometric properties of zonoids.

We do not know an easy characterization of the maximal generators of these order relations. It is clear that the generators mentioned above are not maximal, since they do not form a convex cone. The function $f(\mathbf{x}) = \sum_{i=1}^{n} x_i^{+}$, for example, is not a composition of a convex and a linear function. On the other hand, the maximal generator of $\leq_{lcx}$ does not contain all convex functions as the following example shows.

Example 3.5.4. (Convex order is stronger than linear convex order.)
Let $\mathbf{X}$ and $\mathbf{Y}$ be bivariate random vectors with

$$P(\mathbf{X} = (-1,-1)) = P(\mathbf{X} = (0,1)) = P(\mathbf{X} = (1,0)) = \frac{1}{3},$$

$$P(\mathbf{Y} = (-2,0)) = P(\mathbf{Y} = (0,-2)) = P(\mathbf{Y} = (2,2)) = \frac{1}{6}$$

and

$$P(\mathbf{Y} = (0,0)) = \frac{1}{2}.$$

It can be shown (see Elton and Hill (1992) for details) that $\mathbf{X} \leq_{lcx} \mathbf{Y}$ and thus also $\mathbf{X} \leq_{plcx} \mathbf{Y}$ and $\mathbf{X} \leq_{iplcx} \mathbf{Y}$ hold, but for the increasing convex function $f(x_1, x_2) = \max\{x_1, x_2, 0\}$

$$Ef(\mathbf{X}) = 2/3 > 1/3 = Ef(\mathbf{Y})$$

and hence $\mathbf{X} \not\leq_{cx} \mathbf{Y}$ and $\mathbf{X} \not\leq_{icx} \mathbf{Y}$.

Nevertheless, for parametric families of distributions the orderings $\leq_{lcx}$ and $\leq_{cx}$ often coincide. For the family of multivariate normal distributions this is a consequence of the following result and Theorem 3.4.7.

Theorem 3.5.5. *Let $\mathbf{X}$ and $\mathbf{X}'$ be n-dimensional random vectors with normal distributions $\mathcal{N}(\boldsymbol{\mu}, \boldsymbol{\Sigma})$ and $\mathcal{N}(\boldsymbol{\mu}', \boldsymbol{\Sigma}')$ respectively. Then the following equivalences hold:*
a) $\mathbf{X} \leq_{lcx} \mathbf{X}'$ if and only if $\boldsymbol{\mu} = \boldsymbol{\mu}'$ and $\boldsymbol{\Sigma}' - \boldsymbol{\Sigma}$ is non-negative definite,
b) $\mathbf{X} \leq_{plcx} \mathbf{X}'$ if and only if $\boldsymbol{\mu} = \boldsymbol{\mu}'$ and $\boldsymbol{\Sigma}' - \boldsymbol{\Sigma}$ is copositive,
c) $\mathbf{X} \leq_{iplcx} \mathbf{X}'$ if and only if $\boldsymbol{\mu} \leq \boldsymbol{\mu}'$ and $\boldsymbol{\Sigma}' - \boldsymbol{\Sigma}$ is copositive.

Proof. The assertions follow immediately from the definitions of $\leq_{lcx}$, $\leq_{plcx}$ and $\leq_{iplcx}$, taking into account the fact that for univariate normal distributions [increasing] convex order holds, if and only if $\mu = \mu'$ [$\mu \leq \mu'$] and $\sigma^2 \leq \sigma'^2$. □

Theorem 3.5.5 shows that even for normally distributed vectors the stochastic order relations $\leq_{lcx}$, $\leq_{plcx}$ and $\leq_{iplcx}$ are different.

Example 3.5.6. (A copositive matrix that is neither non-negative nor non-negative definite.) The condition of copositivity of $\boldsymbol{\Sigma}' - \boldsymbol{\Sigma}$, i.e. that $\mathbf{a}^T(\boldsymbol{\Sigma}' - \boldsymbol{\Sigma})\mathbf{a} \geq 0$ for all $\mathbf{a} \geq \mathbf{0}$, is obviously fulfilled, if either $\boldsymbol{\Sigma}' - \boldsymbol{\Sigma}$ is non-negative definite or if $\boldsymbol{\Sigma}$ is elementwise smaller than $\boldsymbol{\Sigma}'$, i.e. if $\sigma_{ij} \leq \sigma'_{ij}$ for all $1 \leq i, j \leq n$. But neither of these two conditions is necessary. Consider the following example in the three-dimensional case. Let $\boldsymbol{\Sigma} = \mathbf{I}$ be the identity matrix and let

$$\boldsymbol{\Sigma}' = \begin{pmatrix} 2 & -1 & 1 \\ -1 & 3 & 0 \\ 1 & 0 & 2 \end{pmatrix}, \quad \text{so that} \quad \boldsymbol{\Sigma}' - \boldsymbol{\Sigma} = \begin{pmatrix} 1 & -1 & 1 \\ -1 & 2 & 0 \\ 1 & 0 & 1 \end{pmatrix}.$$

Then $\boldsymbol{\Sigma}$ and $\boldsymbol{\Sigma}'$ are covariance matrices and

$$\mathbf{a}^T(\boldsymbol{\Sigma}' - \boldsymbol{\Sigma})\mathbf{a} = (a_1 - a_2 + a_3)^2 + a_2^2 + 2a_2a_3 \geq 0$$

for all $\mathbf{a} \geq \mathbf{0}$. However, $\boldsymbol{\Sigma}' - \boldsymbol{\Sigma}$ is neither non-negative nor non-negative definite, since $\det(\boldsymbol{\Sigma}' - \boldsymbol{\Sigma}) = -1 < 0$.

3.6 Componentwise Convex Order

A stronger relation than convex order is obtained by requiring $Ef(\mathbf{X}) \leq Ef(\mathbf{Y})$ for all functions f that are componentwise convex, i.e. convex in each argument, when the other arguments are held fixed. This order was introduced in Mosler (1982), see also Shaked and Shanthikumar (1994), 5.A.6. The formal definition is as follows.

Definition 3.6.1. Let $\mathbf{X}$ and $\mathbf{Y}$ be random vectors. Then
a) $\mathbf{X}$ is less than $\mathbf{Y}$ in *componentwise convex order* (written $\mathbf{X} \leq_{ccx} \mathbf{Y}$), if $Ef(\mathbf{X}) \leq Ef(\mathbf{Y})$ for all componentwise convex functions $f : \mathbb{R}^n \to \mathbb{R}$ such that the expectations exist,
b) $\mathbf{X}$ is less than $\mathbf{Y}$ in *increasing componentwise convex order* (written $\mathbf{X} \leq_{iccx} \mathbf{Y}$), if $Ef(\mathbf{X}) \leq Ef(\mathbf{Y})$ for all increasing componentwise convex functions $f : \mathbb{R}^n \to \mathbb{R}$ such that the expectations exist.

If f is twice differentiable, then f is componentwise convex if and only if

$$\frac{\partial^2}{\partial x_i^2}f(\mathbf{x}) \geq 0 \quad \text{for all } \mathbf{x} \text{ and all } i = 1, \dots, n.$$

It is clear that every convex function is also componentwise convex, but, of course, there are functions which are componentwise convex but not convex. Important examples are the functions $f(\mathbf{x}) = x_i x_j$ and $f(\mathbf{x}) = -x_i x_j$, $1 \leq i < j \leq n$. These functions are componentwise linear and hence componentwise convex. This implies that for $\mathbf{X} \leq_{ccx} \mathbf{Y}$ it is necessary that $\mathbf{X}$ and $\mathbf{Y}$ have the same covariance matrix. We state this result together with some other properties of componentwise convex order in the next theorem.

Theorem 3.6.2. *Let $\mathbf{X}$ and $\mathbf{Y}$ be two random vectors.*
a) If $\mathbf{X} \leq_{ccx} \mathbf{Y}$ then $\mathbf{X} \leq_{cx} \mathbf{Y}$.
b) If $\mathbf{X} \leq_{iccx} \mathbf{Y}$ then $\mathbf{X} \leq_{icx} \mathbf{Y}$.
c) If $\mathbf{X} \leq_{ccx} \mathbf{Y}$ then $X_i \leq_{cx} Y_i$ for all $i = 1, \dots, n$.
d) If $\mathbf{X} \leq_{ccx} \mathbf{Y}$ then $Cov(X_i, X_j) = Cov(Y_i, Y_j)$ for all $1 \leq i < j \leq n$.

For random vectors with independent components we can derive the following characterization of componentwise convex order.

Theorem 3.6.3. *Let $\mathbf{X} = (X_1, \dots, X_n)$ and $\mathbf{Y} = (Y_1, \dots, Y_n)$ be random vectors with independent components. Then $\mathbf{X} \leq_{ccx} \mathbf{Y}$ holds if and only if $X_i \leq_{cx} Y_i$ for all $1 \leq i \leq n$.*

Proof. We show the result by induction on n. The case $n = 1$ is obvious. Hence assume that the assertion holds for $n - 1$, and let $f : \mathbb{R}^n \to \mathbb{R}$ be componentwise convex. Then

$$Ef(\mathbf{X}) = \int Ef(X_1, \dots, X_{n-1}, t) \, P_{X_n}(dt)$$

$$\leq \int Ef(Y_1, \dots, Y_{n-1}, t) \, P_{X_n}(dt)$$

$$\leq \int Ef(Y_1, \dots, Y_{n-1}, t) \, P_{Y_n}(dt)$$

$$= Ef(\mathbf{Y}).$$

Here the first inequality follows from the induction hypothesis, and the second from the fact that $t \mapsto Ef(Y_1, \dots, Y_{n-1}, t)$ is convex for any componentwise convex f. $\square$

From Theorem 2.4.2 and Theorem 3.2.3 the following properties are easily derived.

Theorem 3.6.4. *Componentwise convex order has the properties (M), (C), (MI), (MA), and (IN).*

Notice that $\leq_{ccx}$ does not have the property (ID). In fact, assume that $\mathbf{X} = (X_1, X_2)$ where X_1 and X_2 are i.i.d. with a standard normal distribution. Moreover, let $\mathbf{Y} = 2\mathbf{X}$. Then $\mathbf{X} \leq_{ccx} \mathbf{Y}$, as can be seen from Theorem 3.6.5

below, but $(\mathbf{X}, \mathbf{X}) \not\leq_{ccx} (\mathbf{Y}, \mathbf{Y})$. To see this, take $f(x_1, x_2, x_3, x_4) = -x_1 x_3$. Then f is componentwise convex, but

$$Ef(\mathbf{X}, \mathbf{X}) = -EX_1^2 = -1 > -4 = -EY_1^2 = Ef(\mathbf{Y}, \mathbf{Y}).$$

With the help of the properties in Theorem 3.6.4 we can characterize componentwise convex order of normal distributions.

Theorem 3.6.5. *Let $\mathbf{X}$ and $\mathbf{X}'$ be n-dimensional random vectors with normal distributions $\mathcal{N}(\boldsymbol{\mu}, \boldsymbol{\Sigma})$ and $\mathcal{N}(\boldsymbol{\mu}', \boldsymbol{\Sigma}')$ respectively. Then $\mathbf{X} \leq_{ccx} \mathbf{X}'$ if and only if $\mu = \mu'$, $\sigma_{ii} \leq \sigma'_{ii}$ for all $i = 1, \dots, n$, and $\sigma_{ij} = \sigma'_{ij}$, $1 \leq i < j \leq n$.*

Proof. Let $\mathbf{Y} \equiv (0, \dots, 0)$, i.e. $\mathbf{Y}$ is concentrated at the origin with probability one, and let $\mathbf{Y}' = (Y_1', \dots, Y_n')$ be a vector of independent random variables with $Y_i' \sim \mathcal{N}(0, \sigma'_{ii} - \sigma_{ii})$. Then it follows from Theorem 3.6.3 that $\mathbf{Y} \leq_{ccx} \mathbf{Y}'$. According to Theorem 3.6.4 componentwise convex order has property (C). Hence, if $\mathbf{X}$ is independent of $\mathbf{Y}$ and $\mathbf{Y}'$, then $\mathbf{X} + \mathbf{Y} \leq_{ccx} \mathbf{X} + \mathbf{Y}'$. Since $\mathbf{X} + \mathbf{Y} = \mathbf{X}$ and $\mathbf{X} + \mathbf{Y}' =_{st} \mathbf{X}'$ this proves the if-part. The only-if part follows from Theorem 3.6.2 c) and d). $\square$

3.7 Stochastic Orders Defined by Difference Operators

A general principle of defining stochastic orders, which generalizes the approach of Section 1.6, is based on the use of difference operators. Recall that for a function $f : \mathbb{R}^n \to \mathbb{R}$, arbitrary $i \in \{1, \dots, n\}$ and $\varepsilon > 0$ the difference operator Δ_i^ε is defined as

$$\Delta_i^\varepsilon f(\mathbf{x}) = f(\mathbf{x} + \varepsilon \mathbf{e}_i) - f(\mathbf{x}),$$

where $\mathbf{e}_i = (0, \dots, 0, 1, 0, \dots, 0)$ denotes the i-th unit vector. In the univariate case the simpler notation

$$\Delta^\varepsilon f(x) = f(x + \varepsilon) - f(x)$$

is used.

For a tuple $\mathbf{i} = (i_1, \dots, i_k)$ of k not necessarily different numbers $i_j \in \{1, \dots, n\}$ the generator $\mathcal{F}_{\mathbf{i}+}$ consists of all functions such that

$$\Delta_{i_1}^{\varepsilon_1} \dots \Delta_{i_k}^{\varepsilon_k} f(\mathbf{x}) \geq 0 \quad \text{for all } \mathbf{x} \text{ and all } \varepsilon_1, \dots, \varepsilon_k > 0,$$

and analogously $\mathcal{F}_{\mathbf{i}-}$ is the set of all functions with

$$\Delta_{i_1}^{\varepsilon_1} \dots \Delta_{i_k}^{\varepsilon_k} f(\mathbf{x}) \leq 0 \quad \text{for all } \mathbf{x} \text{ and all } \varepsilon_1, \dots, \varepsilon_k > 0.$$

Given a set $\mathfrak{S}$ of such 'signed indices' let

$$\mathcal{F}_{\mathfrak{S}} = \bigcap_{\mathbf{j} \in \mathfrak{S}} \mathcal{F}_{\mathbf{j}}. \tag{3.7.1}$$

Many integral stochastic orders that have been considered in the literature possess a generator of the form $\mathcal{F}_{\mathfrak{S}}$.

Example 3.7.1. (Univariate examples of difference operator orders.) In the univariate case denote by $\mathbf{i}_k = (1, \ldots, 1)$ the tuple consisting of k repetitions of the number 1. Notice that

$$\Delta^\varepsilon f(x) \geq 0 \quad \text{for all } x \in \mathbb{R} \text{ and all } \varepsilon > 0$$

if and only if f is increasing. Hence $\mathcal{F}_{\mathbf{i}_1^+}$ is the set of all increasing functions and thus the generator of the usual stochastic order $\leq_{st}$. Similarly,

$$\Delta^{\varepsilon_1} \Delta^{\varepsilon_2} f(x) \geq 0 \quad \text{for all } x \in \mathbb{R} \text{ and all } \varepsilon_1, \varepsilon_2 > 0$$

if and only if f is convex. Hence $\mathcal{F}_{\mathbf{i}_2^+}$ generates the convex order $\leq_{cx}$, and taking the intersection yields $\mathcal{F}_{\mathfrak{I}_2}$ with $\mathfrak{I}_2 = \{\mathbf{i}_1^+, \mathbf{i}_2^+\}$, which generates the increasing convex order $\leq_{icx}$. Similarly $\mathcal{F}_{\mathbf{i}_2^-}$ generates the concave order $\leq_{cv}$, and the increasing concave order $\leq_{icv}$ is obtained from $\mathfrak{I}_2^- = \{\mathbf{i}_1^+, \mathbf{i}_2^-\}$. Considering higher differences yields the s-convex orders considered in Section 1.6.

Another order with such a characterization is the Laplace transform order $\leq_{Lt}$. Indeed, it can be shown that the maximal generator of $\leq_{Lt}$ described in Theorem 1.6.6 is equivalent to the set of all functions with

$$(-1)^{n+1} \Delta^{\varepsilon_1} \ldots \Delta^{\varepsilon_n} f(x) \geq 0 \text{ for all } x \in \mathbb{R}, \ \varepsilon_1, \ldots, \varepsilon_n > 0 \text{ and all } n = 1, 2, \ldots$$

Thus it is equivalent to $\mathcal{F}_{\mathfrak{I}}$ with $\mathfrak{I} = \{\mathbf{i}_1^+, \mathbf{i}_2^-, \mathbf{i}_3^+, \ldots\}$.

There are also numerous multivariate stochastic orders that can be characterized in this way. Notice that a function $f : \mathbb{R}^n \to \mathbb{R}$ is increasing if and only if

$$\Delta_i^\varepsilon f(\mathbf{x}) \geq 0 \quad \text{for all } \mathbf{x}, \text{ all } \varepsilon > 0 \text{ and all } i = 1, \ldots, n.$$

Hence the multivariate version of $\leq_{st}$ is generated by $\mathcal{F}_{\mathfrak{I}_{st}}$ for $\mathfrak{I}_{st} = \{i^+ : i = 1, \ldots, n\}$.

In Theorem 3.3.15 it was shown that the maximal generator of the upper orthant order $\leq_{uo}$ is the set of all functions with

$$\Delta_{i_1}^{\varepsilon_1} \ldots \Delta_{i_k}^{\varepsilon_k} f(\mathbf{x}) \geq 0 \quad \text{for all } \mathbf{x}, \text{ all } \varepsilon_1, \ldots, \varepsilon_k > 0$$

and all distinct $i_1, \ldots, i_k$, $1 \leq k \leq n$. Hence $\leq_{uo}$ is generated by $\mathcal{F}_{\mathfrak{I}_{uo}}$, where $\mathfrak{I}_{uo} = \{(i_1, \ldots, i_k)^+ : i_1, \ldots, i_k \text{ distinct}, 1 \leq k \leq n\}$. A similar characterization is possible for $\leq_{lo}$ as it is generated by the class of all functions f such that $-f(-\mathbf{x})$ is in $\mathcal{F}_{\mathfrak{I}_{uo}}$.

More examples like supermodular order and directionally convex order will be considered in separate sections below. Here we show that for such stochastic orders the convolution property (C) always holds.

Theorem 3.7.2. Let $\mathfrak{I}$ be an arbitrary set of signed indices and let $\leq_{\mathcal{F}_{\mathfrak{I}}}$ be the corresponding stochastic order. Then $\leq_{\mathcal{F}_{\mathfrak{I}}}$ has the property (C).

Proof. It is obvious from the definition that for a fixed index $\mathbf{i}$ the sets $\mathcal{F}_{\mathbf{i}+}$ and $\mathcal{F}_{\mathbf{i}-}$ are invariant under translations. Hence $\mathcal{F}_{\mathfrak{J}}$ as an intersection of such sets has the same property. Thus the assertion follows from Theorem 2.4.2 e). $\qquad\square$

Since all of the orders discussed here have generators consisting only of continuous functions, Theorem 2.5.5 can be applied to show that for all these orders generators consisting of infinitely differentiable functions are sufficient. The nice feature of the next theorem is that it says that we can just translate the conditions given in terms of differences into conditions given in terms of derivatives. Therefore define for a given $\mathbf{i} = (i_1, \ldots, i_k)$ the set $\mathcal{F}_{\mathbf{i}+}^{\infty}$ ($\mathcal{F}_{\mathbf{i}-}^{\infty}$) as the set of all infinitely differentiable functions with the property that

$$\frac{\partial^k}{\partial x_{i_1} \ldots \partial x_{i_k}} f(\mathbf{x}) \geq (\leq) \, 0 \quad \text{for all } \mathbf{x},$$

and for a given set $\mathfrak{J}$ of 'signed indices' let

$$\mathcal{F}_{\mathfrak{J}}^{\infty} = \bigcap_{\mathbf{j} \in \mathfrak{J}} \mathcal{F}_{\mathbf{j}}^{\infty}.$$

Note that the elements $\mathbf{j} \in \mathfrak{J}$ may have different signs. This happens, for example, for $\leq_{Lt}$ and $\leq_{lo}$, where alternating sequences of signs occur.

The following theorem contains some results for univariate as well as for multivariate orders. It is taken from Denuit and Müller (2001).

Theorem 3.7.3. *Let $\mathcal{F}_{\mathfrak{J}}$ be the generator of one of the stochastic orders $\leq_{st}, \leq_{s-cx}, \leq_{s-icx}, \leq_{s-cv}, \leq_{s-icv}, \leq_{Lt}, \leq_{uo}$ or $\leq_{lo}$. Then each of these orders $\leq_{\mathcal{F}_{\mathfrak{J}}}$ is also generated by the corresponding $\mathcal{F}_{\mathfrak{J}}^{\infty}$.*

Proof. In view of Theorem 2.5.5 it remains to show that the mentioned orders have property (C) and a generator consisting of continuous functions. The property (C) has been shown in general in Theorem 3.7.2. Hence it remains to be shown that there is always a generator consisting of continuous functions. For $\leq_{st}$ this was shown in Theorem 3.3.4 and for $\leq_{lo}$ and $\leq_{uo}$ it was demonstrated in the proof of Theorem 3.3.19. For all other orders it is obvious that even the maximal generator has this property. $\qquad\square$

3.8 Dependence Orders

Some order relations for multivariate distributions give rise to notions of dependence, when they are restricted to distributions with the same marginals, i.e. $X_i =_{st} Y_i$ for all i. If in this case the generator of an integral stochastic order $\leq_{\mathcal{F}}$ contains the functions $f(x) = x_i x_j$ for all $1 \leq i < j \leq n$, then it is natural to say that this order is a dependence order. This has the following reason: Let $\mathbf{X}$ and $\mathbf{Y}$ be random vectors with the same marginal distributions. Then for such an order $\mathbf{X} \leq_{\mathcal{F}} \mathbf{Y}$ implies $Cov(X_i, X_j) \leq Cov(Y_i, Y_j)$.

The covariance is a measure of linear dependence and is thus scale-dependent. Therefore invariance with respect to scale transformations is another desirable property of a dependence order, i.e.

$$(X_1, \ldots, X_n) \leq_{\mathcal{F}} (Y_1, \ldots, Y_n)$$

should imply

$$(f_1(X_1), \ldots, f_n(X_n)) \leq_{\mathcal{F}} (f_1(Y_1), \ldots, f_n(Y_n))$$

for all increasing functions $f_1, \ldots, f_n$.

In the *bivariate case* this means that the maximal generator should at least contain all separable functions f satisfying $f(x_1, x_2) = f_1(x_1) f_2(x_2)$ with f_1 and f_2 increasing. This leads to a stochastic order that is well known as *concordance order*; see Tchen (1980) or Joe (1997). Usually it is defined as follows.

Definition 3.8.1. Let $\mathbf{X} = (X_1, X_2)$ and $\mathbf{Y} = (Y_1, Y_2)$ be bivariate random vectors with the same marginals. Then $\mathbf{X}$ is said to be smaller than $\mathbf{Y}$ in *concordance order* (written as $\mathbf{X} \leq_c \mathbf{Y}$), if

$$P(X_1 \leq s, X_2 \leq t) \leq P(Y_1 \leq s, Y_2 \leq t) \quad \text{for all } s \text{ and } t.$$

There are many different equivalent characterizations of this order that are summarized in the following result.

Theorem 3.8.2. *Let* $\mathbf{X}$ *and* $\mathbf{Y}$ *be bivariate random vectors with the same marginals. Then the following conditions are equivalent:*

(i) $\mathbf{X} \leq_c \mathbf{Y}$.

(ii) $P(X_1 > s, X_2 > t) \leq P(Y_1 > s, Y_2 > t)$ *for all* s *and* t.

(iii) $E(f_1(X_1) f_2(X_2)) \leq E(f_1(Y_1) f_2(Y_2))$ *for all increasing* f_1 *and* f_2.

(iv) $Cov(f_1(X_1), f_2(X_2)) \leq Cov(f_1(Y_1), f_2(Y_2))$ *for all increasing* f_1 *and* f_2.

(v) $Ef(\mathbf{X}) \leq Ef(\mathbf{Y})$ *for all* Δ-*monotone functions* $f : \mathbb{R}^2 \to \mathbb{R}$.

(vi) $Ef(\mathbf{X}) \leq Ef(\mathbf{Y})$ *for all supermodular functions* $f : \mathbb{R}^2 \to \mathbb{R}$, *i.e. for all functions with*

$$f(x_1 + \varepsilon, x_2 + \delta) + f(x_1, x_2) \geq f(x_1, x_2 + \delta) + f(x_1 + \varepsilon, x_2)$$

for all $x_1, x_2 \in \mathbb{R}$ *and all* $\varepsilon, \delta > 0$.

Proof. The equivalence of parts (i) to (iv) can easily be obtained by combining the results in Tchen (1980), Rüschendorf (1981b) and Bergmann (1991). The equivalences to (v) and (vi) were shown by Tchen (1980), Cambanis and Simons (1982) and Cambanis et al. (1976), but always under some additional regularity conditions. It follows from Müller and Scarsini (2000), however, that these regularity conditions are not necessary. $\square$

Concordance order fulfills all of the following nine properties that have been proposed by Kimeldorf and Sampson (1987, 1989) (see also Joe (1997), p. 38) as axioms for a bivariate dependence order $\preceq$ that compares bivariate distributions F and F' with the same marginals F_1 and F_2.

(P1) (concordance) $F \preceq F'$ implies $F(x_1, x_2) \leq F'(x_1, x_2)$ for all x_1, x_2;

(P2) (transitivity) $F \preceq F'$ and $F' \preceq F''$ imply $F \preceq F''$;

(P3) (reflexivity) $F \preceq F$ for all F;

(P4) (antisymmetry) $F \preceq F'$ and $F' \preceq F$ imply $F = F'$;

(P5) (bound) $F^- \preceq F \preceq F^+$, where $F^-(\mathbf{x}) = \max\{F_1(x_1) + F_2(x_2) - 1, 0\}$ is the lower Fréchet bound and $F^+(\mathbf{x}) = \min\{F_1(x_1), F_2(x_2)\}$ is the upper Fréchet bound;

(P6) (weak convergence) $F_k \preceq F'_k$, $k \in \mathbb{N}$, and $F_k \to F$, $F'_k \to F'$ weakly imply $F \preceq F'$;

(P7) (invariance to order of indices) $(X_1, X_2) \preceq (X'_1, X'_2)$ implies $(X_2, X_1) \preceq (X'_2, X'_1)$;

(P8) (invariance to increasing transforms) $(X_1, X_2) \preceq (X'_1, X'_2)$ implies $(g_1(X_1), g_2(X_2)) \preceq (g_1(X'_1), g_2(X'_2))$ for all increasing functions g_1 and g_2;

(P9) (invariance to decreasing transforms) $(X_1, X_2) \preceq (X'_1, X'_2)$ implies $(g(X'_1), X'_2) \preceq (g(X_1), X_2)$ for all decreasing functions g.

Concordance order is the only integral stochastic order satisfying all these nine properties. In fact, the following much stronger result is true.

Theorem 3.8.3. *Assume that $\leq_{\mathcal{F}}$ is an integral stochastic order for bivariate distributions with the same marginals that has the following properties:*

a) $\mathbf{X} \leq_{\mathcal{F}} \mathbf{Y}$ *implies* $Cov(X_1, X_2) \leq Cov(Y_1, Y_2)$.

b) $\leq_{\mathcal{F}}$ *is invariant to increasing transformations, i.e. (P8) holds.*

c) $F^- \leq_{\mathcal{F}} F^+$ *for arbitrary marginals F_1 and F_2.*

Then $\leq_{\mathcal{F}}$ is necessarily the concordance order.

Proof. It follows from property a) that the maximal generator of $\leq_{\mathcal{F}}$ contains the function $f(x_1, x_2) = x_1 x_2$. Property b) then implies that the maximal generator also contains the functions $f(x_1, x_2) = f_1(x_1) f_2(x_2)$ with f_1, f_2 increasing. According to the equivalence of part (iii) and (vi) in Theorem 3.8.2 the maximal generator of $\leq_{\mathcal{F}}$ thus has to contain all supermodular functions.

On the other hand, assume that F_1 assigns probability $1/2$ to x_1 and $x_1 + \varepsilon$, and that F_2 assigns probability $1/2$ to x_2 and $x_2 + \delta$. In this case $F^- \leq_{\mathcal{F}} F^+$ means

$$f(x_1 + \varepsilon, x_2 + \delta) + f(x_1, x_2) \geq f(x_1, x_2 + \delta) + f(x_1 + \varepsilon, x_2)$$

for all $f \in \mathcal{F}$. According to assumption c) this must hold for arbitrary x_1 and x_2 and all $\varepsilon, \delta > 0$, and thus all $f \in \mathcal{F}$ must be supermodular. Hence the maximal generator of $\leq_{\mathcal{F}}$ consists exactly of the set of all supermodular functions, and thus $\leq_{\mathcal{F}}$ is necessarily $\leq_c$. $\qquad\square$

Notice that assumption a) is weaker than (P1), and that assumption c) is a weakening of (P5).

Of course there are also other interesting bivariate dependence orders, but according to Theorem 3.8.3 they cannot be integral orders. We will not describe them here; instead we refer the interested reader to the survey in Scarsini and Shaked (1997) and to Joe (1997).

Consider now dependence orders in *higher dimensions*. Most of the nine properties suggested for a bivariate dependence order can be extended naturally to higher dimensions. In (P5), however, we have to omit the condition $F^- \preceq F$, since in general the lower Fréchet bound is not a distribution function. Moreover, (P9) does not make sense and is therefore replaced by the natural requirement of closure under marginalization. Thus it is natural to state the following nine desirable properties for a multivariate dependence order, which were suggested by Joe (1997).

Recall that $\Gamma(F_1, \dots, F_n)$ is the set of random vectors which have the same univariate marginals $F_1, \dots, F_n$. A binary relation $\preceq$ on $\Gamma(F_1, \dots, F_n)$ is said to be a *multivariate dependence order* (MPDO), if it fulfills the following properties:

(P1) (bivariate concordance) $F \preceq F'$ implies $F_{ij} \leq_c F'_{ij}$ for all $1 \leq i < j \leq n$, where F_{ij} and F'_{ij} are the (i, j)-bivariate marginals;

(P2) (transitivity) $F \preceq F'$ and $F' \preceq F''$ imply $F \preceq F''$;

(P3) (reflexivity) $F \preceq F$ for all F;

(P4) (antisymmetry) $F \preceq F'$ and $F' \preceq F$ imply $F = F'$;

(P5) (bound) $F \preceq F^+$, where $F^+(\mathbf{x}) = \min_i F_i(x_i)$ is the upper Fréchet bound;

(P6) (weak convergence) $F_k \preceq F'_k$, $k \in \mathbb{N}$, and $F_k \to F$, $F'_k \to F'$ weakly imply $F \preceq F'$;

(P7) (invariance with respect to permutation of indices) $(X_1, \dots, X_n) \preceq (X'_1, \dots, X'_n)$ implies $(X_{i_1}, \dots, X_{i_n}) \preceq (X'_{i_1}, \dots, X'_{i_n})$ for all permutations $(i_1, \dots, i_n)$ of $(1, \dots, n)$;

(P8) (invariance with respect to increasing transforms) $(X_1, \ldots, X_n) \preceq$ $(X_1', \ldots, X_n')$ implies $(g_1(X_1), \ldots, g_n(X_n)) \preceq (g_1(X_1'), \ldots, g_n(X_n'))$ for all increasing functions $g_1, \ldots, g_n$.

(P9) (closure under marginalization) $(X_1, \ldots, X_n) \preceq (X_1', \ldots, X_n')$ implies $(X_{i_1}, \ldots, X_{i_k}) \preceq (X_{i_1}', \ldots, X_{i_k}')$ for all $i_1 < \ldots < i_k$, $2 \le k < n$;

In contrast to the bivariate case, however, there are several integral stochastic orders that fulfill all these properties in the case $n \ge 3$. They can be introduced by generalizing the different equivalent conditions of Theorem 3.8.2. The natural generalizations of characterizations a) and b) lead to the orthant orders $\ge_{lo}$ (it is not a misprint that 'greater equal' is written here) and $\le_{uo}$, which have already been introduced in Definition 3.3.1. Joe (1997) uses the notations $\le_{cL}$ and $\le_{cU}$ for these orders, which stand for *lower concordance order* and *upper concordance order*, respectively. We follow him when we consider the orthant orders restricted to the comparison of random vectors with equal marginals. Formally the orders $\le_{cL}$ and $\le_{cU}$ can be introduced as integral stochastic orders by adding to the respective generators of $\ge_{lo}$ and $\le_{uo}$ all functions $f : \mathbb{R}^n \to \mathbb{R}$ that depend only on one variable. Both these orders fulfill all of the properties (P1) - (P9). Property (P6) has been shown in Theorem 3.3.19, all other properties are easy to see.

Example 3.8.4. ($\le_{cL}$ and $\le_{cU}$ are different) Let $n = 3$ and assume that $\mathbf{X}$ is uniformly distributed in the four points $(0, 0, 0), (1, 1, 0), (1, 0, 1)$ and $(0,1,1)$, and let $\mathbf{Y}$ be uniformly distributed in the four points $(1,0,0), (0,1,0), (0,0,1)$ and $(1,1,1)$. Then $\mathbf{X}$ and $\mathbf{Y}$ have the same marginals (in fact they even have the same bivariate marginals), but it is easy to check that $F_{\mathbf{X}}(\mathbf{t}) \ge F_{\mathbf{Y}}(\mathbf{t})$ for all $\mathbf{t}$ and that $\bar{F}_{\mathbf{X}}(\mathbf{t}) \le \bar{F}_{\mathbf{Y}}(\mathbf{t})$ for all $\mathbf{t}$. Thus we have $\mathbf{X} \le_{cU} \mathbf{Y}$, but not $\mathbf{X} \le_{cL} \mathbf{Y}$. Instead we even have $\mathbf{X} \ge_{cL} \mathbf{Y}$!

The orders $\le_{cL}$ and $\le_{cU}$ are rather weak dependence orders. In fact, in the above example we have $\mathbf{X} \le_{uo} \mathbf{Y}$, but in our opinion these random vectors should not be comparable by any reasonable dependence order, since $\mathbf{Y} =_{st} (1, 1, 1) - \mathbf{X}$, i.e. $\mathbf{Y}$ can be considered as a reflection of $\mathbf{X}$ with respect to the point $(1/2, 1/2, 1/2)$, and hence one should consider them as of the same degree of dependence. Therefore Joe (1990) suggested the following stronger definition as a generalization of bivariate concordance order.

Definition 3.8.5. Let $\mathbf{X}$ and $\mathbf{Y}$ be n-dimensional random vectors. Then $\mathbf{X}$ is said to be less than $\mathbf{Y}$ in *concordance order* (written $\mathbf{X} \le_c \mathbf{Y}$), if $F_{\mathbf{X}}(\mathbf{t}) \le F_{\mathbf{Y}}(\mathbf{t})$ as well as $\bar{F}_{\mathbf{X}}(\mathbf{t}) \le \bar{F}_{\mathbf{Y}}(\mathbf{t})$ hold for all $\mathbf{t}$, i.e. if $\mathbf{X} \le_{cU} \mathbf{Y}$ and $\mathbf{X} \le_{cL} \mathbf{Y}$.

Notice that the requirement of equal marginals is not included in the definition, since it is automatically satisfied if both inequalities $F_{\mathbf{X}}(\mathbf{t}) \le F_{\mathbf{Y}}(\mathbf{t})$ and $\bar{F}_{\mathbf{X}}(\mathbf{t}) \le \bar{F}_{\mathbf{Y}}(\mathbf{t})$ hold for all $\mathbf{t}$.

Example 3.8.6. (Concordance order for normal distribution.) It follows immediately from Slepian's inequality (Theorem 3.3.20) that normal random vectors are ordered with respect to $\leq_c$, $\leq_{cL}$ and $\leq_{cU}$, if and only if they have the same marginals and all their covariances are ordered; the same holds for arbitrary elliptically contoured distributions.

The definition shows that $\leq_c$ is generated by all indicator functions of the form $\mathbf{1}_{(-\infty,\mathbf{t}]}$ and $\mathbf{1}_{(\mathbf{t},\infty)}$ for $\mathbf{t} \in \mathbb{R}^n$. It follows immediately from Theorem 3.3.15 that the maximal generator contains all functions that can be written as a sum of a Δ-monotone and a Δ-antitone function. These functions must have the property that all higher differences of even order up to n are non-negative, but this property does not provide a full characterization of the maximal generator in the case $n \geq 3$, as we will see in Example 3.9.7. We are not aware of a simple description of the maximal generator. Nevertheless we can derive interesting properties of $\leq_c$.

Theorem 3.8.7. *The order $\leq_c$ has all the properties (P1) – (P9), and, moreover, the properties (M), (C), (MI), (MA), (ID), and (IN).*

Proof. The properties (P1) – (P9) hold for $\leq_{cL}$ as well as $\leq_{cU}$, and thus they are inherited by $\leq_c$. The properties (M), (C), (MI), (MA), (ID) and (IN) follow immediately from Theorem 2.4.2 and Theorem 3.2.3, respectively. □

3.9   Supermodular Order

Now we will turn our attention to a dependence order that is stronger than $\leq_c$. It was mentioned above that the functions in the maximal generator of concordance order have the property that all differences of even order are non-negative. We get a larger class of functions if we only require the second differences to be non-negative. These functions are usually called *supermodular*. Sometimes they are also called *L-superadditive* (with L for *lattice*, see Marshall and Olkin (1979), p. 146) or *superadditive* (see Hu (2000)). The name 'superadditive', however, is also used in a different context, see Definition 1.7.9. Therefore we prefer the name 'supermodular'. In Heyman and Sobel (1984) these functions are called functions with *increasing differences*. The formal definition can be given as follows.

Definition 3.9.1. A function $f : \mathbb{R}^n \to \mathbb{R}$ is said to be *supermodular*, if

$$\Delta_i^\varepsilon \Delta_j^\delta f(\mathbf{x}) \geq 0 \qquad (3.9.1)$$

holds for all $\mathbf{x} \in \mathbb{R}^n$, $1 \leq i < j \leq n$ and all $\varepsilon, \delta > 0$. The set of all supermodular functions is denoted by $\mathcal{SM}$, and $\mathcal{ISM}$ shall be the set of all increasing supermodular functions.

Remark 3.9.2. Supermodular functions can alternatively be defined as follows. A function $f : \mathbb{R}^n \to \mathbb{R}$ is supermodular, if

$$f(\mathbf{x} \wedge \mathbf{y}) + f(\mathbf{x} \vee \mathbf{y}) \geq f(\mathbf{x}) + f(\mathbf{y}) \quad \text{for all } \mathbf{x} \text{ and } \mathbf{y}, \tag{3.9.2}$$

where the lattice operators $\wedge$ and $\vee$ are defined as

$$\mathbf{x} \wedge \mathbf{y} = (\min\{x_1, y_1\}, \ldots, \min\{x_n, y_n\})$$

and

$$\mathbf{x} \vee \mathbf{y} = (\max\{x_1, y_1\}, \ldots, \max\{x_n, y_n\}).$$

The equivalence of (3.9.1) and (3.9.2) was shown by Kemperman (1977).

In the following theorem some useful properties of supermodular functions are collected.

Theorem 3.9.3. *a) If f is twice differentiable then $f \in \mathcal{SM}$ if and only if*

$$\frac{\partial^2}{\partial x_i \partial x_j} f(\mathbf{x}) \geq 0 \text{ for all } \mathbf{x} \text{ and all } 1 \leq i < j \leq n.$$

b) If $g_1, \ldots, g_n : \mathbb{R} \to \mathbb{R}$ are increasing and $f \in \mathcal{SM}$ then

$$f(g_1(\cdot), \ldots, g_n(\cdot)) \in \mathcal{SM}.$$

c) If $f, g \in \mathcal{SM}$ then $\alpha f + \beta g \in \mathcal{SM}$ for all $\alpha, \beta \geq 0$.
d) If $f, g \in \mathcal{ISM}$ and $f, g \geq 0$, then $f \cdot g \in \mathcal{ISM}$.
e) If $f \in \mathcal{ISM}$, then $\max\{f, c\} \in \mathcal{ISM}$ for all real constants c.
f) If f is increasing and supermodular and $\phi : \mathbb{R} \to \mathbb{R}$ is increasing and convex then $\phi \circ f \in \mathcal{ISM}$.

For proofs of these properties as well as examples, the reader is referred to Marshall and Olkin (1979), p. 146, and Bäuerle (1997a).

There are interesting stochastic orders based on the notion of supermodularity.

Definition 3.9.4. a) A random vector $\mathbf{X} = (X_1, \ldots, X_n)$ is said to be smaller than the random vector $\mathbf{Y} = (Y_1, \ldots, Y_n)$ in the *supermodular order*, written $\mathbf{X} \leq_{sm} \mathbf{Y}$, if $Ef(\mathbf{X}) \leq Ef(\mathbf{Y})$ for all supermodular functions f such that the expectations exist.
b) A random vector $\mathbf{X} = (X_1, \ldots, X_n)$ is said to be smaller than the random vector $\mathbf{Y} = (Y_1, \ldots, Y_n)$ in the *increasing supermodular order*, written $\mathbf{X} \leq_{ism} \mathbf{Y}$, if $Ef(\mathbf{X}) \leq Ef(\mathbf{Y})$ for all increasing supermodular functions f such that the expectations exist.
c) A random vector $\mathbf{X} = (X_1, \ldots, X_n)$ is said to be smaller than the random vector $\mathbf{Y} = (Y_1, \ldots, Y_n)$ in the *symmetric supermodular order*, written $\mathbf{X} \leq_{symsm} \mathbf{Y}$, if $Ef(\mathbf{X}) \leq Ef(\mathbf{Y})$ for all symmetric supermodular functions f such that the expectations exist.

To the best of our knowledge the definition of $\leq_{sm}$ first appeared in Szekli, Disney, and Hur (1994), and has been considered since then by Bäuerle (1997a,b), Bäuerle and Müller (1998), Müller (1997c), Müller and Scarsini (2000), Shaked and Shanthikumar (1997), and Szekli (1995), among others. Implicitly, however, it has already been used earlier; see Tchen (1980), Rolski (1986) and Meester and Shanthikumar (1993).

Symmetric supermodular order was used in Bäuerle (1997a), Bäuerle and Müller (1998) and Frostig (2001). In contrast to $\leq_{sm}$ it is not a dependence order. It can also be helpful to compare two random vectors of independent components; see Frostig (2001). Therefore we will concentrate here on the order $\leq_{sm}$ that will turn out to be a dependence order.

The results of the following theorem are not difficult to show, but for completeness the proof is given.

Theorem 3.9.5. *The supermodular order has the following properties:*
a) If $\mathbf{X} \leq_{sm} \mathbf{Y}$ *then* $\mathbf{X}$ *and* $\mathbf{Y}$ *have the same marginals.*
b) If $\mathbf{X} \leq_{sm} \mathbf{Y}$ *then* $\mathbf{X} \leq_c \mathbf{Y}$.
c) If $\mathbf{X} \leq_{sm} \mathbf{Y}$ *then* $Cov(X_i, X_j) \leq Cov(Y_i, Y_j)$ *for all* $1 \leq i < j \leq n$.
d) If $\mathbf{X} \leq_{sm} \mathbf{Y}$ *then* $\mathbf{X} \leq_{plcx} \mathbf{Y}$.

Proof. a) follows from the fact that a function $f : \mathbb{R}^n \to \mathbb{R}$ is supermodular if it depends only on one variable. b) holds since the functions $\mathbf{1}_{(-\infty,\mathbf{t}]}$ and $\mathbf{1}_{(\mathbf{t},\infty)}$ are supermodular for all $\mathbf{t}$, and c) is a consequence of a) and the fact that the functions $f(\mathbf{x}) = x_i x_j$ are supermodular for all $1 \leq i < j \leq n$.

d) It is sufficient to show that for all $\mathbf{a} \in \mathbb{R}^n_+$ and all convex functions $f : \mathbb{R} \to \mathbb{R}$ the function $\mathbf{x} \mapsto g(\mathbf{x}) = f(\mathbf{a}^T \mathbf{x})$ is supermodular. Indeed, the convexity of f implies

$$\Delta_i^\varepsilon \Delta_j^\delta g(\mathbf{x}) = f(\mathbf{a}^T\mathbf{x} + \varepsilon a_i + \delta a_j) + f(\mathbf{a}^T\mathbf{x}) - f(\mathbf{a}^T\mathbf{x} + \varepsilon a_i) - f(\mathbf{a}^T\mathbf{x} + \delta a_j)$$
$$= \Delta^{\varepsilon a_i} \Delta^{\delta a_j} f(\mathbf{a}^T\mathbf{x}) \geq 0.$$

$\square$

The increasing supermodular order $\leq_{ism}$ can also hold for random vectors with different marginals. It is necessary, however, that the components are ordered with respect to $\leq_{st}$.

Theorem 3.9.6. *The increasing supermodular order has the following properties:*
a) If $\mathbf{X} \leq_{ism} \mathbf{Y}$ *then* $X_i \leq_{st} Y_i$ *for all* $i = 1, \ldots, n$.
b) If $\mathbf{X} \leq_{ism} \mathbf{Y}$ *then* $\mathbf{X} \leq_{uo} \mathbf{Y}$.
c) If $\mathbf{X} \leq_{ism} \mathbf{Y}$ *then* $\mathbf{X} \leq_{iplcx} \mathbf{Y}$.

Proof. a) follows from the fact that for any increasing function $g : \mathbb{R} \to \mathbb{R}$ and any $i = 1, \ldots, n$ the function $f : \mathbb{R}^n \to \mathbb{R}$ given by $f(\mathbf{x}) = g(x_i)$ is increasing and supermodular. b) holds since the functions $\mathbf{1}_{(\mathbf{t},\infty)}$ are increasing and supermodular for all $\mathbf{t}$, and the proof of c) is similar to that of Theorem 3.9.5 d). □

It was shown in Theorem 3.8.2 that supermodular order coincides with concordance order in the bivariate case. The following example shows that this is no longer true in higher dimensions.

Example 3.9.7. (In dimension $n = 3$ the supermodular order is strictly stronger than the concordance order, Müller and Scarsini (2000).)
Let $\mathbf{X} = (X_1, X_2, X_3)$ and $\mathbf{Y} = (Y_1, Y_2, Y_3)$ be random vectors such that $\mathbf{X}$ is uniformly distributed on the six points

$$(2,2,1),\ (2,1,2),\ (1,2,2),\ (1,1,1),\ (0,0,2)\ \text{and}\ (2,0,0),$$

and let $\mathbf{Y}$ be uniformly distributed on the six points

$$(2,2,2),\ (2,1,1),\ (1,2,1),\ (1,1,2),\ (2,0,2)\ \text{and}\ (0,0,0).$$

It is easy to see that $\mathbf{X}$ and $\mathbf{Y}$ have the same marginal distributions. Furthermore, we claim that $\mathbf{X} \leq_{uo} \mathbf{Y}$ and $\mathbf{X} \leq_{lo} \mathbf{Y}$. As Dyckerhoff and Mosler (1997) have shown, it is sufficient to check the 27 lattice points in $\{0,1,2\}^3$. A straightforward, but tedious, calculation shows that indeed in all these points $\bar{F}_{\mathbf{X}}(\mathbf{t}) \leq \bar{F}_{\mathbf{Y}}(\mathbf{t})$ as well as $F_{\mathbf{X}}(\mathbf{t}) \leq F_{\mathbf{Y}}(\mathbf{t})$. Hence $\mathbf{X} \leq_c \mathbf{Y}$ holds.

On the other hand, there are supermodular functions $f : \mathbb{R}^3 \to \mathbb{R}$ such that $Ef(\mathbf{X}) > Ef(\mathbf{Y})$. As an example, consider the function

$$f(x_1, x_2, x_3) = \max\{x_1 + x_2 + x_3 - 4, 0\}.$$

According to Theorem 3.9.3 this function is supermodular, since it is a composition of an increasing convex function on the real line and an increasing supermodular function. However,

$$Ef(\mathbf{X}) = \frac{1}{2} > \frac{1}{3} = Ef(\mathbf{Y}).$$

Note that Joe (1990) had given an example which shows that $\leq_{sm}$ is strictly stronger than $\leq_c$ for dimension $n \geq 4$, and in Joe (1997), p. 56, the case $n = 3$ was posed as an unsolved problem.

The construction in Example 3.9.7 was already considered by Müller (1997c). There it was used to show that the concordance order of two vectors does not imply the stop-loss order for the sum of the components. This is a question of practical relevance for the calculation of reinsurance premiums for portfolios of dependent risks. We will turn back to this problem in Section 8.3.

Joe (1997), p. 56, also posed the problem whether or not supermodular order fulfills all the nine properties of an MPDO defined on page 110. This question was answered in the affirmative by Müller and Scarsini (2000), where the following material is taken from.

Theorem 3.9.8. *The supermodular order is a multivariate positive depen-*
dence order, i.e. it fulfills all the properties (P1) – (P9).

Proof. The properties (P1) and (P9) follow from the fact that a supermodular
function of $k < n$ variables remains supermodular if it is considered as a
function of n variables. (P2) and (P3) are trivial. (P4) follows from the
fact that $\mathbf{X} \leq_{sm} \mathbf{Y}$ implies $F_{\mathbf{X}} \leq F_{\mathbf{Y}}$ as well as $\bar{F}_{\mathbf{X}} \leq \bar{F}_{\mathbf{Y}}$. Property
(P5) is known as Lorentz inequality and was shown under some additional
regularity conditions in Tchen (1980). It follows from Theorem 3.9.13 that
these regularity conditions can be removed. (P7) is obvious from the definition
of supermodularity and (P8) follows immediately from Theorem 3.9.3 b).
Hence only (P6) remains to be shown. This will be done in Theorem 3.9.12. $\square$

We will now turn to the problem of showing the weak convergence property
for supermodular order. Since it is much easier to show this for the increasing
supermodular order, that case will be treated first.

Due to Theorem 2.4.2 an integral stochastic order is closed with respect to
weak convergence if and only if it is generated by a class of bounded continuous
functions. Therefore we are going to show now that for $\leq_{ism}$ it is sufficient to
consider bounded continuous functions in $\mathcal{ISM}$. The following Lemma will
be helpful in the proof.

Lemma 3.9.9. *Let $f : \mathbb{R} \to \mathbb{R}$ be increasing. Then there is a sequence (f_m)*
of right continuous increasing functions which converges monotonically to f.

Proof. Define $s_{k,m} = \inf\{x : f(x) \geq k/2^m\}$, $k \in \mathbb{Z}$, $m \in \mathbb{N}$, and let
$S_m = \{s_{k,m} : k \in \mathbb{Z}\} \cup \{k/2^m : k \in \mathbb{Z}\}$. Then S_m is a discrete set. Next
define $f_m(x) = \max\{f(t) : t \in S_m, t \leq x\}$. Then f_m is increasing and right-
continuous. Moreover, since $S_m \subset S_{m+1}$ and f is increasing, we also have
$f_m \leq f_{m+1} \leq f$. Hence it remains to show that $\tilde{f} = \lim_{m \to \infty} f_m \geq f$. Fix x
and distinguish the following two cases.
(i) Assume that f is left-continuous in x. For all $m \in \mathbb{N}$ there is some $x_m \in S_m$
with $x - 1/2^m \leq x_m \leq x$. Hence

$$\tilde{f}(x) = \lim_{m \to \infty} f_m(x) \geq \lim_{m \to \infty} f(x_m) = f(x).$$

(ii) If f is not left continuous in x, then there is some m_0 such that

$$\lim_{t \uparrow x} f(t) < f(x) - \frac{1}{2^{m_0}}.$$

But this implies that $x \in S_m$ and hence $\tilde{f}(x) = f_m(x) = f(x)$ for all
$m \geq m_0$. $\square$

Theorem 3.9.10. *The following statements are equivalent:*

(i) $\mathbf{X} \leq_{ism} \mathbf{Y}$,

(ii) $Ef(\mathbf{X}) \leq Ef(\mathbf{Y})$ *for all bounded continuous increasing supermodular functions* f.

Proof. It is clear that (i) implies (ii). Therefore we only have to show that (ii) implies (i). This will be done in several steps.

Step 1: We claim that (ii) implies $Ef(\mathbf{X}) \leq Ef(\mathbf{Y})$ for all bounded *semi-continuous* functions $f \in \mathcal{ISM}$. (Recall that a function is semi-continuous, if it is a monotone limit of continuous functions.)

Verification: Let P_m, $m \in \mathbb{N}$, be the uniform distribution on the segment $[\mathbf{0}, (1/m, \ldots, 1/m)]$ and let $f \in \mathcal{ISM}$ be a bounded lower semi-continuous function. Then the function $f_m(\mathbf{x}) = \int f(\mathbf{x} - \mathbf{y}) P_m(\mathbf{dy})$ is bounded and continuous, and $f_m \in \mathcal{ISM}$. Moreover, for $m \to \infty$ the sequence (f_m) converges to f from below. Hence the assertion follows from the monotone convergence theorem. The case of upper semi-continuous functions can be treated analogously by taking for P_m the uniform distribution on the segment $[(-1/m, \ldots, -1/m), \mathbf{0}]$.

Step 2: We claim that step 1 implies $Ef(\mathbf{X}) \leq Ef(\mathbf{Y})$ for all bounded functions $f \in \mathcal{ISM}$.

Verification: We use an approximation as in Lemma 3.9.9. Define the real-valued functions

$$f_i(t) = \lim_{\substack{x_j \to \infty \\ j \neq i}} f(x_1, \ldots, x_{i-1}, t, x_{i+1}, \ldots, x_n) \quad \text{for } i = 1, \ldots, n.$$

Since f is bounded and supermodular, these limits do exist. Moreover, f_i is a bounded increasing function. Let $S_m(f_i)$, $m \in \mathbb{N}$, $i = 1, \ldots, n$, be the set of points as defined in the proof of Lemma 3.9.9 and let

$$f_{im}(t) = \max\{f_i(s) : s \in S_m(f_i), \ s \leq t\} \quad \text{for } t \in \mathbb{R}, \ 1 \leq i \leq n, \ m \in \mathbb{N}$$

be the corresponding lower approximation of f_i. Now let $S_m = S_m(f_1) \times \ldots \times S_m(f_n)$ and define

$$f_m(x_1, \ldots, x_n) = \max\{f(\mathbf{t}) : \mathbf{t} \in S_m, \mathbf{t} \leq \mathbf{x}\}.$$

Then f_m is increasing and right-continuous (hence semi-continuous) and $f_m \in \mathcal{ISM}$. We claim that f_m converges monotonically to f. It is clear that $m \mapsto f_m$ is monotone, hence only the convergence has to be verified. To do this, fix $\mathbf{x} = (x_1, \ldots, x_n)$ and $\varepsilon > 0$. According to Lemma 3.9.9 there is some $m \in \mathbb{N}$ such that $f_i(x_i) \leq f_{im}(x_i) + \varepsilon/n$ for all $i = 1, \ldots, n$. Let $x_{im} \in S_m(f_i)$ be the corresponding point with $x_{im} \leq x_i$ and $f_{im}(x_i) = f_i(x_{im})$. Hence $\mathbf{x}_m = (x_{1m}, \ldots, x_{nm}) \in S_m$ and $\mathbf{x}_m \leq \mathbf{x}$. Therefore

$$f(\mathbf{x}) - f_m(\mathbf{x}) \leq f(\mathbf{x}) - f(\mathbf{x}_m) \leq \sum_{i=1}^{n} f_i(x_i) - f_i(x_{im}) \leq n \cdot \frac{\varepsilon}{n} = \varepsilon.$$

The second inequality follows from the fact that supermodularity of f implies $f(\mathbf{x} + \delta \mathbf{e}_i) - f(\mathbf{x}) \leq f_i(x_i + \delta) - f_i(x_i)$ for all $i = 1, \ldots, n$, $\delta > 0$, where $\mathbf{e}_i$ denotes the i-th unit vector. Thus we have shown that the sequence (f_m) converges to f from below, and hence the assertion follows from the monotone convergence theorem.

Step 3: Finally we claim that this implies $Ef(\mathbf{X}) \leq Ef(\mathbf{Y})$ also for unbounded functions $f \in \mathcal{ISM}$.

Verification: It suffices to find sequences of upper and lower truncations of unbounded function $f \in \mathcal{ISM}$ such that the truncations are still in $\mathcal{ISM}$ and converge monotonically to the original function f. For a lower truncation we use $f_m(\mathbf{x}) = f(\mathbf{x} \vee -m\mathbf{1})$ and an upper truncation is $f_m(\mathbf{x}) = f(\mathbf{x} \wedge m\mathbf{1})$, where $\mathbf{1} = (1, \ldots, 1)$. It follows immediately from Theorem 3.9.3 b) that both of these approximations are in $\mathcal{ISM}$.   □

Next we will investigate the relationship between $\leq_{ism}$ and $\leq_{sm}$. The result is of interest in its own.

Theorem 3.9.11. *The following statements are equivalent:*

(i) $\mathbf{X} \leq_{sm} \mathbf{Y}$,

(ii) $\mathbf{X}$ *and* $\mathbf{Y}$ *have the same marginals and* $\mathbf{X} \leq_{ism} \mathbf{Y}$,

(iii) $\mathbf{X}$ *and* $\mathbf{Y}$ *have the same expectation and* $\mathbf{X} \leq_{ism} \mathbf{Y}$.

Proof. a) The implication (i) $\Rightarrow$ (ii) is trivial. To show that (ii) implies (i) let us assume that $\mathbf{X}$ and $\mathbf{Y}$ have the same marginals and $\mathbf{X} \leq_{ism} \mathbf{Y}$, and let f be any bounded supermodular function. We have to show that $Ef(\mathbf{X}) \leq Ef(\mathbf{Y})$. To do so define for arbitrary real a

$$f_a(\mathbf{x}) = f(\mathbf{x}) - \sum_{i=1}^{n} f\left((x_i - a) \cdot \mathbf{e}_i + a\mathbf{1}\right) + (n-1)f(a\mathbf{1}).$$

Then f_a is supermodular, and f_a is nonnegative and increasing on $[a\mathbf{1}, \infty)$ since for all $\varepsilon > 0$ and all $\mathbf{x} \geq a\mathbf{1}$

$$\begin{aligned}
&f_a(x_1, \ldots, x_i + \varepsilon, \ldots, x_n) - f_a(x_1, \ldots, x_i, \ldots, x_n) \\
&\geq \quad f_a(a, x_2, \ldots, x_i + \varepsilon, \ldots, x_n) - f_a(a, x_2, \ldots, x_i, \ldots, x_n) \\
&\geq \quad \cdots \\
&\geq \quad f_a(a, \ldots, a, x_i + \varepsilon, a, \ldots, a) - f_a(a, \ldots, a, x_i, a, \ldots, a) \;=\; 0.
\end{aligned}$$

Moreover, since $\mathbf{X}$ and $\mathbf{Y}$ have the same marginals,

$$Ef(\mathbf{Y}) - Ef(\mathbf{X}) = Ef_a(\mathbf{Y}) - Ef_a(\mathbf{X}).$$

Now for every $\varepsilon > 0$ there is some $a_\varepsilon \in \mathbb{R}$ such that

$$P(\mathbf{X} \notin [a_\varepsilon \mathbf{1}, \infty)) + P(\mathbf{Y} \notin [a_\varepsilon \mathbf{1}, \infty)) \;\leq\; \frac{\varepsilon}{\|f\|_\infty}.$$

Define $\tilde{f} = \mathbf{1}_{[a_\varepsilon \mathbf{1}, \infty)} \cdot f_{a_\varepsilon}$. Then $\tilde{f}$ is increasing and supermodular and thus

$$
\begin{aligned}
Ef(\mathbf{Y}) - Ef(\mathbf{X}) &= Ef_{a_\varepsilon}(\mathbf{Y}) - Ef_{a_\varepsilon}(\mathbf{X}) \\
&= \int \mathbf{1}_{[a_\varepsilon \mathbf{1}, \infty)} \cdot f_{a_\varepsilon} \, d(P_{\mathbf{Y}} - P_{\mathbf{X}}) + \int \mathbf{1}_{[a_\varepsilon \mathbf{1}, \infty)^c} \cdot f_{a_\varepsilon} \, d(P_{\mathbf{Y}} - P_{\mathbf{X}}) \\
&\geq \int \tilde{f} \, d(P_{\mathbf{Y}} - P_{\mathbf{X}}) - \|f\|_\infty \cdot [P(\mathbf{X} \notin [a_\varepsilon \mathbf{1}, \infty)) + P(\mathbf{Y} \notin [a_\varepsilon \mathbf{1}, \infty))] \\
&\geq -\varepsilon.
\end{aligned}
$$

Since $\varepsilon > 0$ was arbitrary, this implies $Ef(\mathbf{X}) \leq Ef(\mathbf{Y})$.

If $f \in \mathcal{SM}$ is unbounded, then it can be approximated by bounded functions $f_k \in \mathcal{SM}$ similarly as in Step 3 of the proof of Theorem 3.9.10.

b) It is clear that (ii) implies (iii). Vice versa: assume (iii), then a function $f(x_1, \ldots, x_n) = g(x_i)$ with increasing $g : \mathbb{R} \to \mathbb{R}$ is increasing and supermodular. Therefore for all such functions $Eg(X_i) \leq Eg(Y_i)$, i.e. $X_i \leq_{st} Y_i$, but since $EX_i = EY_i$, this implies $X_i =_{st} Y_i$, see Theorem 1.2.9. $\qquad \square$

Now we immediately get the following result.

Theorem 3.9.12. *The stochastic orders $\leq_{ism}$ and $\leq_{sm}$ are closed with respect to weak convergence.*

Proof. Due to Theorem 2.4.2 an integral stochastic order is closed with respect to weak convergence if and only if it is generated by a class of bounded continuous functions. Therefore for $\leq_{ism}$ the assertion follows immediately from Theorem 3.9.10. Now assume that $\mathbf{X}_k \leq_{sm} \mathbf{Y}_k$ for $k = 1, 2, \ldots$, and that $(\mathbf{X}_k)$ and $(\mathbf{Y}_k)$ converge in distribution to $\mathbf{X}$ and $\mathbf{Y}$, respectively. According to Theorem 3.9.11 this implies that $\mathbf{X}_k$ and $\mathbf{Y}_k$ have the same marginals and that $\mathbf{X}_k \leq_{ism} \mathbf{Y}_k$ for all k. Since $\leq_{ism}$ is closed with respect to weak convergence, this implies $\mathbf{X} \leq_{ism} \mathbf{Y}$. As weak convergence of random vectors implies weak convergence of all marginals, $\mathbf{X}$ and $\mathbf{Y}$ must also have the same marginals. Hence applying Theorem 3.9.11 once more yields $\mathbf{X} \leq_{sm} \mathbf{Y}$. $\qquad \square$

Theorem 3.9.13. *The stochastic order relation $\leq_{sm}$ is generated by the set of all infinite differentiable supermodular functions, and similarly $\leq_{ism}$ is generated by the infinite differentiable increasing supermodular functions.*

Proof. In view of Theorem 3.9.12 this is a consequence of Theorem 2.5.6 and Theorem 3.7.2. $\qquad \square$

In addition to the properties (P1) – (P9) of a multivariate dependence order, the following properties of supermodular order can easily be obtained from Theorem 2.4.2 and Theorem 3.2.3, respectively. Some of these properties were already given in Li and Xu (2000).

Theorem 3.9.14. *The supermodular order $\leq_{sm}$ has the properties (T), (C), (MI), (MA), (ID), and (IN).*

We have seen that there is a maximal element with respect to supermodular order within $\Gamma(F_1, \ldots, F_n)$, namely the upper Fréchet bound. It is therefore natural to ask whether or not there is also a minimal element. We already know that the lower Fréchet bound is not a distribution function in general. However, in Theorem 3.1.2 necessary and sufficient conditions have been stated for the lower Fréchet bound to be a proper distribution function. If this is the case, it is a minimal element with respect to supermodular order.

Theorem 3.9.15. *Assume that the lower Fréchet bound F^- is a proper distribution function. Then for any $F \in \Gamma(F_1, \ldots, F_n)$ it is $F^- \leq_{sm} F$.*

Proof. First suppose that at most two of the distributions $F_i, 1 \leq i \leq n$, are non-degenerate. Then it can be assumed without loss of generality that $n = 2$. But in this case it was already shown in Theorem 3.8.2 that supermodular order is equivalent to concordance order, if the distributions have the same marginals. Hence in this case the result is obvious.

Therefore assume from now on that at least three of the F_i are non-degenerate. Thus one of the conditions a) or b) of Theorem 3.1.2 must hold. If b) holds then all distributions are bounded below by some ξ_i, $1 \leq i \leq n$, and these ξ_is have point masses summing up at least to $n - 1$. Since supermodular order is invariant with respect to translations of the coordinates, we assume without loss of generality that $\boldsymbol{\xi} = \mathbf{0}$. Hence, if $\mathbf{X} \sim F^+$, then almost surely at most one component of $\mathbf{X}$ is positive and all others vanish. Now let $\mathbf{Y}$ be a random vector with distribution function F and let f be an arbitrary supermodular function. Define

$$\tilde{f}(\mathbf{x}) = f(\mathbf{x}) - \sum_{i=1}^{n} f(x_i \mathbf{e}_i) + (n-1)f(\mathbf{0}),$$

where $\mathbf{e}_i$ denotes the i-th unit vector. As $\mathbf{X}$ and $\mathbf{Y}$ have the same marginals, $Eh(\mathbf{X}) = Eh(\mathbf{Y})$ holds for any function that depends only on one component. Hence $Ef(\mathbf{X}) \leq Ef(\mathbf{Y})$ holds if and only if $E\tilde{f}(\mathbf{X}) \leq E\tilde{f}(\mathbf{Y})$. But $\tilde{f}$ is a supermodular function with the property that $\tilde{f}(t\mathbf{e}_i) = 0$ for all $t \in \mathbb{R}$ and all $i = 1, \ldots, n$. Hence it vanishes on the axes. Since it is supermodular this implies

$$\tilde{f}(x_1, \ldots, x_i + \varepsilon, \ldots, x_n) - \tilde{f}(x_1, \ldots, x_i, \ldots, x_n)$$
$$\geq \tilde{f}(0, x_2, \ldots, x_i + \varepsilon, \ldots, x_n) - \tilde{f}(0, x_2, \ldots, x_i, \ldots, x_n)$$
$$\geq \cdots$$
$$\geq \tilde{f}(0, \ldots, 0, x_i + \varepsilon, 0, \ldots, 0) - \tilde{f}(0, \ldots, 0, x_i, 0, \ldots, 0)$$
$$= 0$$

for all $x_i \geq 0$, $1 \leq i \leq n$, and all $\varepsilon > 0$. Hence $\tilde{f}$ is increasing and thus non-negative on the positive orthant. This yields $E\tilde{f}(\mathbf{X}) = 0 \leq E\tilde{f}(\mathbf{Y})$ and therefore also $Ef(\mathbf{X}) \leq Ef(\mathbf{Y})$. Since this holds for all supermodular functions, we have $\mathbf{X} \leq_{sm} \mathbf{Y}$.

When condition a) of Theorem 3.1.2 holds, the proof is similar.   □

3.10   Concepts of Dependence

There is a variety of concepts of positive and negative dependence. Some of them are derived from dependence orders in the following way. Assume that $\preceq$ is a dependence order and that $\mathbf{X}^{\perp}$ is a random vector with the same marginals as $\mathbf{X}$, but with independent components. Then $\mathbf{X}$ is said to be positively dependent if $\mathbf{X}^{\perp} \preceq \mathbf{X}$ and it is said to be negatively dependent if $\mathbf{X} \preceq \mathbf{X}^{\perp}$. When this idea is applied to the concordance orders $\leq_{cU}$ and $\leq_{cL}$ then the well known concepts of positive (respectively negative) upper (respectively lower) orthant dependence are obtained. Alternatively, they can be defined as follows.

Definition 3.10.1. A random vector $\mathbf{X} = (X_1, \ldots, X_n)$ is said to be
a) *positive upper orthant dependent* (PUOD) if

$$P(X_1 > t_1, \ldots, X_n > t_n) \geq \prod_{i=1}^{n} P(X_i > t_i) \qquad (3.10.1)$$

for all $t_1, \ldots, t_n$;
b) *positive lower orthant dependent* (PLOD) if

$$P(X_1 \leq t_1, \ldots, X_n \leq t_n) \geq \prod_{i=1}^{n} P(X_i \leq t_i) \qquad (3.10.2)$$

for all $t_1, \ldots, t_n$;
c) *negative upper orthant dependent* (NUOD) if

$$P(X_1 > t_1, \ldots, X_n > t_n) \leq \prod_{i=1}^{n} P(X_i > t_i) \qquad (3.10.3)$$

for all $t_1, \ldots, t_n$;
b) *negative lower orthant dependent* (NLOD) if

$$P(X_1 \leq t_1, \ldots, X_n \leq t_n) \leq \prod_{i=1}^{n} P(X_i \leq t_i) \qquad (3.10.4)$$

for all $t_1, \ldots, t_n$.

These definitions can be traced back to Lehmann (1966). Similarly, there are notions of dependence based on the supermodular order.

Definition 3.10.2. A random vector $\mathbf{X} = (X_1, \ldots, X_n)$ is said to be

a) *positive supermodular dependent* (PSMD) if

$$\mathbf{X} \geq_{sm} \mathbf{X}^{\perp}; \tag{3.10.5}$$

b) *negative supermodular dependent* (NSMD) if

$$\mathbf{X} \leq_{sm} \mathbf{X}^{\perp}. \tag{3.10.6}$$

It is clear that PSMD implies PUOD and PLOD, and that NSMD implies NUOD and NLOD, where the implications are strict for $n \geq 3$, while in the bivariate case these concepts coincide. The concept of negative supermodular dependence is investigated in Hu (2000) under the name of *negatively superadditive dependence*. To the best of our knowledge there does not exist a detailed investigation of PSMD so far.

Esary, Proschan, and Walkup (1967) introduced the concept of *association*.

Definition 3.10.3. A random vector $\mathbf{X} = (X_1, \ldots, X_n)$ is said to be (positively) *associated* if

$$\mathrm{Cov}(f(\mathbf{X}), g(\mathbf{X})) \geq 0$$

for all increasing functions $f, g : \mathbb{R}^n \to \mathbb{R}$.

Some authors, in particular physicists, use the notion *positively correlated* for associated random variables, see Liggett (1985). We will use the term 'associated' throughout, since positively correlated could also mean the much weaker requirement of $\mathrm{Cov}(X_i, X_j)$ being non-negative for all i and j.

Theorem 3.10.4. *If* $\mathbf{X}$ *is associated, then* $\mathbf{X}$ *is PUOD and PLOD.*

Proof. The definition of association can be stated equivalently as

$$E(f(\mathbf{X})g(\mathbf{X})) \geq Ef(\mathbf{X})Eg(\mathbf{X})$$

for all increasing f and g. Thus in the case of association it follows by induction that

$$E\left(\prod_{i=1}^{n} f_i(X_i)\right) \geq \prod_{i=1}^{n} Ef_i(X_i)$$

for every collection $f_1, \ldots, f_n$ of univariate non-negative increasing functions. Consequently, it follows from Theorem 3.3.16 a) that PUOD holds. Similarly, PLOD follows from Theorem 3.3.16 b). $\qquad\square$

It seems to be an open problem whether or not association implies PSMD. This clearly is true in the bivariate case, where PSMD, PUOD and PLOD coincide, but it is not clear in higher dimensions.

Association has the following properties.

Theorem 3.10.5.

(i) *if $\mathbf{X}$ is associated then $\mathbf{X}_K$ is associated for any $K \subset \{1, \ldots, n\}$;*

(ii) *if $\mathbf{X}$ and $\mathbf{Y}$ are associated and independent, then $(\mathbf{X}, \mathbf{Y})$ is associated;*

(iii) *any univariate random variable X is associated;*

(iv) *if $\mathbf{X}$ is associated and $f_1, \ldots, f_m : \mathbb{R}^n \to \mathbb{R}$ are increasing (decreasing), then $(f_1(\mathbf{X}), \ldots, f_m(\mathbf{X}))$ is associated;*

(v) *if $\mathbf{X} = (X_1, \ldots, X_n)$ has independent components, then it is associated.*

Proof. (i) and (iv) follow immediately from the definition. Notice that in fact (i) is a special case of (iv).

(ii): Assume that $\mathbf{X} = (X_1, \ldots, X_m)$ and $\mathbf{Y} = (Y_1, \ldots, Y_n)$ are associated and independent and that $f, g : \mathbb{R}^{m+n} \to \mathbb{R}$ are increasing. Define $\tilde{f}(\mathbf{y}) = \int f(\mathbf{x}, \mathbf{y}) P_{\mathbf{X}}(\mathrm{d}\mathbf{x})$ and similarly $\tilde{g}(\mathbf{y}) = \int g(\mathbf{x}, \mathbf{y}) P_{\mathbf{X}}(\mathrm{d}\mathbf{x})$. Then $\tilde{f}$ and $\tilde{g}$ are increasing and therefore

$$E(f(\mathbf{X}, \mathbf{Y})g(\mathbf{X}, \mathbf{Y})) - Ef(\mathbf{X}, \mathbf{Y})Eg(\mathbf{X}, \mathbf{Y})$$

$$= \int \left(\int fg\,\mathrm{d}P_{\mathbf{X}} - \int f\,\mathrm{d}P_{\mathbf{X}} \int g\,\mathrm{d}P_{\mathbf{X}} \right) \mathrm{d}P_{\mathbf{Y}} + \left(\int \tilde{f}\tilde{g}\,\mathrm{d}P_{\mathbf{Y}} - \int \tilde{f}\,\mathrm{d}P_{\mathbf{Y}} \int \tilde{g}\,\mathrm{d}P_{\mathbf{Y}} \right)$$

$$\geq 0.$$

(iii) It has to be shown that $E(f(X)g(X)) - Ef(X)Eg(X) \geq 0$ for all increasing $f, g : \mathbb{R} \to \mathbb{R}$. This is Harris' inequality, see Theorem 4.3.17. Combining (ii) and (iii) yields (v). $\qquad\square$

The definition of association can obviously be extended to Polish spaces with closed partial orders. A thorough investigation for that case was carried out by Lindqvist (1988). We need this extension for the investigation of association properties of stochastic processes in Chapter 4. Therefore the most important results of Lindqvist (1988) are briefly reported here.

A random element $\mathbf{X}$ or its distribution P_X on a partially ordered Polish space S is said to be *associated* if

$$\mathrm{Cov}(f(\mathbf{X}), g(\mathbf{X})) \geq 0$$

for all increasing $f, g : S \to \mathbb{R}$. By a straightforward approximation argument as in the proof of Theorem 3.3.4 it can be shown that this holds if and only if

$$P_{\mathbf{X}}(A \cap B) \geq P_{\mathbf{X}}(A)P_{\mathbf{X}}(B) \qquad (3.10.7)$$

for all measurable upper sets $A, B \subset S$. The properties (ii) – (iv) in Theorem 3.10.5 have generalizations to this more general situation. Property (ii) remains valid as stated. As an extension of property (iii) the following result can be shown. Recall that a partially ordered space S is said to be *totally ordered* if for any $x, y \in S$ either $x \leq y$ or $x \geq y$.

Theorem 3.10.6. *On a Polish space S with a closed partial order the following statements are equivalent:*

(i) *Every probability measure on S is associated;*

(ii) *S is totally ordered.*

Proof. If S is totally ordered and $A, B \subset S$ are upper then either $A \cap B = A$ or $A \cap B = B$ and thus (3.10.7) holds. On the other hand, if S is not totally ordered then there are x and y which are not comparable. Define $C_x = \{a \in S : a \geq x\}$ and let P be the probability measure assigning probability $1/2$ to each of x and y. Then C_x and C_y are obviously increasing and $P(C_x \cap C_y) = 0$ whereas $P(C_x)P(C_y) = 1/4$. Hence P is not associated. $\square$

Property (iv) in Theorem 3.10.5 has the following generalization. The proof easily follows from (3.10.7).

Theorem 3.10.7. *Suppose that S_1 and S_2 are Polish spaces with a closed partial order. If $\mathbf{X}$ is an associated random element on S_1 and $f : S_1 \to S_2$ is increasing then $f(\mathbf{X})$ is an associated random element on S_2.*

Important instances of infinite dimensional partially ordered spaces are infinite products of the real line or subsets of such spaces. The most prominent examples are ℓ^∞, the space of all sequences of real numbers endowed with the product topology and pointwise order, and $D[0, \infty)$, the space of all real functions on $[0, \infty)$ which are cadlag, i.e. they are right-continuous and have left-hand limits everywhere. This space is a partially ordered Polish space if it is endowed with the Skorokhod topology and the pointwise order; see Billingsley (1999). In these two cases the following characterization holds; see Lindqvist (1988).

Theorem 3.10.8. *Let $\mathbf{X} = (X_t)_{t \in T}$ be a real valued stochastic process, where either $T = \mathbb{N}$ or $T = [0, \infty)$ and the paths of (X_t) are cadlag almost surely. Then (X_t) is associated if and only if*

$$(X_{t_1}, \dots, X_{t_n}) \quad \text{is associated}$$

for all $t_1, \dots, t_n \in T$ and all n.

Cohen, Sackrowitz, and Samuel-Cahn (1995) consider association on the Euclidean space $\mathbb{R}^n$ endowed with a general cone order. Association on products of general partially ordered spaces was also considered in Ahmed, Leon, and Proschan (1981).

There also exist attempts to define *negative association*. Note that the condition

$$\text{Cov}(f(\mathbf{X}), g(\mathbf{X})) \leq 0$$

for all increasing functions $f, g : \mathbb{R}^n \to \mathbb{R}$ does not yield a useful concept of negative dependence. In fact, it would imply $\text{Cov}(X_i, X_i) = \text{Var}(X_i) \leq 0$ for all $i = 1, \ldots, n$ and thus $\mathbf{X}$ must be degenerated to satisfy this condition. Therefore the concept of negative association is typically defined as follows. A random vector $\mathbf{X}$ is said to be *negatively associated* if

$$\text{Cov}(f(\mathbf{X}_K), g(\mathbf{X}_L)) \leq 0$$

for all disjoint subsets K and L of $\{1, \ldots, n\}$ and all increasing $f : \mathbb{R}^{|K|} \to \mathbb{R}$ and $g : \mathbb{R}^{|L|} \to \mathbb{R}$. We refer the reader to Joag-Dev and Proschan (1983) for a study of this concept.

Some other concepts of dependence are based on the idea that a random vector should be considered as positively dependent if the conditional distribution of a subset of the components, given the knowledge of some others, is stochastically increasing in the known components. We use the notation $\mathbf{X} \uparrow_{st} \mathbf{Y}$ if $\mathbf{X}$ is *stochastically increasing* in $\mathbf{Y}$, i.e. if the conditional distribution of $\mathbf{X}$ given $\mathbf{Y} = \mathbf{y}$ is $\leq_{st}$-increasing in $\mathbf{y}$.

Definition 3.10.9. a) A random vector $\mathbf{X} = (X_1, \ldots, X_n)$ is said to be *positive dependent through stochastic ordering* (PDS) if $(X_i : i \neq j) \uparrow_{st} X_j$ for all $j = 1, \ldots, n$.
b) A random vector $\mathbf{X} = (X_1, \ldots, X_n)$ is said to be *conditionally increasing in sequence* (CIS) if $X_i \uparrow_{st} (X_1, X_2, \ldots, X_{i-1})$ for all $i = 2, \ldots, n$.
c) A random vector $\mathbf{X} = (X_1, \ldots, X_n)$ is said to be *conditionally increasing* (CI) if $X_i \uparrow_{st} \mathbf{X}_J$ for all $J \subset \{1, \ldots, n\}$.

Note that CI holds if and only if $\mathbf{X}_\pi = (X_{\pi(1)}, \ldots, X_{\pi(n)})$ is CIS for all permutations π. Consequently, CI implies CIS.

The notion of CIS is closely related to the standard construction introduced in (3.1.7). The following result can be found implicitly already in Barlow and Proschan (1975). It is stated explicitly in Rubinstein, Samorodnitsky, and Shaked (1985). The proof is obvious.

Lemma 3.10.10. *Let F be the distribution function of the random vector $\mathbf{X}$. Then $\mathbf{X}$ is CIS if and only if the function Ψ_F^* in (3.1.7) is increasing.*

This can be used to show that CIS is stronger than association.

Theorem 3.10.11. *If $\mathbf{X}$ is CIS then it is associated.*

Proof. The proof is given by induction on the dimension n. For $n = 1$ there is nothing to show. Therefore assume that the assertion holds for $n - 1$. According to Lemma 3.10.10 it can be assumed without loss of generality that there are independent random variables $U_1, \ldots, U_n$ such that $X_i = g_i(X_1, \ldots, X_{i-1}, U_i)$ for increasing functions g_i for all $i = 1, \ldots, n$. According to the induction hypothesis $(X_1, \ldots, X_{n-1})$ are associated and $X_n = g_n(X_1, \ldots, X_{n-1}, U_n)$ with U_n independent of $(X_1, \ldots, X_{n-1})$. Hence it follows from properties (ii) and (iv) in Theorem 3.10.5 that $(X_1, \ldots, X_n)$ is associated, too. $\square$

The following result is contained in Meester and Shanthikumar (1993).

Theorem 3.10.12. *CIS implies PSMD.*

The notion of PDS was introduced by Block, Savits, and Shaked (1985). They show that it is weaker than CI and stronger than PUOD and PLOD. However, it is neither implied by nor implies association.

Karlin and Rinott (1980a) investigated the concept of *multivariate total positivity of order 2* (MTP$_2$), which they defined as follows: Assume that the random vector $\mathbf{X} = (X_1, \ldots, X_n)$ with distribution P has a density f with respect to some σ-finite product measure $\mu = \mu_1 \times \cdots \times \mu_n$. Typically μ is either the Lebesgue measure (for continuous distributions) or the counting measure (for discrete distributions). Then $\mathbf{X}$ (or f) is said to be MTP$_2$, if

$$f(\mathbf{x})f(\mathbf{y}) \le f(\mathbf{x} \wedge \mathbf{y})f(\mathbf{x} \vee \mathbf{y}) \quad \text{for all } \mathbf{x}, \mathbf{y} \in \mathbb{R}^n. \tag{3.10.8}$$

In the discrete case this concept appears in the context of the FKG inequality, see (7.5.11). Notice that the assumption of a density with respect to a product measure is crucial, since every distribution has a density fulfilling (3.10.8), namely the density $f \equiv 1$ with respect to itself.

It is possible to extend this definition to probability measures that do not have a density with respect to a σ-finite product measure. This has been done in the literature on economics by Milgrom and Weber (1982), where this concept is called *affiliation*. For the definition we need the notion of a sublattice. A subset $L \subset \mathbb{R}^n$ is called a *sublattice*, if $\mathbf{x}, \mathbf{y} \in L$ implies $\mathbf{x} \vee \mathbf{y} \in L$ and $\mathbf{x} \wedge \mathbf{y} \in L$. Notice that L is a sublattice if and only if its indicator function $\mathbf{1}_L$ is MTP$_2$.

Definition 3.10.13. A probability measure P on $\mathbb{R}^n$ is called *affiliated*, if

$$P(A \cap B | L) \ge P(A|L)P(B|L) \tag{3.10.9}$$

for all increasing subsets A and B and all sublattices $L \subset \mathbb{R}^n$.

Thus a distribution is affiliated if and only if all conditional distributions obtained by constrictions to sublattices are associated. This reveals a strong relationship between affiliation and association. In particular, it shows that affiliation implies association.

In the next result we use the following notation. For arbitrary sets $A, B \subset \mathbb{R}^n$ it is

$$A \vee B = \{\mathbf{a} \vee \mathbf{b} : \mathbf{a} \in A, \mathbf{b} \in B\}$$

and

$$A \wedge B = \{\mathbf{a} \wedge \mathbf{b} : \mathbf{a} \in A, \mathbf{b} \in B\}.$$

Theorem 3.10.14. *For a distribution P on $\mathbb{R}^n$ having a density with respect to a product measure the following statements are equivalent:*

(i) *P is MTP$_2$,*

(ii) *$P(A \vee B)P(A \wedge B) \geq P(A)P(B)$ for all $A, B \subset \mathbb{R}^n$,*

(iii) *P is affiliated.*

Proof. For the proof of the implication (i) $\Rightarrow$ (ii) we refer to Karlin and Rinott (1980a) and for (iii) $\Rightarrow$ (i) to Milgrom and Weber (1982). Only the implication (ii) $\Rightarrow$ (iii) will be shown here. Notice that the inequality in (iii) can be rewritten as

$$P(ABL)P(L) \geq P(AL)P(BL),$$

where as usual AB is an abbreviation for $A \cap B$. Now assume that (ii) holds, and let A and B be increasing sets. Then $A \vee B = AB$ and therefore, if L is a sublattice, $AL \vee BL = ABL$. Since obviously $AL \wedge BL \subset L$ this yields

$$P(ABL)P(L) \geq P(AL \vee BL)P(AL \wedge BL) \geq P(AL)P(BL),$$

and hence P is affiliated. $\qquad\square$

Characterization (ii) of Theorem 3.10.14 is mentioned in Block, Savits, and Shaked (1982), where also the counterpart for negative dependence is treated.

Theorem 3.10.15. *For any marginals $F_1, \ldots, F_n$ the upper Fréchet bound $F^+ = \min\{F_1, \ldots, F_n\}$ is affiliated.*

Proof. Since the support of F^+ is a totally ordered set, it is for increasing A and B and a sublattice $L \subset \mathbb{R}^n$ either $P(A \cap B|L) = P(A|L)$ or $P(A \cap B|L) = P(B|L)$. Thus (3.10.9) holds. $\qquad\square$

Notice that for continuous marginals $F_1, \ldots, F_n$ the upper Fréchet bound does not have a density with respect to a σ-finite product measure. Therefore in general we cannot say that the upper Fréchet bound is MTP$_2$.

Karlin and Rinott (1980a) showed that MTP$_2$ implies CI. Since the MTP$_2$ property is invariant under permutations of the components, this implies that MTP$_2$ is also stronger than CI; see Müller and Scarsini (2001). We show now that this result remains true for affiliation.

Theorem 3.10.16. *Affiliation implies CI.*

Proof. Assume that $\mathbf{X}$ is affiliated and has a discrete distribution. According to Theorem 3.10.14 this is equivalent to $P_{\mathbf{X}}(A \vee B)P_{\mathbf{X}}(A \wedge B) \geq P_{\mathbf{X}}(A)P_{\mathbf{X}}(B)$

for all $A, B \subset \mathbb{R}^n$. Choosing $A = \{\mathbf{y} : y_i > t, \mathbf{y}_J = \mathbf{x}_J\}$ and $B = \{\mathbf{y} : \mathbf{y}_J = \mathbf{x}'_J\}$ for some $J \subset \{1, \ldots, n\}$, $i \notin J$, and $\mathbf{x}_J \leq \mathbf{x}'_J$ yields

$$P(X_i > t, \mathbf{X}_J = \mathbf{x}'_J)P(\mathbf{X}_J = \mathbf{x}_J) \geq P(X_i > t, \mathbf{X}_J = \mathbf{x}_J)P(\mathbf{X}_J = \mathbf{x}'_J),$$

or equivalently, $P(X_i > t | \mathbf{X}_J = \mathbf{x}_J) \leq P(X_i > t | \mathbf{X}_J = \mathbf{x}'_J)$. Hence $\mathbf{X}$ is CI. The case of a general distribution follows by an approximation argument. $\square$

Example 3.10.17. (The implications $\text{MTP}_2 \Rightarrow \text{CI} \Rightarrow \text{CIS}$ are strict.)

1. Assume that $P((X_1, X_2) = (x_1, x_2)) = 1/6$ for all

$$(x_1, x_2) \in \{(1,1), (1,2), (2,1), (2,3), (3,2), (3,3)\}.$$

Then it is straightforward to show that $X_2 \uparrow_{\text{st}} X_1$ and $X_1 \uparrow_{\text{st}} X_2$, hence $\mathbf{X}$ is CI. However, $\mathbf{X}$ is not MTP_2 since it holds for $\mathbf{x} = (1,2)$ and $\mathbf{y} = (2,1)$ that

$$f(\mathbf{x})f(\mathbf{y}) = \frac{1}{36} > 0 = f(\mathbf{x} \wedge \mathbf{y})f(\mathbf{x} \vee \mathbf{y}) = f(1,1)f(2,2).$$

2. Assume that $P((X_1, X_2) = (x_1, x_2)) = 1/4$ for all

$$(x_1, x_2) \in \{(1,1), (1,2), (2,1), (2,3)\}.$$

Then $\mathbf{X}$ is CIS, but $P(X_1 \geq 2 | X_2 = 1) = 1/2 > 0 = P(X_1 \geq 2 | X_2 = 2)$. Hence $X_1 \not\uparrow_{\text{st}} X_2$ and therefore $\mathbf{X}$ is not CI.

The concept MTP_2 has the advantage that it is easy to check. Notice that a density f is MTP_2 if and only if $\ln f$ is supermodular. Hence it follows from Theorem 3.9.3 that a twice differentiable density (with respect to Lebesgue measure) is MTP_2 if and only if

$$\frac{\partial^2}{\partial x_i \partial x_j} \ln f(\mathbf{x}) \geq 0 \text{ for all } \mathbf{x} \text{ and all } 1 \leq i < j \leq n.$$

This can be used, for example, to characterize MTP_2 in the case of multivariate normal distributions. The following theorem is due to Rüschendorf (1981a).

Theorem 3.10.18. *If the random vector* $\mathbf{X}$ *has a multivariate normal distribution with an invertible covariance matrix* $\boldsymbol{\Sigma}$, *then the following statements are equivalent:*

(i) $\mathbf{X}$ *is* MTP_2,

(ii) $\mathbf{X}$ *is CI,*

(iii) $\boldsymbol{\Sigma}^{-1}$ *is an M-matrix, i.e. its off-diagonal elements are non-positive.*

For further examples of MTP_2 densities see Karlin and Rinott (1980a). The counterpart of negative dependence corresponding to MTP_2 is called *multivariate reverse rule of order 2* (MRR_2). A detailed study of this concept is given in Karlin and Rinott (1980b).

It is easy to show that all concepts of positive dependence considered in this section are preserved under monotone transformations of the marginals. This means that they are properties of copulas and do not depend on the marginals.

Theorem 3.10.19. *If* $\mathbf{X} = (X_1, \ldots, X_n)$ *has one of the properties PUOD, PLOD, NUOD, NLOD, PSMD, NSMD, association, PDS, CIS, CI or MTP_2, and if $f_1, \ldots, f_n$ are increasing functions, then* $\mathbf{Y} = (f_1(X_1), \ldots, f_n(X_n))$ *has this property, too.*

3.11 Multivariate Likelihood Ratio Orders

There are several possibilities of defining a multivariate likelihood ratio order. The most intuitive one is obtained by requiring the ratio of densities to be increasing.

Definition 3.11.1. The random vector $\mathbf{Y}$ is said to be larger than $\mathbf{X}$ in *weak likelihood ratio order* (written $\mathbf{X} \leq_r \mathbf{Y}$), if $\mathbf{X}$ and $\mathbf{Y}$ have densities with respect to some dominating measure μ such that

$$f_{\mathbf{X}}(\mathbf{t}) f_{\mathbf{Y}}(\mathbf{s}) \leq f_{\mathbf{X}}(\mathbf{s}) f_{\mathbf{Y}}(\mathbf{t}) \quad \text{for all } \mathbf{s} \leq \mathbf{t}. \tag{3.11.1}$$

It turns out, however, that there is a stronger concept, which is more useful. It is derived from the MTP_2 concept considered in Theorem 3.10.14.

Definition 3.11.2. The random vector $\mathbf{Y}$ is said to be larger than $\mathbf{X}$ in *strong likelihood ratio order* or *tp_2-order* (written $\mathbf{X} \leq_{tp} \mathbf{Y}$), if $\mathbf{X}$ and $\mathbf{Y}$ have densities with respect to some dominating σ-finite product measure $\mu = \mu_1 \times \cdots \times \mu_n$ such that

$$f_{\mathbf{X}}(\mathbf{s}) f_{\mathbf{Y}}(\mathbf{t}) \leq f_{\mathbf{X}}(\mathbf{s} \wedge \mathbf{t}) f_{\mathbf{Y}}(\mathbf{s} \vee \mathbf{t}) \quad \text{for all } \mathbf{s}, \mathbf{t} \in \mathbb{R}^n. \tag{3.11.2}$$

Sometimes only this stronger tp_2-order is called multivariate likelihood ratio order and then it is denoted as $\leq_{lr}$ (as in Shaked and Shanthikumar (1994)). We will not use the character $\leq_{lr}$ in the multivariate case. To avoid confusion we instead use the symbols $\leq_r$ and $\leq_{tp}$ as in Whitt (1982).

Notice that in the definition of $\leq_{tp}$ the dominating measure must be a product measure, whereas this is not necessary in the definition of $\leq_r$. The reason is that if (3.11.1) holds for some dominating measure then it also holds (almost surely) for any other dominating measure. In (3.11.2), however, this is only true for densities with respect to product measures.

The relation $\leq_{tp}$ is not an order on the set of all probability measures. Even reflexivity does not hold in general.

Theorem 3.11.3. *The relation $\leq_{tp}$ is a partial order on the set of all MTP$_2$ distributions with the same lattice as their support.*

Proof. Obviously $P \leq_{tp} P$ holds if and only if P is MTP$_2$. If P_1, P_2, P_3 are distributions with densities f_1, f_2, f_3 and $P_1 \leq_{tp} P_2$ and $P_2 \leq_{tp} P_3$ then the MTP$_2$ property of f_2 implies

$$f_1(\mathbf{x})f_2(\mathbf{x} \wedge \mathbf{y})f_2(\mathbf{x} \vee \mathbf{y})f_3(\mathbf{y}) \leq f_1(\mathbf{x} \wedge \mathbf{y})f_2(\mathbf{x})f_2(\mathbf{y})f_3(\mathbf{x} \vee \mathbf{y})$$
$$\leq f_1(\mathbf{x} \wedge \mathbf{y})f_2(\mathbf{x} \wedge \mathbf{y})f_2(\mathbf{x} \vee \mathbf{y})f_3(\mathbf{x} \vee \mathbf{y})$$

for all $\mathbf{x}$ and $\mathbf{y}$ in the common support and hence $P_1 \leq_{tp} P_3$. Antisymmetry follows from the fact that $P_1 \leq_{tp} P_2$ and $P_2 \leq_{tp} P_1$ together imply that f_2/f_1 is constant. $\qquad\square$

Nevertheless $\leq_{tp}$ can be useful, since it has the following properties.

Theorem 3.11.4.
a) $\mathbf{X} \leq_{tp} \mathbf{Y}$ *implies* $\mathbf{X} \leq_{st} \mathbf{Y}$.
b) $\mathbf{X} \leq_{tp} \mathbf{Y}$ *implies* $[\mathbf{X}|\mathbf{X} \in L] \leq_{tp} [\mathbf{Y}|\mathbf{Y} \in L]$ *for all sublattices L.*

Proof. For a) we refer to Karlin and Rinott (1980a). b) is obvious. $\qquad\square$

Remark 3.11.5. In the case of distributions on a discrete sublattice of $\mathbb{R}^n$ part a) of Theorem 3.11.4 is known as 'Holley's inequality'. Under this name it is well known in statistical physics, see p. 258.

In the univariate case it has been shown in Theorem 1.4.6 that likelihood ratio order is equivalent to uniform conditional stochastic order, meaning that $[\mathbf{X}|\mathbf{X} \in A] \leq_{st} [\mathbf{Y}|\mathbf{Y} \in A]$ for all measurable subsets A. In the multivariate case the uniform conditional stochastic order is stronger than likelihood ratio order. It can be shown that $\mathbf{X} \leq_{ucso} \mathbf{Y}$ essentially holds only if $\mathbf{X}$ and $\mathbf{Y}$ are identical except on a totally ordered set, where they must be ordered by $\leq_r$; see Rüschendorf (1991).

The main disadvantage of $\leq_r$ is that in contrast to $\leq_{tp}$ it in general does not imply $\leq_{st}$.

Example 3.11.6. ($\leq_r$ does not imply $\leq_{st}$, Whitt (1982).) Consider $\mathbf{X}$ and $\mathbf{Y}$ concentrated on $\{0,1\}^2$ with $P(\mathbf{X} = (0,0)) = 0.1$,

$$P(\mathbf{X} = (0,1)) = P(\mathbf{X} = (1,0)) = P(\mathbf{X} = (1,1)) = 0.3,$$

$$P(\mathbf{Y} = (0,0)) = 0.01, \ P(\mathbf{Y} = (1,0)) = 0.09$$

and

$$P(\mathbf{Y} = (0,1)) = P(\mathbf{Y} = (1,1)) = 0.45.$$

Then $\mathbf{X} \leq_r \mathbf{Y}$, but for the increasing set $A = \{(x_1, x_2) : x_1 \geq 1\}$

$$P(\mathbf{X} \in A) = 0.6 > 0.54 = P(\mathbf{Y} \in A).$$

Hence it is not $\mathbf{X} \leq_{st} \mathbf{Y}$.

Also the hazard rate order has a multivariate generalization, which was introduced by Shaked and Shanthikumar (1987b). Naturally, it is based on a definition of a multivariate hazard rate.

Let $\mathbf{X}$ be a non-negative random vector having a density with respect to Lebesgue measure. For positive t the *hazard rate* depends on those components of $\mathbf{X}$ which have failed till time t, X_i with $i \in I$, and which survive t, X_i with $i \in I^c$. Thus I is the subset of $\{1, \dots, n\}$ of failed components and I^c its complement.

$$r_{X,k|I}(t|\mathbf{x}_I) = \lim_{\varepsilon \to 0} \frac{1}{\varepsilon} P(t < X_k \leq t + \varepsilon | \mathbf{X}_I = \mathbf{x}_I, \mathbf{X}_{I^c} > t\mathbf{1}),$$

where $\mathbf{1} = (1, \dots, 1)$ is a vector of 1s and $k \notin I$. Thus $r_{X,k|I}(t|\mathbf{x}_I)$ is the failure rate of component k evaluated at time t, under the condition that the components in I have failed at the times $\mathbf{x}_I$, and the components in I^c have survived time t.

Definition 3.11.7. The random vector $\mathbf{Y}$ is said to be larger than $\mathbf{X}$ in *multivariate hazard rate order* (written $\mathbf{X} \leq_{hr} \mathbf{Y}$), if $\mathbf{X}$ and $\mathbf{Y}$ have hazard rates r_X and r_Y satisfying

$$r_{X,k|J}(t|\mathbf{x}_J) \geq r_{Y,k|I}(t|\mathbf{y}_I) \quad \text{for all } k \in J^c$$

for all positive t whenever $J \supseteq I, 0 \leq \mathbf{x}_I \leq \mathbf{y}_I \leq t\mathbf{1}$ and $0 \leq \mathbf{x}_J \leq t\mathbf{1}$.

Interesting applications of $\leq_{hr}$ exist for point processes, see Section 5.4.

Theorem 3.11.8. $\mathbf{X} \leq_{tp} \mathbf{Y} \Rightarrow \mathbf{X} \leq_{hr} \mathbf{Y} \Rightarrow \mathbf{X} \leq_{st} \mathbf{Y}$.

A proof of this result as well as other properties of this and related orders can be found in chapter 4 of Shaked and Shanthikumar (1994).

3.12   Directionally Convex Order

In Section 3.9 it was shown that supermodular order is a useful tool for the comparison of the dependence structure of random vectors with fixed marginals. The present section is devoted to a generalization of this order which in addition takes into account the variability of the marginals. This can be achieved by requiring the functions of the generator to be supermodular and additionally to have some convexity property.

Definition 3.12.1. A function $f : \mathbb{R}^n \to \mathbb{R}$ is said to be *directionally convex* if it is supermodular and componentwise convex.

The reason for this name becomes clearer in the next theorem, where some characterizations of these functions are collected. The proof is omitted. For details see Shaked and Shanthikumar (1990) and Müller and Scarsini (2001).

Theorem 3.12.2. *a) A function $f : \mathbb{R}^n \to \mathbb{R}$ is directionally convex if and only if one of the following equivalent characterizations holds.*

(i)

$$\Delta_i^\varepsilon \Delta_j^\delta f(\mathbf{x}) \geq 0 \qquad\qquad (3.12.1)$$

for all $\mathbf{x}$, *all* $1 \leq i,j \leq n$ *and all* $\varepsilon, \delta > 0$.

(ii) *For all* $\mathbf{x}_i \in \mathbb{R}^n$, $i = 1,2,3,4$, *with* $\mathbf{x}_1 \leq \mathbf{x}_2 \leq \mathbf{x}_4$, $\mathbf{x}_1 \leq \mathbf{x}_3 \leq \mathbf{x}_4$ *and* $\mathbf{x}_1 + \mathbf{x}_4 = \mathbf{x}_2 + \mathbf{x}_3$

$$f(\mathbf{x}_2) + f(\mathbf{x}_3) \ \leq \ f(\mathbf{x}_1) + f(\mathbf{x}_4). \qquad\qquad (3.12.2)$$

(iii) *For all* $\mathbf{x}_1$ *and* $\mathbf{x}_2$ *with* $\mathbf{x}_1 \leq \mathbf{x}_2$ *and all* $\mathbf{y} \geq \mathbf{0}$

$$f(\mathbf{x}_1 + \mathbf{y}) - f(\mathbf{x}_1) \ \leq \ f(\mathbf{x}_2 + \mathbf{y}) - f(\mathbf{x}_2). \qquad\qquad (3.12.3)$$

(iv) *For all* $\mathbf{x} \in \mathbb{R}^n$, $1 \leq i \leq j \leq n$, *and all* $\varepsilon, \delta > 0$,

$$f(\mathbf{x} + \varepsilon\mathbf{e}_i + \delta\mathbf{e}_j) - f(\mathbf{x} + \delta\mathbf{e}_j) - f(\mathbf{x}) + f(\mathbf{x} - \varepsilon\mathbf{e}_i) \geq 0. \qquad (3.12.4)$$

(v) *For all* $\alpha, \beta > 0$, $i \in \{1,\dots,n\}$ *and all* $\mathbf{x}_1, \mathbf{x}_2 \in \mathbb{R}^n$ *with* $\mathbf{x}_1 \leq \mathbf{x}_2$ *and* $\mathbf{x}_1 + \alpha\mathbf{e}_i \leq \mathbf{x}_2 + \beta\mathbf{e}_i$,

$$\beta(f(\mathbf{x}_1 + \alpha\mathbf{e}_i) - f(\mathbf{x}_1)) \leq \alpha(f(\mathbf{x}_2 + \beta\mathbf{e}_i) - f(\mathbf{x}_2)). \qquad (3.12.5)$$

b) A twice differentiable function f is directionally convex if and only if

$$\frac{\partial^2}{\partial x_i \partial x_j} f(\mathbf{x}) \ \geq \ 0$$

for all $\mathbf{x}$ *and all* $1 \leq i,j \leq n$.

Notice that usual convexity neither implies nor is implied by directional convexity, though especially the characterization in (3.12.3) seems to be a very natural extension of univariate convexity to higher dimensions.

In the following theorem some useful properties of directionally convex functions are collected. We will write $f \in \mathcal{DCX}$ if f is directionally convex, and $f \in \mathcal{IDCX}$ if f is increasing and directionally convex.

Theorem 3.12.3. *a) If $g_1, \dots, g_n : \mathbb{R} \to \mathbb{R}$ are increasing convex and $f \in \mathcal{IDCX}$ then*

$$f(g_1(\cdot), \dots, g_n(\cdot)) \in \mathcal{IDCX}.$$

b) If $f,g \in \mathcal{DCX}$ then $\alpha f + \beta g \in \mathcal{DCX}$ for all $\alpha, \beta \geq 0$.
c) If $f,g \in \mathcal{IDCX}$ and $f,g \geq 0$, then $f \cdot g \in \mathcal{IDCX}$.
d) If $f \in \mathcal{IDCX}$ then $\max\{f,c\} \in \mathcal{IDCX}$ for all real constants c.
e) If $f \in \mathcal{IDCX}$ and $\phi : \mathbb{R} \to \mathbb{R}$ is increasing and convex then $\phi \circ f \in \mathcal{IDCX}$.
f) If $f : \mathbb{R}^2 \to \mathbb{R}$ and $g : \mathbb{R}^n \to \mathbb{R}$ are in $\mathcal{IDCX}$, then $f(\cdot, g(\cdot)) \in \mathcal{IDCX}$.

The easy proof is omitted.

Meester and Shanthikumar (1993) introduced stochastic order relations generated by directionally convex functions.

Definition 3.12.4. Let $\mathbf{X}$ and $\mathbf{Y}$ be random vectors.

a) $\mathbf{X}$ precedes $\mathbf{Y}$ in *directionally convex order* (written $\mathbf{X} \leq_{dcx} \mathbf{Y}$), if $Ef(\mathbf{X}) \leq Ef(\mathbf{Y})$ for all directionally convex functions f.

b) $\mathbf{X}$ precedes $\mathbf{Y}$ in *increasing directionally convex order* (written $\mathbf{X} \leq_{idcx} \mathbf{Y}$), if $Ef(\mathbf{X}) \leq Ef(\mathbf{Y})$ for all increasing directionally convex functions f.

The next result describes the relation of directionally convex order to other stochastic orders.

Theorem 3.12.5.

a) $\mathbf{X} \leq_{sm} \mathbf{Y}$ *implies* $\mathbf{X} \leq_{dcx} \mathbf{Y}$ *and* $\mathbf{X} \leq_{ism} \mathbf{Y}$ *implies* $\mathbf{X} \leq_{idcx} \mathbf{Y}$.

b) $\mathbf{X} \leq_{ccx} \mathbf{Y}$ *implies* $\mathbf{X} \leq_{dcx} \mathbf{Y}$ *and* $\mathbf{X} \leq_{iccx} \mathbf{Y}$ *implies* $\mathbf{X} \leq_{idcx} \mathbf{Y}$.

c) $\mathbf{X} \leq_{dcx} \mathbf{Y}$ *implies* $\mathbf{X} \leq_{plcx} \mathbf{Y}$ *and* $\mathbf{X} \leq_{idcx} \mathbf{Y}$ *implies* $\mathbf{X} \leq_{iplcx} \mathbf{Y}$.

d) $\mathbf{X} \leq_{dcx} (\leq_{idcx}) \mathbf{Y}$ *implies* $X_i \leq_{cx} (\leq_{icx}) Y_i$ *for all* $i = 1, \ldots, n$.

Theorem 2.4.2 and Theorem 3.2.3 imply the following properties.

Theorem 3.12.6.

a) *The order* $\leq_{dcx}$ *has the properties (M), (C), (MI), (MA), (ID), and (IN).*

b) *The order* $\leq_{idcx}$ *has the properties (R), (E), (M), (T), (C), (MI), (MA), (ID), and (IN).*

Neither $\leq_{dcx}$ nor $\leq_{idcx}$ are closed with respect to weak convergence. This follows immediately from Theorem 2.4.2 g) since a directionally convex function is bounded and continuous only if it is constant. It can be shown, however, that directionally convex order is generated by functions with a bounded growth behavior at infinity.

Theorem 3.12.7. *If* $Ef(\mathbf{X}) \leq Ef(\mathbf{Y})$ *for all directionally convex f such that $f(\mathbf{x}) = O(\|\mathbf{x}\|)$ at infinity then* $\mathbf{X} \leq_{dcx} \mathbf{Y}$.

Proof. It will be shown that any directionally convex function f can be approximated monotonically from below by directionally convex functions f_k such that $f_k(\mathbf{x}) = O(\|\mathbf{x}\|)$ at infinity. The assertion then follows from the monotone convergence theorem. The idea of the construction is as follows. Let f_k coincide with f in a cube with side length $2k$, i.e. $f_k(\mathbf{x}) = f(\mathbf{x})$ for all $\mathbf{x} \in \mathbb{R}^n$ with $\|\mathbf{x}\|_\infty = \max_{i=1}^n |x_i| \leq k$, and linearly extrapolate it outside of this cube. Formally this means that we define $\psi_k : \mathbb{R} \to \mathbb{R}$ as

$$\psi_k(x) = \begin{cases} k, & x > k, \\ x, & -k \leq x \leq k, \\ -k, & x < -k, \end{cases}$$

and $\boldsymbol{\Psi}_k : \mathbb{R}^n \to \mathbb{R}^n$ as $\boldsymbol{\Psi}_k(\mathbf{x}) = (\psi_k(x_1), \ldots, \psi_k(x_n))$. Since f is convex in each variable when the others are held fixed, it has left and right partial derivatives. Hence we can define $g_{k,i} : \mathbb{R}^n \to \mathbb{R}$, $i = 1, \ldots, n$ as

$$g_{k,i}(\mathbf{x}) = \begin{cases} \dfrac{\partial^-}{\partial x_i} f(\boldsymbol{\Psi}_k(\mathbf{x})), & x_i > k, \\ \dfrac{\partial^+}{\partial x_i} f(\boldsymbol{\Psi}_k(\mathbf{x})), & x_i < -k, \\ 0, & \text{otherwise.} \end{cases}$$

Finally the function f_k is given as

$$f_k(\mathbf{x}) = f(\boldsymbol{\Psi}_k(\mathbf{x})) + \sum_{i=1}^{n} g_{k,i}(\mathbf{x})(x_i - \psi_k(x_i)).$$

It is clear that $f_k(\mathbf{x}) = O(\|\mathbf{x}\|)$ at infinity. Thus it remains to show that the sequence f_k converges monotonically to f and that each f_k is directionally convex. We will assume for simplicity that f is twice differentiable (the general case is similar, but more technical). In this case the definition of f_k reduces to

$$f_k(\mathbf{x}) = f(\boldsymbol{\Psi}_k(\mathbf{x})) + \nabla f(\boldsymbol{\Psi}_k(\mathbf{x}))^T (\mathbf{x} - \boldsymbol{\Psi}_k(\mathbf{x})),$$

where ∇f denotes the gradient of f. Obviously (f_k) converges to f, since $f_k(\mathbf{x})$ and $f(\mathbf{x})$ coincide for $k \geq \|\mathbf{x}\|_\infty$. It is also clear that $f_k(\mathbf{x}) \leq f_{k+1}(\mathbf{x}) \leq f(\mathbf{x})$, since f is convex in each variable when the others are held fixed, and we have replaced $f(\mathbf{x})$ outside the cube by the corresponding supporting line (or plane) at $f(\boldsymbol{\Psi}_k(\mathbf{x}))$. Directional convexity of f_k follows from the fact that

$$\frac{\partial^2}{\partial x_i \partial x_j} f_k(\mathbf{x}) \;=\; \mathbf{1}_{[-k \leq x_i \leq k] \cup [-k \leq x_j \leq k]} \cdot \frac{\partial^2}{\partial x_i \partial x_j} f(\boldsymbol{\Psi}_k(\mathbf{x})) \;\geq\; 0$$

for all $1 \leq i, j \leq n$ and all $\mathbf{x} \in \mathbb{R}^n$. $\square$

The following result is an immediate consequence of Theorem 3.12.7, taking into account Lemma 3.4.5.

Theorem 3.12.8. *Let* $(\mathbf{X}^{(k)})$ *and* $(\mathbf{Y}^{(k)})$ *be sequences of random vectors with* $\mathbf{X}^{(k)} \leq_{dcx} \mathbf{Y}^{(k)}$ *for all* $k \in \mathbb{N}$. *If* $\mathbf{X}^{(k)} \to \mathbf{X}$ *and* $\mathbf{Y}^{(k)} \to \mathbf{Y}$ *in distribution, and if moreover* $E\mathbf{X}^{(k)} \to E\mathbf{X}$ *and* $E\mathbf{Y}^{(k)} \to E\mathbf{Y}$, *then* $\mathbf{X} \leq_{dcx} \mathbf{Y}$.

Another small generator is given in the next result.

Theorem 3.12.9. *The stochastic order relation* $\leq_{dcx}$ *is generated by the set of all infinite differentiable directionally convex functions.*

Proof. A directionally convex function is convex in each argument and hence continuous. Therefore the assertion follows from Theorem 2.5.6 in combination with Theorem 3.7.2. □

It will be shown below that directionally convex order is the order which is most suitable for the comparison of random vectors with a common copula but different variability in the marginals. Convex order is not helpful in that situation, as is demonstrated for the case of normal distributions in the following result.

Theorem 3.12.10. *Let* $\mathbf{X} = (X_1, X_2)$ *and* $\mathbf{X}' = (X_1', X_2')$ *be normally distributed with the same copula (i.e.* $Corr(X_1, X_2) = Corr(X_1', X_2')$*) and* $EX = EX' = \mathbf{0}$*, and assume that* $\mathrm{Var}(X_1) < \mathrm{Var}(X_1')$ *and* $\mathrm{Var}(X_2) = \mathrm{Var}(X_2') > 0$*. Then* $\mathbf{X} \leq_{cx} \mathbf{X}'$ *holds if and only if the components are independent.*

Proof. We have

$$\Sigma_{\mathbf{X}} = \begin{bmatrix} \sigma_1^2 & \rho\sigma_1\sigma_2 \\ \rho\sigma_1\sigma_2 & \sigma_2^2 \end{bmatrix} \quad \text{and} \quad \Sigma_{\mathbf{X}'} = \begin{bmatrix} {\sigma_1'}^2 & \rho\sigma_1'\sigma_2 \\ \rho\sigma_1'\sigma_2 & \sigma_2^2 \end{bmatrix}$$

with $\sigma_1 < \sigma_1'$ and correlation coefficient $-1 \leq \rho \leq 1$. It has been shown in Theorem 3.4.7 that $\mathbf{X} \leq_{cx} \mathbf{X}'$ holds if and only if $\Sigma_{\mathbf{X}'} - \Sigma_{\mathbf{X}}$ is non-negative definite. However,

$$\det(\Sigma_{\mathbf{X}'} - \Sigma_{\mathbf{X}}) = -\left((\sigma_1' - \sigma_1)\sigma_2\rho\right)^2 \geq 0$$

only if $\rho = 0$. □

According to Theorem 3.12.10 we cannot expect any reasonable result for convex ordering of random vectors with a common copula. We will have to look for other orders for random vectors. In Theorem 3.6.3 it was shown that for the independent copula, convex ordering of the components implies componentwise convex order for the vectors. Thus we have to look for a weakening of the componentwise convex order. It is of interest (especially in financial and actuarial applications, see Chapter 8) to find a weakening which implies univariate convex order for the sum of the components. Since $\mathbf{X} \leq_{ccx} \mathbf{X}'$ implies $\mathbf{X} \leq_{dcx} \mathbf{X}'$ one possibility is to consider the directionally convex order.

Moreover, it is clear that we cannot expect that convex order of the marginals leads to convex order of the sum of the components if the components are negatively dependent. In the financial literature this phenomenon is known under the name of 'hedging' risks. Consider the owner of some stock which yields a random return of X_1. The risk can be diminished by investing in some other risky asset which is negatively correlated with X_1, for example, a put option for that stock position. Assume that this put option yields $X_2 = \max\{K - X_1, 0\}$ and that EX_2 has to be paid for this put option. Then we can compare the situation of holding the stock and an amount of EX_2 in cash (i.e. the portfolio is $\mathbf{X} = (X_1, EX_2)$) to the situation where the put option is bought (i.e. the portfolio is $\mathbf{X}' = (X_1, X_2)$). The portfolio $\mathbf{X}'$ is less risky than $\mathbf{X}$, namely, $X_1 + EX_2 \geq_{cx} X_1 + X_2$, though $EX_2 \leq_{cx} X_2$.

Thus it is clear that we need some notion of positive dependence. In fact, below (see Theorem 3.12.14) it will be shown that if a common conditionally increasing copula is assumed for the two vectors and if the marginals are ordered by the convex order, then the directionally convex order holds for the vectors. This implies as a corollary that the convex order holds for all positive linear combinations. For the proof of that result we need some lemmata. Recall that a random vector is said to be comonotone if its distribution is the upper Fréchet bound, and for the definition of a mean preserving local spread we refer to Definition 1.5.28.

Lemma 3.12.11. *Let all marginals $F_1, \ldots, F_n$ of the comonotone random vector $\mathbf{X}$ have finite support, and assume that there is a distribution function G_1 which can be obtained from F_1 through a mean preserving local spread. Let $\mathbf{Y}$ be a comonotone random vector with distribution function $G = \min\{G_1, F_2, \ldots F_n\}$. Then $\mathbf{X} \leq_{dcx} \mathbf{Y}$.*

Proof. Since F is comonotone and all marginals have finite support the support of F is given by a finite set $\{\mathbf{x}^{(1)}, \mathbf{x}^{(2)}, \ldots, \mathbf{x}^{(m)}\} \subset \mathbb{R}^n$ with $\mathbf{x}^{(1)} \leq \cdots \leq \mathbf{x}^{(m)}$. Since G_1 is obtained from F_1 by a mean preserving local spread, there are p and β in $(0,1)$ and $\varepsilon > 0$ such that G_1 is obtained from F_1 by removing all mass, say p, from the point $s \in \mathbb{R}$, and moving mass βp to the point $s - (1 - \beta)\varepsilon$ and mass $(1 - \beta)p$ to $s + \beta\varepsilon$. Moreover, there are $1 \leq i \leq j \leq m$, such that $x_1^{(i-1)} < x_1^{(i)} = s = x_1^{(j)} < x_1^{(j+1)}$. To avoid technicalities we assume without loss of generality that $j = i+1$. (The general proof for $j > i + 1$ can then be derived from that by induction, and the case $j = i$ is easy.) Hence there is some $\alpha \in (0,1)$ such that $P(\mathbf{X} = \mathbf{x}^{(j)}) = \alpha p$ and $P(\mathbf{X} = \mathbf{x}^{(i)}) = (1 - \alpha)p$. Since $P(Y_1 = s) = 0$, the random vector $\mathbf{Y}$ has no mass in these two points. Instead, some mass is added to two or three other points. The following two cases have to be distinguished:

Case 1: $\beta \leq 1 - \alpha$: In this case a portion βp of the mass is moved to $\mathbf{x}^{(i)} - (1 - \beta)\varepsilon\mathbf{e}_1$, a portion $(1 - \alpha - \beta)p$ is moved to $\mathbf{x}^{(i)} + \beta\varepsilon\mathbf{e}_1$, and a portion αp is moved to $\mathbf{x}^{(j)} + \beta\varepsilon\mathbf{e}_1$. Hence

$$Ef(\mathbf{Y}) - Ef(\mathbf{X}) = p \left(\beta f(\mathbf{x}^{(i)} - (1 - \beta)\varepsilon\mathbf{e}_1) + (1 - \alpha - \beta)f(\mathbf{x}^{(i)} + \beta\varepsilon\mathbf{e}_1) \right.$$
$$\left. + \alpha f(\mathbf{x}^{(j)} + \beta\varepsilon\mathbf{e}_1) - (1 - \alpha)f(\mathbf{x}^{(i)}) - \alpha f(\mathbf{x}^{(j)}) \right).$$

Since f is directionally convex it is convex in the first variable and therefore

$$f(\mathbf{x}^{(i)}) \leq \frac{1 - \alpha - \beta}{1 - \alpha} f(\mathbf{x}^{(i)} + \beta\varepsilon\mathbf{e}_1) + \frac{\beta}{1 - \alpha} f(\mathbf{x}^{(i)} - (1 - \alpha - \beta)\varepsilon\mathbf{e}_1).$$

This implies

$$Ef(\mathbf{Y}) - Ef(\mathbf{X}) \geq p \Big(\beta f(\mathbf{x}^{(i)} - (1-\beta)\varepsilon\mathbf{e}_1) + \alpha f(\mathbf{x}^{(j)} + \beta\varepsilon\mathbf{e}_1)$$
$$-\beta f(\mathbf{x}^{(i)} - (1-\alpha-\beta)\varepsilon\mathbf{e}_1) - \alpha f(\mathbf{x}^{(j)}) \Big).$$
$$= p\alpha \Big(f(\mathbf{x}^{(j)} + \beta\varepsilon\mathbf{e}_1) - f(\mathbf{x}^{(j)}) \Big)$$
$$- p\beta \Big(f(\mathbf{x}^{(i)} - (1-\alpha-\beta)\varepsilon\mathbf{e}_1) - f(\mathbf{x}^{(i)} - (1-\beta)\varepsilon\mathbf{e}_1) \Big)$$
$$\geq 0.$$

Here the last inequality follows from Theorem 3.12.2 (v).

Case 2: $\beta > 1 - \alpha$: In this case a portion $(1-\alpha)p$ of the mass is moved to $\mathbf{x}^{(i)} - (1-\beta)\varepsilon\mathbf{e}_1$, a portion $(\alpha + \beta - 1)p$ is moved to $\mathbf{x}^{(j)} - (1-\beta)\varepsilon\mathbf{e}_1$, and a portion $(1-\beta)p$ is moved to $\mathbf{x}^{(j)} + \beta\varepsilon\mathbf{e}_1$. Hence

$$Ef(\mathbf{Y}) - Ef(\mathbf{X}) = p \Big((1-\alpha)f(\mathbf{x}^{(i)} - (1-\beta)\varepsilon\mathbf{e}_1) + (1-\beta)f(\mathbf{x}^{(j)} + \beta\varepsilon\mathbf{e}_1)$$
$$(\alpha + \beta - 1)f(\mathbf{x}^{(j)} - (1-\beta)\varepsilon\mathbf{e}_1) - (1-\alpha)f(\mathbf{x}^{(i)}) - \alpha f(\mathbf{x}^{(j)}) \Big).$$

Since f is directionally convex and hence convex in the first variable,

$$f(\mathbf{x}^{(j)}) \leq \frac{\alpha + \beta - 1}{\alpha} f(\mathbf{x}^{(j)} + (\beta - 1)\varepsilon\mathbf{e}_1) + \frac{1-\beta}{\alpha} f(\mathbf{x}^{(j)} + (\alpha + \beta - 1)\varepsilon\mathbf{e}_1).$$

This implies

$$Ef(\mathbf{Y}) - Ef(\mathbf{X}) \geq p(1-\beta) \Big(f(\mathbf{x}^{(j)} + \beta\varepsilon\mathbf{e}_1) - f(\mathbf{x}^{(j)} + (\beta + \alpha - 1)\varepsilon\mathbf{e}_1) \Big)$$
$$- p(1-\alpha) \Big(f(\mathbf{x}^{(i)}) - f(\mathbf{x}^{(i)} - (1-\beta)\varepsilon\mathbf{e}_1) \Big)$$
$$\geq 0.$$

Here the last inequality again follows from Theorem 3.12.2 (v). □

Lemma 3.12.12. *Let* $\mathbf{X}^{(k)} = (X_1^{(k)}, \ldots, X_n^{(k)})$, $k \in \mathbb{N}$, *be a sequence of random vectors, and let the marginals* $F_i^{(k)}$ *converge weakly to some* F_i, $i = 1, \ldots, n$. *Moreover, assume that the distribution of* $\mathbf{X}^{(k)}$ *can be written as*

$$F^{(k)}(\mathbf{x}) = C(F_1^{(k)}(x_1), \ldots, F_n^{(k)}(x_n))$$

for some fixed copula C. *Then* $\mathbf{X}^{(k)}$ *converges weakly to some* $\mathbf{X}$ *with distribution*

$$F(x_1, \ldots, x_n) = C(F_1(x_1), \ldots, F_n(x_n)).$$

Proof. This follows from the fact that each copula is Lipschitz continuous, see Sempi (1982). □

Lemma 3.12.13. *Let* $\mathbf{X}$ *and* $\mathbf{Y}$ *be comonotone random vectors and assume that* $X_i \leq_{cx} Y_i$ *for all* $i = 1, \ldots, n$. *Then* $\mathbf{X} \leq_{dcx} \mathbf{Y}$.

Proof. By Lemma 3.12.11, if there are two discrete comonotone random vectors $\mathbf{X}$ and $\mathbf{Y}$ such that the second is obtained from the first by a local mean preserving spread in just one coordinate, then $\mathbf{X} \leq_{dcx} \mathbf{Y}$.

By Theorem 1.5.29, if X_i and Y_i have finite support and $X_i \leq_{cx} Y_i$, then Y_i can be obtained from X_i by a finite sequence of local mean preserving spreads.

Theorem 1.5.30 ensures that for any pair of random variables (X_i, Y_i) such that $X_i \leq_{cx} Y_i$ there exists a sequence of random pairs $(X_i^{(k)}, Y_i^{(k)})$ with finite support such that $X_i^{(k)} \leq_{cx} Y_i^{(k)}$, $X_i^{(k)} \to X_i$, and $Y_i^{(k)} \to Y_i$ in distribution, and such that the means converge.

Theorem 3.12.8 and Lemma 3.12.12 then yield the result. □

Theorem 3.12.14. *Let* $\mathbf{X}$ *and* $\mathbf{Y}$ *be random vectors with a common conditionally increasing copula* C *and assume that* $X_i \leq_{cx} Y_i$ *for* $i = 1, 2, \ldots, n$. *Then* $\mathbf{X} \leq_{dcx} \mathbf{Y}$.

Proof. First assume that $\mathbf{X}$ and $\mathbf{Y}$ differ only in their first marginal and that the copula is CIS. According to Lemma 3.10.10 there are i.i.d. random variables $U_1, \ldots, U_n \sim U(0, 1)$ and monotone functions $h_i : (0, 1)^i \to (0, 1)$ such that $X_1 = F_1^{-1}(U_1)$, $Y_1 = G_1^{-1}(U_1)$ and

$$X_i = Y_i = F_i^{-1}(W_i) \quad \text{for } i = 2, \ldots, n,$$

where $W_i = h_i(U_1, \ldots, U_i)$, $i = 2, \ldots, n$. Define $\Theta = (U_2, \ldots, U_n)$ and condition on $\Theta = \theta = (\theta_2, \ldots, \theta_n)$. Then $[X_1 | \Theta = \theta] = F_1^{-1}(U_1)$ and $[Y_1 | \Theta = \theta] = G_1^{-1}(U_1)$ are independent of θ, and

$$[X_i | \Theta = \theta] = [Y_i | \Theta = \theta] = F_i^{-1}(h_i(U_1, \theta_2, \ldots, \theta_i)) \quad \text{for } i = 2, \ldots, n.$$

Hence the vectors $[\mathbf{X} | \Theta = \theta]$ and $[\mathbf{Y} | \Theta = \theta]$ are comonotone for all θ and $[X_i | \Theta = \theta] \leq_{cx} [Y_i | \Theta = \theta]$ for all $i = 1, \ldots, n$. Thus it follows from Lemma 3.12.13 that $[\mathbf{X} | \Theta = \theta] \leq_{dcx} [\mathbf{Y} | \Theta = \theta]$ for all θ. Theorem 3.12.6 yields that $\leq_{dcx}$ is closed with respect to mixtures, and therefore unconditioning yields $\mathbf{X} \leq_{dcx} \mathbf{Y}$.

Now assume that $\mathbf{X}$ and $\mathbf{Y}$ differ only in their ith marginal. Choose a permutation π with $\pi(1) = i$. Since the permuted vectors $\mathbf{X}_\pi$ and $\mathbf{Y}_\pi$ are CIS for all permutations π, the vectors $\mathbf{X}_\pi$ and $\mathbf{Y}_\pi$ fulfill the conditions of the first part of the proof, and therefore $\mathbf{X} \leq_{dcx} \mathbf{Y}$. The general case now follows by induction on i, taking into account the transitivity of $\leq_{dcx}$. □

Corollary 3.12.15. *Let* **X** *and* **Y** *fulfill the conditions of Theorem 3.12.14. Then, for all non-negative* $\alpha_1, \ldots, \alpha_n$,

$$\sum_{i=1}^{n} \alpha_i X_i \leq_{cx} \sum_{i=1}^{n} \alpha_i Y_i.$$

Proof. This follows immediately from Theorem 3.12.14 and the fact that for any convex function $g : \mathbb{R} \to \mathbb{R}$, and for any non-negative $\alpha_1, \ldots, \alpha_n$,

$$f(x_1, \ldots, x_n) = g(\alpha_1 x_1 + \cdots + \alpha_n x_n)$$

is directionally convex. $\square$

Example 3.12.16. (CIS is not sufficient for Theorem 3.12.14.) Consider the following copula C in the bivariate case. Let $U_2 \sim U(0,1)$, $U_1 = 2U_2 \bmod 1$ and $(U_1, U_2) \sim C$. This means that the distribution of (U_1, U_2) is concentrated on the two lines connecting $(0,0)$ with $(1, 1/2)$ and $(0, 1/2)$ with $(1, 1)$, on which it is uniformly distributed. Formally we have

$$C(u_1, u_2) = \begin{cases} \frac{1}{2} \min\{u_1, 2u_2\} & \text{for } u_2 \leq \frac{1}{2}, \\ \frac{1}{2} u_1 + \frac{1}{2} \min\{u_1, 2u_2 - 1\} & \text{for } u_2 > \frac{1}{2}, \end{cases}$$

$0 \leq u_1, u_2 \leq 1$. It is easy to see that C is CIS, since

$$[U_2 | U_1 = u] \sim \frac{1}{2} \left(\delta_{u/2} + \delta_{(u+1)/2} \right) \quad \text{for } u \in (0,1).$$

Now assume that **X** and **Y** have this copula C and that

$$X_1 =_{st} Y_1 \sim \frac{1}{2} (\delta_0 + \delta_1)$$

and that

$$X_2 \sim \frac{1}{3} (\delta_0 + \delta_1 + \delta_2),$$

whereas

$$Y_2 \sim \frac{1}{3} \delta_0 + \frac{1}{6} \delta_{1/2} + \frac{1}{6} \delta_{3/2} + \frac{1}{3} \delta_2.$$

Then clearly $X_1 \leq_{cx} Y_1$ and $X_2 \leq_{cx} Y_2$. However,

$$\mathbf{X} \sim \frac{1}{4} \delta_{(0,0)} + \frac{1}{12} \delta_{(1,0)} + \frac{1}{6} \delta_{(1,1)} + \frac{1}{6} \delta_{(0,1)} + \frac{1}{12} \delta_{(0,2)} + \frac{1}{4} \delta_{(1,2)}$$

and

$$\mathbf{Y} \sim \frac{1}{4} \delta_{(0,0)} + \frac{1}{12} \delta_{(1,0)} + \frac{1}{6} \delta_{(1,1/2)} + \frac{1}{6} \delta_{(0,3/2)} + \frac{1}{12} \delta_{(0,2)} + \frac{1}{4} \delta_{(1,2)},$$

hence

$$X_1 + X_2 \sim \frac{1}{4}\left(\delta_0 + \delta_1 + \delta_2 + \delta_3\right)$$

and

$$Y_1 + Y_2 \sim \frac{1}{4}\delta_0 + \frac{1}{12}\delta_1 + \frac{1}{3}\delta_{3/2} + \frac{1}{12}\delta_2 + \frac{1}{4}\delta_3,$$

and thus $X_1 + X_2 \not\leq_{cx} Y_1 + Y_2$. Instead, it even holds that $Y_1 + Y_2 \leq_{cx} X_1 + X_2$.

This result is perhaps surprising. Since CIS is a quite strong notion of positive dependence, stronger than association and orthant dependence, the example shows that even in the situation of rather strong positive dependence a higher variability of the marginals may lead to a lower variability of the sum of the components. Thus the assumptions of Theorem 3.12.14 cannot be weakened to, for instance, association.

Now we will see that a sort of inverse of Theorem 3.12.14 holds for comonotone random vectors. Since comonotonicity is the strongest possible notion of positive dependence, this demonstrates that the directionally convex order is the strongest conclusion that can be expected in Theorem 3.12.14.

Theorem 3.12.17. *If* $X_i \leq_{cx} Y_i$ *for* $1, 2, \ldots, n$ *implies* $Ef(\mathbf{X}) \leq Ef(\mathbf{Y})$ *whenever* $\mathbf{X}$ *and* $\mathbf{Y}$ *are comonotone, then* f *is directionally convex.*

Proof. Fix $\mathbf{x} = (x_1, \ldots, x_n)$, $1 \leq i, j \leq n$ and $\varepsilon, \gamma \geq 0$, and assume that $X_j \sim 1/2(\delta_{x_j} + \delta_{x_j + \gamma})$ and that $X_k \sim \delta_{x_k}$ for all $k \neq j$. Moreover, let $Y_i \sim 1/2(\delta_{x_i - \varepsilon} + \delta_{x_i + \varepsilon})$ in case $i \neq j$ and $Y_i \sim 1/2(\delta_{x_i - \varepsilon} + \delta_{x_i + \gamma + \varepsilon})$ in case $i = j$, and $Y_k =_{st} X_k$ for all $k \neq i$. Then $X_k \leq_{cx} Y_k$ for $k = 1, \ldots, n$, and for comonotone $\mathbf{X}$ and $\mathbf{Y}$ it holds that $Ef(\mathbf{Y}) - Ef(\mathbf{X}) \geq 0$ if and only if

$$f(\mathbf{x} + \varepsilon \mathbf{e}_i + \gamma \mathbf{e}_j) - f(\mathbf{x} + \gamma \mathbf{e}_j) - f(\mathbf{x}) + f(\mathbf{x} - \varepsilon \mathbf{e}_i) \geq 0.$$

Since $\mathbf{x} \in \mathbb{R}^n$, $1 \leq i, j \leq n$ and $\varepsilon, \gamma \geq 0$ are arbitrary, it follows from Theorem 3.12.2 (iv) that f is directionally convex. $\qquad\square$

For normally distributed random vectors there is a complete characterization of directionally convex order. It can be shown that $\mathcal{N}(\boldsymbol{\mu}, \boldsymbol{\Sigma}) \leq_{dcx} \mathcal{N}(\boldsymbol{\mu}', \boldsymbol{\Sigma}')$, if and only if $\boldsymbol{\mu} = \boldsymbol{\mu}'$ and $\sigma_{ij} \leq \sigma'_{ij}$ for all $1 \leq i, j \leq n$. This will be demonstrated below in Theorem 3.13.6.

Applications of directionally convex order are given in Meester and Shanthikumar (1993, 1999), Bäuerle and Rolski (1998) and Müller and Scarsini (2001), among others.

3.13   Stochastic Ordering of Multivariate Normal Distributions

A very important multivariate distribution is the normal distribution. Recall that a random vector $\mathbf{X}$ is said to have a normal distribution $\mathcal{N}(\boldsymbol{\mu}, \boldsymbol{\Sigma})$ with expectation $\boldsymbol{\mu} = E\mathbf{X}$ and covariance matrix $\boldsymbol{\Sigma} = E((\mathbf{X} - \boldsymbol{\mu})(\mathbf{X} - \boldsymbol{\mu})^T)$ if it has the characteristic function

$$\Psi_{\mathbf{X}}(\mathbf{t}) = Ee^{i\mathbf{t}^T\mathbf{X}} = \exp\left(i\mathbf{t}^T\boldsymbol{\mu} - \frac{1}{2}\mathbf{t}^T\boldsymbol{\Sigma}\mathbf{t}\right).$$

If $\boldsymbol{\Sigma}$ is positive definite and hence invertible then $\mathbf{X}$ has the density

$$\phi_{\mathbf{X}}(\mathbf{x}) = \frac{1}{\sqrt{(2\pi)^n \det(\boldsymbol{\Sigma})}} \exp\left(-\frac{1}{2}(\mathbf{x} - \boldsymbol{\mu})^T\boldsymbol{\Sigma}^{-1}(\mathbf{x} - \boldsymbol{\mu})\right).$$

The following identity is very useful, and follows from the Fourier inversion theorem, see lemma 9.5.4 in Dudley (1989).

Lemma 3.13.1. *If* $\mathbf{X} \sim \mathcal{N}(\boldsymbol{\mu}, \boldsymbol{\Sigma})$ *and* $\boldsymbol{\Sigma}$ *is positive definite then the following relation holds between the characteristic function* $\Psi_{\mathbf{X}}$ *and the density function* $\phi_{\mathbf{X}}$:

$$\phi_{\mathbf{X}}(\mathbf{x}) = \frac{1}{(2\pi)^n}\int e^{-i\mathbf{t}^T\mathbf{x}}\Psi_{\mathbf{X}}(\mathbf{t})d\mathbf{t} = \frac{1}{(2\pi)^n}\int \exp\left(-i\mathbf{t}^T(\mathbf{x}-\boldsymbol{\mu}) - \frac{1}{2}\mathbf{t}^T\boldsymbol{\Sigma}\mathbf{t}\right)d\mathbf{t}.$$

To characterize integral stochastic order relations for normally distributed vectors we have to consider the difference $Ef(\mathbf{X}') - Ef(\mathbf{X})$ for normally distributed random vectors $\mathbf{X}$ and $\mathbf{X}'$ and suitable functions f. Therefore we will derive now an identity for this expression under some weak regularity conditions for f. If $f : \mathbb{R}^n \to \mathbb{R}$ is twice continuously differentiable, then as usual

$$\nabla f(\mathbf{x}) = \left(\frac{\partial}{\partial x_i}f(\mathbf{x})\right)_{i=1}^n \quad \text{and } \mathbf{H}_f(\mathbf{x}) = \left(\frac{\partial^2}{\partial x_i \partial x_j}f(\mathbf{x})\right)_{i,j=1}^n$$

are the gradient and the Hesse matrix of f. Recall also that for a matrix $\mathbf{A} = (a_{ij})$ the trace is defined as $tr(\mathbf{A}) = \sum_{i=1}^n a_{ii}$, and hence

$$tr(\mathbf{AB}) = \sum_{i,j=1}^n a_{ij}b_{ij},$$

if $\mathbf{A}$ and $\mathbf{B}$ are symmetric matrices.

The following result is shown in Müller (2001b).

Theorem 3.13.2. *Let* $\mathbf{X} \sim \mathcal{N}(\boldsymbol{\mu}, \boldsymbol{\Sigma})$ *and* $\mathbf{X}' \sim \mathcal{N}(\boldsymbol{\mu}', \boldsymbol{\Sigma}')$, *where* $\boldsymbol{\Sigma}$ *and* $\boldsymbol{\Sigma}'$ *are positive definite, and let* ϕ_λ *be the density of*

$$\mathcal{N}(\lambda\boldsymbol{\mu}' + (1 - \lambda)\boldsymbol{\mu}, \lambda\boldsymbol{\Sigma}' + (1 - \lambda)\boldsymbol{\Sigma}), \quad 0 \leq \lambda \leq 1.$$

Moreover, assume that $f : \mathbb{R}^n \to \mathbb{R}$ is twice continuously differentiable with $f(\mathbf{x}) = O(\|\mathbf{x}\|)$ and $\nabla f(\mathbf{x}) = O(\|\mathbf{x}\|)$ at infinity. Then

$$Ef(\mathbf{X}') - Ef(\mathbf{X})$$

$$= \int_0^1 \int \left((\boldsymbol{\mu}' - \boldsymbol{\mu})^T \nabla f(\mathbf{x}) + \frac{1}{2} tr((\boldsymbol{\Sigma}' - \boldsymbol{\Sigma})\mathbf{H}_f(\mathbf{x})) \right) \phi_\lambda(\mathbf{x}) \, d\mathbf{x} \, d\lambda. \quad (3.13.1)$$

Proof. Define

$$g(\lambda) = \int f(\mathbf{x})\phi_\lambda(\mathbf{x})d\mathbf{x}.$$

Then $Ef(\mathbf{X}') - Ef(\mathbf{X}) = g(1) - g(0)$. Hence it is sufficient to show that the derivative of g is equal to the expression inside the outer integral in (3.13.1). This can be seen as follows. Lemma 3.13.1 implies

$$
\begin{aligned}
\frac{\partial}{\partial\lambda}\phi_\lambda(\mathbf{x}) &= \frac{\partial}{\partial\lambda}\frac{1}{(2\pi)^n}\int e^{-i\mathbf{t}^T\mathbf{x}}\Psi_\lambda(\mathbf{t})d\mathbf{t} \\
&= \frac{1}{(2\pi)^n}\int \frac{\partial}{\partial\lambda}e^{-i\mathbf{t}^T\mathbf{x}}\Psi_\lambda(\mathbf{t})d\mathbf{t} \\
&= \frac{1}{(2\pi)^n}\int e^{-i\mathbf{t}^T\mathbf{x}}\Psi_\lambda(\mathbf{t})\left(-i\mathbf{t}^T(\boldsymbol{\mu}'-\boldsymbol{\mu}) - \frac{1}{2}\mathbf{t}^T(\boldsymbol{\Sigma}'-\boldsymbol{\Sigma})\mathbf{t}\right)d\mathbf{t} \\
&= -\sum_{i=1}^n (\mu_i' - \mu_i)\frac{\partial}{\partial x_i}\phi_\lambda(\mathbf{x}) + \frac{1}{2}\sum_{i,j=1}^n (\sigma_{ij}' - \sigma_{ij})\frac{\partial^2}{\partial x_i\partial x_j}\phi_\lambda(\mathbf{x}).
\end{aligned}
$$

This yields

$$
\begin{aligned}
g'(\lambda) &= \int f(\mathbf{x})\frac{\partial}{\partial\lambda}\phi_\lambda(\mathbf{x})d\mathbf{x} \\
&= \sum_{i=1}^n (\mu_i - \mu_i')\int f(\mathbf{x})\frac{\partial\phi_\lambda(\mathbf{x})}{\partial x_i}d\mathbf{x} + \sum_{i,j=1}^n \frac{\sigma_{ij}' - \sigma_{ij}}{2}\int f(\mathbf{x})\frac{\partial^2\phi_\lambda(\mathbf{x})}{\partial x_i\partial x_j}d\mathbf{x} \\
&= \sum_{i=1}^n (\mu_i' - \mu_i)\int \phi_\lambda(\mathbf{x})\frac{\partial f(\mathbf{x})}{\partial x_i}d\mathbf{x} + \sum_{i,j=1}^n \frac{\sigma_{ij}' - \sigma_{ij}}{2}\int \phi_\lambda(\mathbf{x})\frac{\partial^2 f(\mathbf{x})}{\partial x_i\partial x_j}d\mathbf{x} \\
&= \int \left((\boldsymbol{\mu}' - \boldsymbol{\mu})^T\nabla f(\mathbf{x}) + \frac{1}{2}tr\left((\boldsymbol{\Sigma}' - \boldsymbol{\Sigma})\mathbf{H}_f(\mathbf{x})\right)\right)\phi_\lambda(\mathbf{x})\,d\mathbf{x}.
\end{aligned}
$$

Here the third equality follows form a twofold application of partial integration, taking into account the growth conditions on f. □

From Theorem 3.13.2 the following sufficient condition for non-negativity of $Ef(\mathbf{X}') - Ef(\mathbf{X})$ can be derived.

Corollary 3.13.3. *Let* $\mathbf{X} \sim \mathcal{N}(\boldsymbol{\mu}, \boldsymbol{\Sigma})$ *and* $\mathbf{X}' \sim \mathcal{N}(\boldsymbol{\mu}', \boldsymbol{\Sigma}')$ *and assume that* f *satisfies the conditions of Theorem 3.13.2. Then* $Ef(\mathbf{X}') - Ef(\mathbf{X}) \geq 0$ *if the following two conditions hold:*

$$\sum_{i=1}^{n} (\mu_i' - \mu_i) \frac{\partial f(\mathbf{x})}{\partial x_i} \geq 0 \quad \text{for all } \mathbf{x} \tag{3.13.2}$$

and

$$\sum_{i,j=1}^{n} (\sigma_{ij}' - \sigma_{ij}) \frac{\partial^2 f(\mathbf{x})}{\partial x_i \partial x_j} \geq 0 \quad \text{for all } \mathbf{x}. \tag{3.13.3}$$

Proof. If $\boldsymbol{\Sigma}$ and $\boldsymbol{\Sigma}'$ are positive definite, then the assertion follows immediately from Theorem 3.13.2. If one or both of them are only non-negative definite, then the following approximation argument can be used. Let $\mathbf{I}$ be the identity matrix. Then for arbitrary $\varepsilon > 0$ the matrices $\boldsymbol{\Sigma} + \varepsilon\mathbf{I}$ and $\boldsymbol{\Sigma}' + \varepsilon\mathbf{I}$ are positive definite. Hence Theorem 3.13.2 can be applied to this perturbed matrices. Finally let ε approach zero to get the assertion. □

Many of the sufficient conditions for stochastic ordering of multivariate normal distributions which have been described in the previous sections can alternatively be derived from this result.

Example 3.13.4. (Stochastic and convex order of multivariate normal distributions.)
a) If $\boldsymbol{\mu} \leq \boldsymbol{\mu}'$ then $\mathcal{N}(\boldsymbol{\mu}, \boldsymbol{\Sigma}) \leq_{st} \mathcal{N}(\boldsymbol{\mu}', \boldsymbol{\Sigma})$. This follows from the fact that a differentiable function is increasing if and only if $\nabla f(\mathbf{x}) \geq \mathbf{0}$ for all $\mathbf{x}$, and hence (3.13.2) holds. Equation (3.13.3) holds trivially when $\boldsymbol{\Sigma} = \boldsymbol{\Sigma}'$.
b) If $\boldsymbol{\Sigma}' - \boldsymbol{\Sigma}$ is non-negative definite, then $\mathcal{N}(\boldsymbol{\mu}, \boldsymbol{\Sigma}) \leq_{cx} \mathcal{N}(\boldsymbol{\mu}, \boldsymbol{\Sigma}')$. Here equation (3.13.2) holds trivially, and (3.13.3) can be shown as follows: Since $\boldsymbol{\Sigma}' - \boldsymbol{\Sigma}$ is non-negative definite, it has the canonical representation $\boldsymbol{\Sigma}' - \boldsymbol{\Sigma} = \sum_{k=1}^{n} \lambda_k \mathbf{a}^{(k)} \mathbf{a}^{(k)T}$, where $\mathbf{a}^{(k)}$ are the eigenvectors and $\lambda_k \geq 0$ are the corresponding eigenvalues of $\boldsymbol{\Sigma}' - \boldsymbol{\Sigma}$. Hence

$$\sum_{i,j=1}^{n} (\sigma_{ij}' - \sigma_{ij}) \frac{\partial^2 f(\mathbf{x})}{\partial x_i \partial x_j} = \sum_{k=1}^{n} \lambda_k \sum_{i,j=1}^{n} a_i^{(k)} a_j^{(k)} \frac{\partial^2 f(\mathbf{x})}{\partial x_i \partial x_j}$$

$$= \sum_{k=1}^{n} \lambda_k \cdot \mathbf{a}^{(k)T} \mathbf{H}_f(\mathbf{x}) \mathbf{a}^{(k)} \geq 0,$$

since a twice differentiable function f is convex if and only if its Hesse matrix $\mathbf{H}_f$ is non-negative definite.
c) If $\boldsymbol{\mu} \leq \boldsymbol{\mu}'$ and $\boldsymbol{\Sigma}' - \boldsymbol{\Sigma}$ is non-negative definite, then $\mathcal{N}(\boldsymbol{\mu}, \boldsymbol{\Sigma}) \leq_{icx} \mathcal{N}(\boldsymbol{\mu}', \boldsymbol{\Sigma}')$. In this case (3.13.2) follows as in part a) and (3.13.3) follows as in part b).

Theorem 3.13.2 also enables the proof of new results, for example, for supermodular and directionally convex order.

Theorem 3.13.5. *If* $\mathbf{X} \sim \mathcal{N}(\boldsymbol{\mu}, \boldsymbol{\Sigma})$ *and* $\mathbf{X}' \sim \mathcal{N}(\boldsymbol{\mu}', \boldsymbol{\Sigma}')$*, then the following conditions are equivalent:*

(i) $\mathbf{X} \leq_{sm} \mathbf{X}'$*;*

(ii) $\mathbf{X}$ *and* $\mathbf{X}'$ *have the same marginals and* $\sigma_{ij} \leq \sigma'_{ij}$ *for all* i, j*.*

Proof. The implication (i) $\Rightarrow$ (ii) is an immediate consequence of Theorem 3.9.5. To see the converse notice that a twice differentiable function f is supermodular if and only if

$$\frac{\partial^2 f(\mathbf{x})}{\partial x_i \partial x_j} \geq 0 \quad \text{for all } \mathbf{x} \in \mathbb{R}^n,\ 1 \leq i < j \leq n$$

(see Theorem 3.9.3 a). Thus the assertion follows from Corollary 3.13.3, taking into account Theorem 3.9.13. $\square$

Theorem 3.13.6. *Let* $\mathbf{X} \sim \mathcal{N}(\boldsymbol{\mu}, \boldsymbol{\Sigma})$ *and* $\mathbf{X}' \sim \mathcal{N}(\boldsymbol{\mu}', \boldsymbol{\Sigma}')$*. Then:*
a) $\mathbf{X} \leq_{dcx} \mathbf{X}'$ *if and only if* $\boldsymbol{\mu} = \boldsymbol{\mu}'$ *and* $\sigma_{ij} \leq \sigma'_{ij}$ *for all* $1 \leq i, j \leq n$*,*
b) if $\boldsymbol{\mu} \leq \boldsymbol{\mu}'$ *and* $\sigma_{ij} \leq \sigma'_{ij}$ *for all* $1 \leq i, j \leq n$ *then* $\mathbf{X} \leq_{idcx} \mathbf{X}'$*.*

Proof. a) A twice differentiable function f is directionally convex if and only if

$$\frac{\partial^2 f(\mathbf{x})}{\partial x_i \partial x_j} \geq 0 \quad \text{for all } \mathbf{x} \text{ and all } 1 \leq i, j \leq n$$

(see Theorem 3.12.2 b). Hence the sufficiency of the mentioned condition follows from Corollary 3.13.3, taking into account Theorem 3.12.9. But this condition is also necessary, since all of the following functions are directionally convex: $f(\mathbf{x}) = x_i$, $f(\mathbf{x}) = -x_i$ and $f(\mathbf{x}) = x_i x_j$, $1 \leq i, j \leq n$.

b) This follows as part a) from Corollary 3.13.3, or alternatively by combining part a) and Theorem 3.3.13, taking into account the transitivity of $\leq_{idcx}$. $\square$

Remark 3.13.7. It follows from Theorem 3.13.6 that for normal distributions with the same marginals there is no difference between supermodular and directional convex order. Both orders hold if the off-diagonal elements of the covariance matrix are ordered. The advantage of $\leq_{dcx}$ lies in the fact that it is able to compare random vectors with different marginals. Indeed, it follows from Theorem 3.13.6 that for normal distributions directional convex order can be decomposed in a supermodular order part (i.e. a dependence order part) and a convex order part (i.e. a variability order part). Assume that $\sigma_{ij} \leq \sigma'_{ij}$ for all $1 \leq i, j \leq n$. Define the matrix $\tilde{\boldsymbol{\Sigma}}$ by

$$\tilde{\sigma}_{ij} = \begin{cases} \sigma_{ij} & \text{for } i = j, \\ \sigma'_{ij} & \text{for } i \neq j. \end{cases}$$

Then $\boldsymbol{\Sigma}' - \tilde{\boldsymbol{\Sigma}}$ is non-negative definite, since it is a diagonal matrix with non-negative entries. Moreover, $\tilde{\boldsymbol{\Sigma}} - \boldsymbol{\Sigma}$ has non-negative entries, and the two matrices have the same diagonal elements. Hence it can be deduced from Theorem 3.6.5 that $\mathcal{N}(\boldsymbol{\mu}, \boldsymbol{\Sigma}) \leq_{dcx} \mathcal{N}(\boldsymbol{\mu}, \boldsymbol{\Sigma}')$ holds if and only if there is some $\tilde{\boldsymbol{\Sigma}}$ such that

$$\mathcal{N}(\boldsymbol{\mu}, \boldsymbol{\Sigma}) \leq_{sm} \mathcal{N}(\boldsymbol{\mu}, \tilde{\boldsymbol{\Sigma}}) \leq_{ccx} \mathcal{N}(\boldsymbol{\mu}, \boldsymbol{\Sigma}').$$

The cover of this book shows (the maximum of the) density functions of two bivariate normal distributions with the same means. For the first density the variances are $\sigma_1^2 = \sigma_2^2 = 3/5$ and the correlation coefficient is $\rho = -5/6$, whereas the other function depicts the density of two independent random variables with standard normal distributions. Thus the corresponding random vectors $\mathbf{X}$ and $\mathbf{X}'$ satisfy $\mathbf{X} \leq_{dcx} \mathbf{X}'$ but neither $\mathbf{X} \leq_{cx} \mathbf{X}'$ nor $\mathbf{X} \leq_{sm} \mathbf{X}'$.

3.14   Relationships and Comparison Criteria for Multivariate Stochastic Orders

This section presents a graph which describes relationships between multivariate stochastic orders and another which shows relationships between various concepts of dependence.

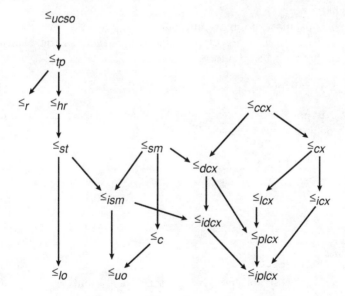

Figure 3.1 Relations between various multivariate stochastic orders. $\preceq \rightarrow \precneqq$ means that $\mathbf{X} \preceq \mathbf{Y}$ implies $\mathbf{X} \precneqq \mathbf{Y}$.

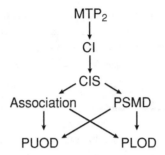

Figure 3.2 Relations between various concepts of positive dependence.

There are many families of multivariate distributions which are of interest in various branches of applied probability. Many of them are described in Kotz, Balakrishnan, and Johnson (2000). We are not aware of any systematic studies of comparison properties for multivariate distributions which are comparable to those mentioned on p. 60. Also the present book gives results only for particular cases.

The case of multivariate normal distributions is considered in Section 3.13, and in the following, two particular multivariate exponential distributions are discussed. For a few other distributions results are given in the literature, for example for Liouville distributions in Gupta and Richards (1992) and Kotz et al. (2000).

The perhaps most popular multivariate exponential distribution is the *Marshall–Olkin multivariate exponential distribution*. In the bivariate case, its survival function has the form

$$\bar{F}_{\mathbf{X}}(x_1, x_2) = \exp(-\lambda_1 x_1 - \lambda_2 x_2 - \lambda_{12} \max\{x_1, x_2\}) \quad \text{for } x_1, x_2 \geq 0.$$

The λ_i are positive parameters and λ_{12} is non-negative. In the multivariate case, the definition uses the sets $\Im_k$ given by

$$\Im_k = \{(i_1, \ldots, i_k) : 1 \leq i_1 < i_2 < \cdots < i_k \leq n\}$$

for $k = 1, \ldots, n$.

Then

$$\bar{F}_{\mathbf{X}}(x_1, \ldots, x_n) = \exp(-\sum_{k=1}^{n} \sum_{(i_1, \ldots, i_k) \in \Im_k} \lambda_{i_1 \ldots i_k} \max\{x_{i_1}, \ldots, x_{i_k}\})$$

for $x_1, \ldots, x_n \geq 0$ and positive parameters λ_i and $\lambda_{i_1 \ldots i_k} \geq 0$ for all $(i_1, \ldots, i_k)$ for $k \geq 2$.

The one-dimensional marginal distributions are exponential distributions with parameters μ_i which are equal to the sums of all $\lambda_{i_1, \ldots, i_k}$ where one of the i_j is equal to i. If at least one of the parameters $\lambda_{i_1, \ldots, i_k}$ with $k \geq 2$ is strictly positive then the Marshall–Olkin multivariate exponential distribution is not absolutely continuous with respect to the Lebesgue measure. It has positive masses on linear subspaces of $\mathbb{R}^n$ of the form $\{\mathbf{x} : x_{i_1} = \ldots = x_{i_k}\}$.

The Marshall–Olkin multivariate exponential distribution can be obtained as the result of the following construction; see also Olkin and Tong (1994) for interesting relations to reliability theory.

For all $\lambda_{i_1,\ldots,i_k} > 0$ let there be given independent random variables $Z_{i_1,\ldots,i_k}$ with exponential distributions with parameters $\lambda_{i_1,\ldots,i_k}$ for $(i_1,\ldots,i_k) \in \Im_k$, $k = 1,\ldots,n$. Define a random vector $\mathbf{X}$ with components X_i given by

$$X_i = \text{minimum of all } Z_{i_1\ldots i_k}\text{with } i \in \{i_1,\ldots,i_k\}$$

for $i = 1,\ldots,n$ (see p. 150 for the bivariate case). This construction makes it easy to derive the following distributional properties.

- The Marshall–Olkin multivariate exponential distribution is associated, see p. 158. In the bivariate case it is also conditionally increasing, see Nelsen (1999), p. 159. However, it is not MTP$_2$ for $\lambda_{12} > 0$. Indeed, let A be the square with sides parallel to the axes, side length $a < 1$ and center $(2,2)$, and let B be the analogous square with center $(3,1)$. Then $P(A)P(B) > P(A \cap B)P(A \cup B)$, i.e. property (ii) in Theorem 3.10.14 cannot hold.

- Comparison with respect to $\leq_{st}$ is possible:

$$\lambda'_{i_1\ldots i_k} \geq \lambda''_{i_1\ldots i_k} \quad \text{for all } (i_1,\ldots,i_k) \in \Im_k \text{ and all } k \text{ implies } \mathbf{X}' \leq_{st} \mathbf{X}''.$$

This is a consequence of Theorem 4.3.3.

- In the particular case of

$$\lambda_{i_1\ldots i_k} \equiv \mu_k \quad \text{for all } (i_1,\ldots,i_k) \in \Im_k \text{ and all } k = 1,\ldots,n$$

the weaker condition

$$\sum_{i=1}^{m} \mu'_i \geq \sum_{i=1}^{m} \mu''_i \quad \text{for } m = 1,\ldots,n \tag{3.14.1}$$

ensures

$$\mathbf{X}' \leq_{uo} \mathbf{X}'',$$

see Olkin and Tong (1994).

- In the case of a symmetric bivariate Marshall–Olkin exponential distribution conditions for the likelihood ratio order $\leq_r$ can be given. A straightforward but tedious calculation yields the following result. If $\lambda'_1 = \lambda'_2$, $\lambda''_1 = \lambda''_2$ and

$$\lambda'_{12}\lambda''_1(\lambda''_1 + \lambda''_{12}) = \lambda''_{12}\lambda'_1(\lambda'_1 + \lambda'_{12}),$$

then $\lambda'_1 \geq \lambda''_1$ and $\lambda'_2 \geq \lambda''_2$ implies $\mathbf{X}' \geq_{uo} \mathbf{X}''$.

An alternative distribution is *Gumbel's bivariate exponential distribution* (model I) (for which there is no multivariate generalization):

$$\bar{F}_{\mathbf{X}}(x_1, x_2) = \exp(-\lambda_1 x_1 - \lambda_2 x_2 - \lambda_{12} x_1 x_2) \quad \text{for } x_1, x_2 \geq 0$$

with non-negative λ_1, λ_2 and λ_{12} satisfying $\lambda_1 \cdot \lambda_2 > \lambda_{12}$. The corresponding correlation coefficient is negative. Thus it is not surprising that the distribution has the negative dependence property NUOD. A simple comparison property is that

$$\lambda_1' \leq \lambda_1'', \lambda_2' \leq \lambda_2'', \lambda_{12}' \leq \lambda_{12}'' \tag{3.14.2}$$

implies

$$\mathbf{X}' \geq_{uo} \mathbf{X}'',$$

but this criterion does not always ensure comparison with respect to $\leq_{lo}$ and $\leq_{st}$. In the case

$$\lambda_1' = \lambda_1'', \lambda_2' = \lambda_2'', \lambda_{12}' \leq \lambda_{12}''$$

it follows

$$\mathbf{X}' \geq_{sm} \mathbf{X}''$$

and if $\lambda_{12}' = \lambda_{12}'' = 0$ then clearly $\lambda_1' \leq \lambda_1''$ and $\lambda_2' \leq \lambda_2''$ imply

$$\mathbf{X}' \geq_{st} \mathbf{X}''.$$

4

Stochastic Models, Comparison and Monotonicity

4.1 General Considerations Concerning Stochastic Models

Stochastic models are the mathematical models studied in applied probability, statistical physics, mathematical economics and other areas. Two large classes of such models are finite and iterative models.

Finite models. Suppose we are given n random variables $Z_1, \dots, Z_n$, which are sufficient to describe the behaviour of the elements of a system studied. Depending on the origin and structure of the model, these Z_i may denote quantities of a quite different nature. For example, in the reliability theory context, the Z_i may be lifetimes of elements or components that run down simultaneously, or one after the other, or in some other fashion. The quantity of interest is the lifetime of the whole system.

The general problem is to consider the properties of a random variable X describing the behaviour of the system and having the form

$$X = \Phi(Z_1, \dots, Z_n) \qquad (4.1.1)$$

for some suitable mapping Φ. Here are four examples for such models.

1) In a *reliability theory model* without repair, Z_i denotes the lifetime of the ith component and X the lifetime of the system. (See Section 7.2 for concrete examples.)

2) The *Marshall–Olkin exponential distribution* (see Section 3.14, page 147) is given by a mapping. The following repeats the construction for the bivariate case.

Let Z_1, Z_2 and Z_{12} be three independent real-valued random variables with exponential distribution with parameters λ_1, λ_2 and λ_{12}. The random vector $\mathbf{X} = (X_1, X_2)$ defined as

$$X_1 = \min\{Z_1, Z_{12}\}$$

and

$$X_2 = \min\{Z_2, Z_{12}\}$$

has a Marshall–Olkin exponential distribution. It can be written in the form

$$\mathbf{X} = \Phi(Z_1, Z_2, Z_{12}),$$

where Φ maps from $\mathbb{R}^3$ into $\mathbb{R}^2$.

3) In a *planning network*, Z_i may denote the duration of the ith activity and X the duration of the full project. Again, X can be written as a function $X = \Phi(Z_1, Z_2, \dots, Z_n)$, see Section 7.3.

4) In a portfolio optimization problem, Z_i may denote the return of investment opportunity i and X the value of the portfolio. Clearly, X can be written as a function $X = \Phi(Z_1, Z_2, \dots, Z_n)$, see Section 8.2.2 for details.

Iterative models. Suppose an infinite sequence of random variables (Z_n) describes random influences on the state of a system considered at the (deterministic or random) instances t_n, with $t_n < t_{n+1}$ for $n = 0, 1, 2, \dots$. The system state X_{n+1} at time t_{n+1} is determined by the recurrence equation

$$X_{n+1} = \varphi_n(X_n, Z_n) \quad \text{for } n = 0, 1, \dots \tag{4.1.2}$$

where the φ_n are suitable mappings. Such processes have been called 'chains with complete connections'; see Iosifescu and Theodorescu (1969). The case of fixed $\varphi_n = \varphi$ has been systematically studied by Brandt, Franken, and Lisek (1990). These authors spoke about 'recursive stochastic equations' and studied in particular existence and continuity (or robustness) problems in the case of stationary sequences (Z_n). A survey paper on this topic is Borovkov and Foss (1992), see also Baccelli and Brémaud (1994) and Duflo (1997). In these works, such processes are called 'random iterative models', 'stochastic recursive sequences', 'stochastic recurrent sequences' or '*stochastic recurrences*', the term used in this book.

Clearly, if the Z_n are i.i.d. and the φ_n are the same for all n, then a stochastic recurrence is a homogeneous Markov chain. It is well known that conversely any homogeneous Markov chain (X_n) can be written in the form

$$X_{n+1} = \varphi(X_n, Z_n)$$

with i.i.d. Z_n; see Borovkov and Foss (1992) and Kifer (1986). In the case of a real-valued process, the Z_n can be taken as uniformly distributed on $[0,1]$ and φ can be expressed in terms of the transition kernel $Q(x, B)$ as

$$\varphi(x, z) = \inf\{y : Q(x, (-\infty, y]) \geq z\}, \tag{4.1.3}$$

which means in the case of a discrete Markov chain with states $i \in \{1, 2, \dots\}$ and transition matrix $((p_{ij}))$ that

$$\varphi(i, z) = j \quad \text{if and only if} \quad \sum_{k=1}^{j-1} p_{ik} \leq z < \sum_{k=1}^{j} p_{ik}. \tag{4.1.4}$$

Three simple examples are as follows.

1) In a queueing model, Z_n may be, for example, the vector of the nth inter-arrival time and the nth service time, and X_n the waiting time associated with the arrival of the nth customer. See Sections 6.1 and 6.6 for concrete definitions.

2) In an inventory model, Z_n may be the demand during the nth period and X_n the inventory at the beginning of the nth period.

3) In economics, physics and other fields, sequences appear, sometimes called *perpetuities*, of the form

$$X_{n+1} = Z_n X_n + Y_n. \tag{4.1.5}$$

In an economic setting, X_n may represent an accumulated stock at time n, Y_n a return added just before time $n + 1$, and the discount factor Z_n denotes the intrinsic decay or increase of the stock X_n between times n and $n + 1$, see Vervaat (1979), Brandt (1986) and, for financial applications, Embrechts, Klüppelberg, and Mikosch (1997), section 8.4.

In what follows the distributions F_i of the Z_i for $i = 0, 1, \ldots$ (and of analogous random variables appearing in stochastic models that do not lie within the framework of (4.1.1) and (4.1.2)) will be called *constituent distributions*. Whitt (1974) used the term 'initial data' for quantities having the nature of the Z_i. In the case of the queueing example above, if the Z_i are independent and identically distributed, then it suffices to consider as the constituent distribution just that of Z_1. If the Z_i are dependent then of course not only their marginal distributions are of interest; also the form of the multivariate distribution, for example characterized by the corresponding copula, plays a role. In such a case the constituent distribution is the joint distribution F_i of a vector or a sequence of dependent random variables. Thus the character F_i does not only stand for a real valued distribution *function*; it may also denote a distribution on some more general space.

The mappings Φ or φ_n are determined for the system under consideration by its structure, i.e. by the rules for transitions between the states of the system. Just as the term 'stochastic model' has not been described precisely, neither has the term 'structure': What is important in the sequel is the convention of referring to two systems as being identical in structure if the relevant mappings Φ or φ_n are the same, while differences in their constituent distributions are allowed, in particular, different forms of stochastic dependency of the Z_i. For example, two queueing systems with the same numbers of servers, the same queue discipline, the same waiting room capacity etc., have the same structure notwithstanding any differences in constituent distributions or any different stochastic dependencies in those distributions. The behaviour of a stochastic model of given structure is unambiguously determined once the constituent distributions (including their dependency properties) and, in the case of iterative models, the initial distribution, are known.

The principal aim of studying stochastic models mathematically is to obtain statements about quantities which, for a given model Σ consisting of specified structure and constituent distribution(s), describe its behaviour as a whole. These may be called *system quantities*, a term we use to cover real-valued quantities like means and probabilities (for example, mean lifetime or loss probability) as well as distribution functions or distributions (for example, of waiting time or output process of a queueing system).

Let c_Σ be a system quantity of Σ and let C_Σ denote the set of possible values of c_Σ. For given structure and initial distribution, c_Σ depends only on the distributions F_i, symbolically,

$$c_\Sigma = c_\Sigma(F_1, F_2, \dots) \in C_\Sigma. \qquad (4.1.6)$$

In simpler models we may be able to deduce an explicit expression for c_Σ, but in many situations such mathematical work may not be feasible or may lead to complicated formulae that are not useful from a practical viewpoint.

Such circumstances lead to the aim to seek for qualitative properties of c_Σ with respect to the F_i which characterize, roughly speaking, the way in which c_Σ is affected by variations or changes in the F_i. Important qualitative properties of stochastic models include *insensitivity* (meaning that c_Σ depends only on mean value characteristics of some of the F_i such as expectation or intensity) *comparability* or *monotonicity* (if the F_i 'increase' in some sense, so does c_Σ), and *robustness* (small perturbations in the F_i lead to small perturbations in the c_Σ).

A famous example for insensitivity is Erlang's formula in queueing theory. It says that in an $M/G/s$ loss system the stationary state probabilities follow a truncated Poisson distribution the parameter of which depends only on the intensity of the arrival process and the mean service time.

For many reliability systems, system lifetimes are 'comparable' in the sense that they increase monotonically if the element lifetimes increase (see Section 7.2). In many queueing systems, system characteristics such as waiting times increase if the service times increase; they are 'comparable' in this sense. Frequently in such systems with fixed constituent distributions, characteristics tend to increase if at the beginning the system is empty. In the present book the term 'monotonicity' is restricted to this and similar situations, in contrast to 'comparability'. Of great interest also are properties such as 'association' or 'positive correlation', yielding that covariances of pairs of vector components, members of sequences or random fields are positive.

For example, 'small changes' of durations of activities in a planning network lead to 'small changes' of project duration, in the sense of a suitable metric for distributions weak convergence. Such robustness properties are systematically studied in Stoyan (1983) and Rachev (1991), while insensitivity properties are treated in Franken, König, Arndt, and Schmidt (1982) and Schassberger (1977, 1978a,b). The present book concentrates on comparability and monotonicity properties.

The existence of relations such as (4.1.1) and (4.1.2) with tractable Φ and φ_n is an ideal starting point for the proof of comparability and monotonicity properties of stochastic models. Particular properties such as monotonicity, convexity, supermodularity and so on, of the mappings Φ and φ_n determine which qualitative properties can be proved and which stochastic order relations can be successfully used. The stochastic order relation $\leq_{st}$ is not always the only or best choice; there are more interesting relations which may be particularly adapted to given stochastic models. In reliability theory, for example, the hazard rate order $\leq_{hr}$, which is stronger than $\leq_{st}$, plays an important role. In some other situations it is possible to apply relations weaker than $\leq_{st}$ which enable the comparison of distributions with the same mean value characteristics (such as expectation or intensity); on the other hand, the study of the effects of different dependence structures (for example comparability with respect to supermodular order $\leq_{sm}$) on system characteristics is an important topic, too. Therefore this book contains much material on stochastic orders of distributions.

The authors emphasize the possibility of constructing new order relations for proving new comparability and monotonicity properties. In the past the aim of better understanding comparability and extremality properties of stochastic models has led to such relations or to new applications of already known relations. An example is the increasing convex order $\leq_{icx}$, which was 'constructed' (or rediscovered) in Stoyan and Stoyan (1969) in order to formulate comparability properties of the queueing system $GI/G/1$ (see Section 6.2). Similarly, the aim of proving qualitative properties of density operators led Uhlmann (1972) to partial orders of such operators and finally to a comparison theorem for Gibbs distributions (see Theorem 7.5.1), in which the randomness order $\leq_{rand}$ plays an important and natural role.

It is by means of monotonicity and comparability properties that bounding formulae can be obtained for characteristics of otherwise intractable systems and that approximations are given a rigorous basis. For example, for many stochastic models it is easy to determine system quantities if the constituent distributions are exponential distributions. If such a model has a comparison property saying that the system quantities are monotone in the constituent distributions and in a given application they are 'smaller' than exponential distributions, then the values obtained for those distributions are upper bounds. (Of course, such bounds can be obtained also by other methods, see, for example, Coyle and Taylor (1995).) Further, often in practice constituent distributions may not be known exactly but only through statistical data or through predictions of experts; in particular, their full dependence structure may be unknown. This may lead to the statement that some constituent distributions F_k belong to a set $\mathbb{F}_k$ of distributions, for example, the set of all distributions with mean m_k and support $[a_k, b_k]$. Then there is good reason to study systematically qualitative properties of system quantities in their own right: Comparability properties may lead to bounds on system quantities suitable to the given 'incomplete information', based on extremal elements of $\mathbb{F}_k$.

Monotonicity and comparability properties play also a role in the study of asymptotic properties of stochastic processes. If, for example, for a given non-negative process there is a majorizing process which is stable, then the same is true for the process under analysis.

4.2   Monotonicity and Comparability

4.2.1   Monotonicity

Let Σ be a stochastic model with constituent distributions $F_1, F_2, \ldots$, and let c_Σ be a system quantity which is time dependent, being defined either for all real t or for all positive t (for example, with Σ a queueing system, c_Σ could be the mean virtual waiting time at t, or the distribution function of the number of customers in Σ at t) or for a sequence of (deterministic or random) instances t_n ($n = 1, 2, \ldots$) (for example, with the same Σ, the waiting time distribution function or the loss probability of the nth customer). Write then c_Σ more precisely as $c_\Sigma(t)$ or $c_\Sigma(n)$. Let $\preceq$ denote a partial order on the set C_Σ of possible values of c_Σ.

Definition 4.2.1. The system quantity $c_\Sigma(\cdot)$ is *increasing (decreasing)* if always

$$t \leq t' \quad \text{implies} \quad c_\Sigma(t) \preceq (\succeq) c_\Sigma(t'). \tag{4.2.1}$$

This property is called *monotonicity*.

An essential feature is that monotonicity is defined for fixed constituents F_i, with the possible exception of an initial distribution. If $c_\Sigma(t)$ converges to some limit c_Σ (dependent or independent of the initial distribution) for $t \to \infty$, then it follows in the case of (4.2.1) that for all t

$$c_\Sigma(t) \preceq (\succeq) c_\Sigma \tag{4.2.2}$$

provided $\preceq$ is closed under the mode of convergence used. Thus, bounds on c_Σ may follow from the choice of suitable initial distributions for monotone system quantities.

4.2.2   Comparability

With Σ, F_i and c_Σ as above, write $\mathbb{F}_k$ for sets of constituent distributions F_k partially ordered by $\preceq_k$, and let $\preceq_c$ be a partial order on the set C_Σ of all possible c_Σ. Typically the relation $\preceq_c$ stands for $\leq$ for real numbers; however, it may also be some stochastic order (like $\leq_{icx}$) for distribution functions.

Definition 4.2.2. The system quantity c_Σ is *increasing* on $\mathbb{F}_k$ with respect to $\preceq_k$ if for any F'_k and F''_k in $\mathbb{F}_k$ and all other constituent distributions fixed, $F'_k \preceq_k F''_k$ implies

$$c_\Sigma(F_1, \ldots, F_{k-1}, F'_k, F_{k+1}, \ldots) \preceq_c c_\Sigma(F_1, \ldots, F_{k-1}, F''_k, F_{k+1}, \ldots), \quad (4.2.3)$$

In this case we will speak about *comparability*.

Comparability is of great value for constructing bounds on system quantities. If, for example, a given constituent distribution F_k can be bounded by distributions F^l_k and F^u_k in $\mathbb{F}_k$ for which

$$F^l_k \preceq_k F_k \preceq_k F^u_k, \quad (4.2.4)$$

then the corresponding system quantities c_Σ satisfy

$$c_\Sigma(F^l_k) \preceq_c c_\Sigma(F_k) \preceq_c c_\Sigma(F^u_k) \quad (4.2.5)$$

if (4.2.3) holds.

Comparability is also of interest when stochastic models are studied with respect to non-distributional system parameters such as the number of servers in queueing systems or the chemical potential in Gibbs distributions. Many examples of comparability are given in the following chapters. The next section and Chapter 5 present general methods for proving comparability properties.

4.3   Methods for Establishing Monotonicity and Comparability Properties

4.3.1   The Functional Method

Monotonicity and comparability properties of stochastic models can often be proved on the basis of formulae for system characteristics c_Σ resulting from preceding calculations. Such a formula may contain a functional v_f defined over a set of distribution functions F or distributions P as

$$v_f(F) = \int\limits_{-\infty}^{\infty} f(t)F(\mathrm{d}t) \quad \text{or} \quad v_f(P) = \int f(x)P(\mathrm{d}x)$$

for a suitable function f. If c_Σ has the form

$$c_\Sigma(F_1, F_2, \ldots) = g(F_1, \ldots, F_{k-1}, v_f(F_k), F_{k+1}, \ldots) \quad (4.3.1)$$

where the function g is real-valued and increasing in its kth argument for all other F_j $(j \neq k)$ fixed, then the following is true.

Theorem 4.3.1. *Let $\leq_{\mathcal{F}}$ be an integral stochastic order on the set of all possible constituent distributions F_k and let f be an element of the maximal generator $\mathcal{R}_{\mathcal{F}}$ of $\leq_{\mathcal{F}}$. Then $F^l_k \leq_{\mathcal{F}} F^u_k$ implies*

$$c_\Sigma(F_1, \ldots, F_{k-1}, F^l_k, \ldots) \leq c_\Sigma(F_1, \ldots, F_{k-1}, F^u_k, \ldots). \quad (4.3.2)$$

To apply the functional method it is clear that a formula for g is needed. This restricts its application to those models for which formulae for c_Σ are already known, where, in principle, an exact computation of the system quantity concerned is possible. However, those formulae may be rather complicated, and both then and in cases of 'incomplete information' (see Section 4.4) the method may be of considerable value. An application of the method is given in Chapter 6 in the proof of Theorem 6.3.5, where

$$v_f(F_A) = \int_0^\infty \exp(-\mu(1-z)t)F_A(\mathrm{d}t).$$

Also the proof of Theorem 4.3.6 below uses the functional method, with

$$v_f(F) = \sum_{n=0}^\infty P(N=n)\pi_n(t),$$

where f corresponds to the function $n \mapsto \pi_n(t)$ and F to the distribution of N.

4.3.2   The Mapping Method

The idea of the mapping method consists in using relations of the form (4.1.1) or (4.1.2) and to exploit qualitative properties of the mappings Φ and φ_n. The aim is to transform comparability of the Z_i to comparability of X. Because (4.1.2) can be reduced to (4.1.1) with an appropriate function Φ, the discussion below, which is oriented more towards (4.1.1), covers both cases.

In the case of iterative models, comparison properties for characteristics of stationary distributions can be obtained in the following way: First non-stationary system quantities (for example, in a queueing system, the distribution function of the waiting time of the nth customer) are considered, and then, if the corresponding stationary system quantities exist and provided the non-stationary quantities converge to their stationary counterparts in a manner that is suitably 'consistent' with the stochastic order used, comparability properties for the stationary quantities follow.

We start by considering generally the comparison of random variables X of the form

$$X = \Phi(Z)$$

as regards their dependence on the random variable Z with values in S and on the measurable mapping $\Phi : S \to T$. Here S and T are abstract spaces, for example, Polish spaces with closed partial order and (if necessary) a linear structure, and $\mathcal{S}$ and $\mathcal{T}$ are (Borel) σ-algebras of subsets of S and T, such that it makes sense to define integral stochastic orders for distributions on

S and $\mathcal{T}$. Let $\mathcal{F}_S$ and $\mathcal{F}_T$ be generators of such orders and $\mathcal{R}_S$ and $\mathcal{R}_T$ the corresponding maximal generators.

Theorem 4.3.2. *If for $\Phi', \Phi'' : S \to T$ there exists a measurable $\Phi : S \to T$ with $f(\Phi(\cdot)) \in \mathcal{R}_S$ for all $f \in \mathcal{R}_T$ and*

$$f(\Phi'(z)) \leq f(\Phi(z)) \leq f(\Phi''(z)) \quad \text{for all } f \in \mathcal{F}_T \text{ and all } z \in S \quad (4.3.3)$$

then $Z' \leq_{\mathcal{F}_S} Z''$ implies

$$\Phi'(Z') \leq_{\mathcal{F}_T} \Phi''(Z''). \quad (4.3.4)$$

Proof. Since $f(\Phi(\cdot)) \in \mathcal{R}_S$ for all $f \in \mathcal{R}_T$, it is $\Phi(Z') \leq_{\mathcal{F}_T} \Phi(Z'')$, and (4.3.3) yields $\Phi'(Z') \leq_{\mathcal{F}_T} \Phi(Z')$ and $\Phi(Z'') \leq_{\mathcal{F}_T} \Phi''(Z'')$. Thus the transitivity of $\leq_{\mathcal{F}_T}$ yields the result.   □

An example for the application of Theorem 4.3.2 is the proof of comparison properties for point processes (see Section 5.4). There $S = \mathbb{R}_+^{\mathbb{N}}$, the space of all infinite sequences of positive numbers, $T = D[0, \infty)$, and $\Phi' = \Phi'' = \Phi$ is the mapping transforming the sequence (T_n) of event epochs into the counting process (N_t). Sequences comparable with respect to $\leq_{st}$ are transformed into $\leq_{st}$-completely comparable counting processes.

Now mappings Φ are considered such as occurring for example in (4.1.1); these mappings may be more general than functionals on S. For some integer n and for each $i = 1, \ldots, n$, let the Z_i be $(S_i, \mathcal{S}_i)$-valued random variables on the probability space $(\Omega, \mathcal{A}, P)$ and let Φ be a mapping from $S_1 \times \cdots \times S_n$ into some other space $(S, \mathcal{S})$ with appropriate measurability properties such that

$$X \equiv \Phi(Z_1, \ldots, Z_n) \quad (4.3.5)$$

is an $(S, \mathcal{S})$-valued random variable on $(\Omega, \mathcal{A}, P)$. Assume that S and the S_i are vector spaces with partial orders $\leq_S$ and $\leq_{S_i}$ respectively. Then monotonicity and convexity properties of Φ facilitate the deduction of analogous properties of X with respect to the Z_i, according as $\leq_{st}, \leq_{icx}$ or $\leq_{icv}$ is used. For other integral stochastic orders other properties ensure similar results. It should be noted that the statements below remain true if X is a function of a possibly infinite sequence of random variables.

Monotonicity and convexity are here defined for mappings with values in spaces different from $\mathbb{R}$ with respect to the partial orders $\leq_S$ and $\leq_{S_i}$. These concepts have already been defined for general Polish spaces in Chapter 2. On product spaces, however, it is useful to introduce componentwise monotonicity and convexity. The mapping Φ is called *increasing in z_i* if for all fixed z_j for $j \neq i$ and all z_i' and $z_i'' \in S_i$, $z_i' \leq_{S_i} z_i''$ implies $\Phi(z_1, \ldots, z_i', \ldots, z_n) \leq_S \Phi(z_1, \ldots, z_i'', \ldots, x_n)$. Notice that Φ is increasing if and only if it is increasing in z_i for all $i = 1, \ldots, n$.

Φ is *convex in z_i* if for all fixed z_j for $j \neq i$, and all z_i' and $z_i'' \in S_i$ and $0 < \lambda < 1$,

$$\Phi(z_1, \dots, \lambda z_i' + (1-\lambda)z_i'', \dots, z_n) \leq_S \lambda \Phi(z_1, \dots, z_n)$$
$$+ (1-\lambda)\Phi(z_1, \dots, z_i'', \dots, z_n).$$

Since the spaces S_i are vector spaces, so also are $S_{i_1} \times \cdots \times S_{i_m}$ and *convexity* and *concavity* can be defined *in terms of the vector argument* $z_{i_1}, \dots, z_{i_m}$ of Φ. Notice, however, that a componentwise convex function is not necessarily convex in terms of the vector argument (see Section 3.6).

According to the properties of Φ the following theorem holds.

Theorem 4.3.3. *Let the random variables* $(Z_{i_1}', \dots, Z_{i_m}')$ *and* $(Z_{i_1}'', \dots, Z_{i_m}'')$ *be independent of the Z_j for $j \neq i_1, \dots, i_m$. Then if Φ is increasing in* $(z_{i_1}, \dots, z_{i_m})$

$$(Z_{i_1}', \dots, Z_{i_m}') \leq_{st} (Z_{i_1}'', \dots, Z_{i_m}'') \quad \text{implies} \quad X' \leq_{st} X'', \tag{4.3.6}$$

while for Φ additionally convex in $(z_{i_1}, \dots, z_{i_m})$,

$$(Z_{i_1}', \dots, Z_{i_m}') \leq_{icx} (Z_{i_1}'', \dots, Z_{i_m}'') \quad \text{implies} \quad X' \leq_{icx} X''. \tag{4.3.7}$$

Proof. If Φ is increasing (and convex) in $(x_{i_1}, \dots, x_{i_m})$ then $f(\Phi(\cdot))$ has the same property if $f : S \to \mathbb{R}$ is increasing (and convex). $\qquad\square$

Example 4.3.4. (Comparability and association of the multivariate Marshall–Olkin exponential distribution.)

A simple application is the following result. Consider two random vectors with bivariate Marshall–Olkin exponential distributions with parameters λ_1', λ_1'', λ_2', λ_2'', λ_{12}', and λ_{12}''. If $\lambda_1' \leq \lambda_1''$, $\lambda_2' \leq \lambda_2''$ and $\lambda_{12}' \leq \lambda_{12}''$ then $\mathbf{X}' \leq_{st} \mathbf{X}''$, since the mapping Φ in $\mathbf{X} = \Phi(Z_1, Z_2, Z_{12})$ on page 150 is monotone and $Z_1' \leq_{st} Z_1''$ etc. An analogous relation is true for the multivariate Marshall–Olkin exponential distribution. Also by using the mapping Φ one obtains that the random vector $\mathbf{X}$ is associated; this is an application of Theorem 3.10.5 (iv).

As a further example, *sums* and *compounds* of random variables are considered. While for this case comparability properties can be proved for several stochastic orders, first the case of the usual stochastic order $\leq_{st}$ is considered, generalizing Theorem 1.2.17.

Theorem 4.3.5. *Let (X_n') and (X_n'') be sequences of non-negative random variables and let N' and N'' be non-negative integer-valued random variables (X_n') and (X_n''). Then*

$$(N', X_1', X_2', \dots) \leq_{st} (N'', X_1'', X_2'', \dots)$$

implies

$$\sum_{n=1}^{N'} X'_n \leq_{st} \sum_{n=1}^{N''} X''_n \tag{4.3.8}$$

where the empty sum equals zero.

Proof. Write

$$X = \Phi(N, X_1, X_2, \dots) = \sum_{i=1}^{N} X_i.$$

Since the terms X_i are non-negative, the mapping is increasing in all variables. Consequently, Theorem 4.3.3 yields the result.   □

Now as an example for an application of some other stochastic order relation the increasing convex order $\leq_{icx}$ is considered. Here a combination of mapping and functional method yields the following result; see also Szekli (1995) for more information on compounding.

Theorem 4.3.6. *Let (X'_n) and (X''_n) be sequences of non-negative stochastically independent random variables for which*

$$X'_n \leq_{icx} X'_{n+1} \quad and \quad X''_n \leq_{icx} X''_{n+1} \quad for \ n = 1, 2, \dots, \tag{4.3.9}$$

and let N' and N'' be non-negative integer-valued random variables independent of (X'_n) and (X''_n). Then

$$N' \leq_{icx} N'' \quad and \quad X'_n \leq_{icx} X''_n \quad for \ n = 1, 2, \dots$$

imply

$$\sum_{n=1}^{N'} X'_n \leq_{icx} \sum_{n=1}^{N''} X''_n \tag{4.3.10}$$

where the empty sum equals zero.

Proof. Since $\leq_{icx}$ has the convolution property (C), it follows from the assumptions that $S'_n = \sum_{i=1}^{n} X'_i \leq_{icx} \sum_{i=1}^{n} X''_i = S''_n$ for any fixed integer n, i.e. for any real t

$$\pi'_n(t) \equiv E(S'_n - t)_+ \leq E(S''_n - t)_+ \equiv \pi''_n(t).$$

Clearly, (4.3.10) is equivalent to having

$$\sum_{n=0}^{\infty} P(N' = n)\pi'_n(t) \leq \sum_{n=0}^{\infty} P(N'' = n)\pi''_n(t) \quad \text{for all } t. \tag{4.3.11}$$

For fixed t, the sequence $(\pi'_n(t))$ is increasing in n. Furthermore, it is also convex, which can be shown as follows, using first the non-negativity of X'_n and then that $X'_n \leq_{icx} X'_{n+1}$:

$$
\begin{aligned}
\pi'_{n+1}(t) - \pi'_n(t) &= E(X'_{n+1} + S'_n - t)_+ - E(S'_n - t)_+ \\
&\geq E(X'_{n+1} + S'_{n-1} - t)_+ - E(S'_{n-1} - t)_+ \\
&\geq E(X'_n + S'_{n-1} - t)_+ - E(S'_{n-1} - t)_+ \\
&= \pi'_n(t) - \pi'_{n-1}(t).
\end{aligned}
$$

Consequently, $N' \leq_{icx} N''$ yields

$$
\sum_{n=0}^{\infty} P(N' = n)\pi'_n(t) \leq \sum_{n=0}^{\infty} P(N'' = n)\pi'_n(t) \leq \sum_{n=0}^{\infty} P(N'' = n)\pi''_n(t).
$$

□

For arbitrary stochastic orders, it is clear that for independent X'_i and X''_i

$$
X'_i \preceq X''_i \quad \text{for } i = 1, \ldots, n \quad \text{implies} \quad \sum_{i=1}^{n} X'_i \preceq \sum_{i=1}^{n} X''_i
$$

if $\preceq$ has property (C) and for independent N

$$
X'_i \preceq X''_i \quad \text{for } i = 1, \ldots, n \quad \text{implies} \quad \sum_{i=0}^{N} X'_i \preceq \sum_{i=0}^{N} X''_i
$$

if $\preceq$ has additionally property (MI).

For dependent X'_i and X''_i let be $\mathbf{X}' = (X'_1, \ldots, X'_n)$ and $\mathbf{X}'' = (X''_1, \ldots, X''_n)$. Then

$$
\mathbf{X}' \leq_{lcx} \mathbf{X}'' \quad \text{implies} \quad \sum_{i=1}^{n} X'_i \leq_{cx} \sum_{i=1}^{n} X''_i,
$$

$$
\mathbf{X}' \leq_{plcx} \mathbf{X}'' \quad \text{implies} \quad \sum_{i=1}^{n} X'_i \leq_{cx} \sum_{i=1}^{n} X''_i,
$$

$$
\mathbf{X}' \leq_{iplcx} \mathbf{X}'' \quad \text{implies} \quad \sum_{i=1}^{n} X'_i \leq_{icx} \sum_{i=1}^{n} X''_i.
$$

Analogous relations are true for compounds with independent N. Lefèvre and Utev (1996) gave criteria for the comparison of sums of exchangeable Bernoulli random variables with respect to $\leq_{s-cx}$.

As a cautionary example consider the scheduling problem considered in Theorem 1.9.5. It is not always true that for independent X and Y the inequality $X \leq_{st} Y$ implies $2X + Y \leq_{st} X + 2Y$. The mapping method is not able to prove such an inequality; it only yields that

$$X_1' \leq_{st} X_1'' \quad \text{and} \quad X_2' \leq_{st} X_2''$$

implies

$$2X_1' + X_2' \leq_{st} 2X_1'' + X_2'',$$

an inequality which involves four random variables and not only two.

Another application of the mapping method is related to recursions in the so called (max, +)-algebra. These recursions appear very often in discrete event systems. A comprehensive treatment of ordering properties of such systems is given in Glasserman and Yao (1994). Typical examples are queueing systems where event epochs can be written as functions of random variables built by addition and maximization operations (see Section 6.3).

For other mappings $\Phi : \mathbb{R}^n \to \mathbb{R}$ also the weaker stochastic orders of Chapter 3 are of interest. Examples are

$$X = \Phi(Z_1, \ldots, Z_n) = \min\{a_1 Z_1, \ldots, a_n Z_n\}$$

where

$$(Z_1', \ldots Z_n') \leq_{uo} (Z_1'', \ldots Z_n'') \quad \text{implies } X' \leq_{st} X''$$

and

$$X = \Phi(Z_1, \ldots, Z_n) = \max\{a_1 Z_1, \ldots, a_n Z_n\}$$

where

$$(Z_1', \ldots Z_n') \leq_{lo} (Z_1'', \ldots Z_n'') \quad \text{implies } X' \leq_{st} X''$$

for positive a_i by Theorem 3.3.17.

In the case of iterative models or stochastic recurrences it is of course desirable to express the monotonicity and/or convexity properties of $X_{n+1} = \Phi_n(X_0, Z_0, \ldots, Z_n) = \varphi_n(\varphi_{n-1}(\ldots), Z_n)$ in terms of the properties of the φ_n. For simplicity, only the case of real-valued random variables is considered.

Write $\mathbf{z}_n = (z_0, \ldots, z_n)$ and

$$\Phi_0(x_0, z_0) = \varphi_0(x_0, z_0),$$
$$\Phi_n(x_0, \mathbf{z}_n) = \varphi_n(\Phi_{n-1}(x_0, \mathbf{z}_{n-1}), z_n) \quad \text{for } n = 1, 2, 3, \ldots.$$

Theorem 4.3.7. *(a) If all φ_n are increasing in both variables, then all Φ_n are increasing in $(x_0, \mathbf{z}_n)$.*
(b) If all φ_n are increasing in the first variable x and convex in the vector argument (x, z), then all Φ_n are convex in $(x_0, \mathbf{z}_n)$.

(c) If all φ_n are increasing in the first variable x and componentwise convex in x and z, then all Φ_n are componentwise convex in x_0 and all z_i for $i = 0, \dots, n$.

(d) If all φ_n are convex in the first variable x and increasing and supermodular in the vector argument (x, z), then all Φ_n are increasing and supermodular in $(x_0, \mathbf{z}_n)$.

Proof. (a) Φ_0 is increasing in z_0 by the monotonicity of φ_0 in z. Since

$$\Phi_{k+1}(x, z_0, \dots, z_{k+1}) = \varphi_{k+1}(\Phi_k(x, z_0, \dots, z_k), z_{k+1}),$$

an easy induction step establishes the monotonicity of Φ_k for all $k = 0, 1, 2, \dots$.

(b) Again, Φ_0 is convex in (x_0, z_0) by the convexity of φ_0 in (x, z). For $0 < \lambda < 1$, assuming Φ_k is convex in $(x_0, \mathbf{z}_k)$, it is

$$
\begin{aligned}
&\Phi_{k+1}(\lambda x_0' + (1-\lambda)x_0'', \lambda \mathbf{z}_{k+1}' + (1-\lambda)\mathbf{z}_{k+1}'') \\
&= \varphi_{k+1}(\Phi_k(\lambda x_0' + (1-\lambda)x_0'', \lambda \mathbf{z}_k' + (1-\lambda)\mathbf{z}_k''), \lambda z_{k+1}' + (1-\lambda)z_{k+1}'') \\
&\leq \varphi_{k+1}(\lambda \Phi_k(x_0, \mathbf{z}_k') + (1-\lambda)\Phi_k(x_0, \mathbf{z}_k''), \lambda z_{k+1}' + (1-\lambda)z_{k+1}'') \\
&\leq \lambda \varphi_{k+1}(\Phi_k(x_0, \mathbf{z}_k'), z_{k+1}') + (1-\lambda)\varphi_{k+1}(\Phi_k(x_0'', \mathbf{z}_k''), z_{k+1}'') \\
&= \lambda \varphi_{k+1}(x_0', \mathbf{z}_{k+1}') + (1-\lambda)\Phi_{k+1}(x_0'', \mathbf{z}_{k+1}'')
\end{aligned}
$$

by the properties assumed for Φ_k and φ_{k+1}. Part (c) is proved similarly as (b), and (d) is a consequence of Theorem 3.9.3 f). $\square$

Two important applications of Theorem 4.3.7 are given in the following comparison theorems.

Theorem 4.3.8. *Suppose that (X_n') and (X_n'') are stochastic processes of the form (4.1.2) which are generated by the same φ_n and sequences (Z_n') and (Z_n''). Then*

$$(X_0', Z_0', \dots, Z_n') \leq_{\mathcal{F}} (X_0'', Z_0'', \dots, Z_n'') \tag{4.3.12}$$

implies

$$(X_0', \dots, X_n') \leq_{\mathcal{F}} (X_0'', \dots, X_n'') \quad \text{for all } n \tag{4.3.13}$$

in the following cases:

(a) if $\leq_{\mathcal{F}}$ is $\leq_{st}$ and all φ_n are increasing in both variables;

(b) if $\leq_{\mathcal{F}}$ is $\leq_{icx}$ and all φ_n are increasing in the first variable x and convex in the vector argument (x, z);

(c) if $\leq_{\mathcal{F}}$ is $\leq_{idcx}$ and all φ_n are convex in the first variable x and increasing and supermodular in the vector argument (x, z).

Theorem 4.3.9. *(a) Let (non-homogeneous) Markov chains (X_n') and (X_n'') of the form (4.1.2) be given which are defined by the same φ_n and sequences (Z_n') and (Z_n'') of independent random variables. Assume that $X_0' \leq_{\mathscr{F}} X_0''$ and $Z_n' \leq_{\mathscr{F}} Z_n''$ (for all n) for some integral stochastic order $\leq_{\mathscr{F}}$. If*

$$f(\varphi_n(\cdot, z)) \in \mathcal{R}_{\mathscr{F}} \quad and \quad f(\varphi_n(x, \cdot)) \in \mathcal{R}_{\mathscr{F}}$$

for all $f \in \mathscr{F}$ and all n, z and x, then

$$X_n' \leq_{\mathscr{F}} X_n'' \quad for\ all\ n. \tag{4.3.14}$$

(b) If, in addition, $\leq_{\mathscr{F}}$ is a multivariate stochastic order defined for all n and having properties (ID) and (IN), then

$$(X_0', X_1', \ldots, X_n') \leq_{\mathscr{F}} (X_0'', X_1'', \ldots, X_n'') \tag{4.3.15}$$

for all n.

Proof. (a) The argument goes by induction. Assume that $X_n' \leq_{\mathscr{F}} X_n''$. Under the stated assumptions this implies $\varphi_n(X_n', z) \leq_{\mathscr{F}} \varphi_n(X_n'', z)$ for all z and $\varphi_n(x, Z_n') \leq_{\mathscr{F}} \varphi_n(x, Z_n'')$ for all x. According to Theorem 2.4.2 any integral stochastic order has property (MI). Hence

$$X_{n+1}' = \varphi_n(X_n', Z_n') \leq_{\mathscr{F}} \varphi_n(X_n'', Z_n') \leq_{\mathscr{F}} \varphi_n(X_n'', Z_n'') = X_{n+1}''.$$

(b) For $n = 0$, (4.3.15) is trivial. Let (4.3.15) be true for some $n \geq 0$. For any fixed vector $(x_0, \ldots, x_n)$ it is

$$(x_0, \ldots, x_n, \varphi(x_n, Z_n')) \leq_{\mathscr{F}} (x_0, \ldots, x_n, \varphi(x_n, Z_n'')).$$

This follows from property (IN) using that $\varphi(x_n, Z_n') \leq_{\mathscr{F}} \varphi(x_n, Z_n'')$ according to the assumptions and that trivially $(x_0, \ldots, x_n) \leq_{\mathscr{F}} (x_0, \ldots, x_n)$. Property (MI) of $\leq_{\mathscr{F}}$ ensures

$$(X_0', \ldots, X_n', \varphi(X_n', Z_n')) \leq_{\mathscr{F}} (X_0', \ldots, X_n', \varphi(X_n', Z_n'')). \tag{4.3.16}$$

The induction hypothesis and property (ID) of $\leq_{\mathscr{F}}$ yield

$$(X_0', \ldots, X_n', X_n') \leq_{\mathscr{F}} (X_0'', \ldots, X_n'', X_n'').$$

As above this leads for any x to

$$(X_0', \ldots, X_n', \varphi(X_n', x)) \leq_{\mathscr{F}} (X_0'', \ldots, X_n'', \varphi(X_n'', x)),$$

and (MI) yields

$$(X_0', \ldots, X_n', \varphi(X_n', Z_n'')) \leq_{\mathscr{F}} (X_0'', \ldots, X_n'', \varphi(X_n'', Z_n'')).$$

Together with (4.3.16) this leads to (4.3.15) for $n + 1$. $\qquad \square$

If $\leq_{\mathcal{F}}$ is the usual stochastic order $\leq_{st}$ then the assumptions of Theorem 4.3.9 are fulfilled for φ_n increasing, and for $\leq_{icx}$ they hold whenever φ_n is increasing and componentwise convex. An example for $\leq_{sm}$ and the orthant orders is given below.

A similar argument yields the following monotonicity statement.

Theorem 4.3.10. *Let (X_n) be a (homogeneous) Markov chain of the form (4.1.2) with independent identically distributed Z_n and $\varphi_n = \varphi$. If*

$$f(\varphi(\cdot, z)) \in \mathcal{R}_{\mathcal{F}} \quad \text{for all } f \in \mathcal{F} \quad \text{and all } z,$$

then $X_0 \leq_{\mathcal{F}} X_1$ implies

$$X_n \leq_{\mathcal{F}} X_{n+1} \quad \text{for all } n. \tag{4.3.17}$$

Proof. The argument again goes by induction. Assume that $X_{n-1} \leq_{\mathcal{F}} X_n$. Under the stated assumptions this implies $\varphi(X_{n-1}, z) \leq_{\mathcal{F}} \varphi(X_n, z)$ for all z and thus it follows from property (MI) that

$$X_n = \varphi(X_{n-1}, Z_{n-1}) =_{st} \varphi(X_{n-1}, Z_n) \leq_{\mathcal{F}} \varphi(X_n, Z_n) = X_{n+1}.$$

$\square$

Example 4.3.11. (Perpetuities.) An example with a two-dimensional second variable in φ is the Markov chain given by (4.1.5),

$$X_{n+1} = Z_n X_n + Y_n,$$

with independent random vectors (Z_n, Y_n). (We allow dependence between the two components Z_n and Y_n for fixed n.) Here $\varphi(x, (y, z)) = zx + y$. Assuming that both Z_n and Y_n only assume non-negative values, the function φ is increasing, supermodular and convex. If $X_0 = 0$, then the sequence (X_n) is increasing with respect to $\leq_{st}$ (for any distributions of Z_n and Y_n), while

$$(Z'_n, Y'_n) \leq_{\mathcal{F}} (Z''_n, Y''_n) \quad \text{implies } X'_n \leq_{\mathcal{F}} X''_n \quad \text{for all } n$$

with $\leq_{\mathcal{F}}$ equal to $\leq_{st}$ or $\leq_{icx}$. Moreover, it follows from Theorem 4.3.8 that

$$(Z'_n, Y'_n) \leq_{\mathcal{F}} (Z''_n, Y''_n) \quad \text{imply } (X'_0, \dots, X'_n) \leq_{\mathcal{F}} (X''_0, \dots, X''_n) \quad \text{for all } n,$$

when $\leq_{\mathcal{F}}$ is $\leq_{st}$, $\leq_{icx}$ or $\leq_{idcx}$. This implies for example that $(Z'_n, Y'_n) \leq_{sm} (Z''_n, Y''_n)$ for all n yields $X'_n \leq_{icx} X''_n$. Thus, Z_n and Y_n of a higher variability lead to X_n which are also of a higher variability, and also a stronger dependence (in the $\leq_{sm}$-sense) between Z_n and Y_n leads to a higher variability of the X_n.

Example 4.3.12. (Separable Markov chains.) Li and Xu (2000) consider a 'separable' Markov chain with state space $S = \mathbb{R}^2$

$$\mathbf{X}_{n+1} = \varphi(\mathbf{X}_n, \mathbf{Z}_n) \quad \text{for } n = 0, 1, \ldots$$

with

$$\mathbf{X}_n = (X_{1,n}, X_{2,n}), \quad \mathbf{Z}_n = (Z_{1,n}, Z_{2,n})$$

and

$$X_{i,n+1} = \varphi_{i,n}(X_{i,n}, Z_{i,n}) \quad \text{for } i = 1, 2 \text{ and } n = 0, 1, \ldots,$$

where the $\varphi_{i,n}$ are increasing in both variables. The $\mathbf{Z}_n$ are independent two-dimensional random vectors (with possibly dependent components). Clearly $(X_{1,n})$ and $(X_{2,n})$ are real-valued Markov chains.

Then the function

$$\varphi_n(\mathbf{x}, \mathbf{z}) = (\varphi_{1,n}(x_1, z_1), \varphi_{2,n}(x_2, z_2))$$

fulfills the assumptions of Theorem 4.3.9 (b) if $\leq_{\mathcal{F}}$ is one of the stochastic orders $\leq_{st}, \leq_{cx}, \leq_{icx}, \leq_{sm}, \leq_{lo}$ or $\leq_{uo}$.

Therefore for any of these orders

$$\mathbf{Z}_n' \leq_{\mathcal{F}} \mathbf{Z}_n''$$

implies

$$(\mathbf{X}_0', \mathbf{X}_1', \ldots, \mathbf{X}_n') \leq_{\mathcal{F}} (\mathbf{X}_0'', \mathbf{X}_1'', \ldots, \mathbf{X}_n'') \tag{4.3.18}$$

for all n if $\mathbf{X}_0' \leq_{\mathcal{F}} \mathbf{X}_0''$.

Finally, following Lindqvist (1988) association of stochastic processes of the form (4.1.2) can be proved.

Theorem 4.3.13. *Let $(\mathbf{X}_n)$ be a stochastic process with state space $S = \mathbb{R}^d$ of the form*

$$\mathbf{X}_{n+1} = \varphi(\mathbf{X}_n, \mathbf{Z}_n) \quad \text{for } n = 0, 1, \ldots$$

where the mapping φ is monotone in both variables $\mathbf{x}$ and $\mathbf{z}$. If $(\mathbf{Z}_n)$ and $\mathbf{X}_0$ form a set of associated random variables, then $(\mathbf{X}_n)$ is associated. If the $\mathbf{Z}_n$ are mutually independent and independent on $\mathbf{X}_0$, then $(\mathbf{X}_n)$ is associated if φ is monotone in $\mathbf{x}$.

Proof. The proof is made by induction. Assume that the set of random variables $(\mathbf{X}_0, \mathbf{X}_1, \ldots, \mathbf{X}_k)$ and $(\mathbf{Z}_n)$ is associated for some k. Then property Theorem 3.10.7 yields the association of the set of random variables $(\mathbf{X}_0, \mathbf{X}_1, \ldots, \mathbf{X}_{k+1})$ and $(\mathbf{Z}_n)$ since $\mathbf{X}_{k+1}$ depends monotonically on $(\mathbf{Z}_n)$ and $(\mathbf{X}_0, \mathbf{X}_1, \ldots, \mathbf{X}_k)$.

In the independent case the fact that $(\mathbf{X}_0, \mathbf{X}_1, \ldots, \mathbf{X}_k, \varphi(\mathbf{X}_k, \mathbf{z}))$ is associated for every $\mathbf{z}$ is used. The result finally follows from Theorem 3.10.8. $\square$

4.3.3   The Coupling Method

The comparison methods described until now lead to inequalities for distributional characteristics such as distribution functions or moments. Such quantities are probably completely sufficient for most applications, since such information is closely related to statistical measurement which is possible for compared stochastic models working in separation. The corresponding probability spaces are often only artificial objects and may have little to do with the comparison problem; any information on the joint distribution of stochastic models that are compared can be inessential.

However, there are situations where it makes sense to consider suitable probability spaces explicitly and to consider two stochastic models over the same space. In some situations it is a helpful trick to introduce a common probability space on which both models are defined and to compare directly the corresponding random variables. In both cases it is possible to obtain inequalities for distributional characteristics, which can be of a better quality than those obtained by the functional or mapping method. The corresponding method is called *coupling method* and is described in detail by Lindvall (1992) and Thorisson (2000). Until now it has been used mainly in the context of the stochastic order relation $\leq_{st}$, but it is useful for orderings like $\leq_{cx}$, $\leq_{icx}$ and $\leq_{icv}$ as well.

We describe here the coupling method by means of three examples from the comparison context.

Example 4.3.14. (Comparison of $G/G/s'$ and $G/G/s''$.) Let Σ' and Σ'' be two queueing systems of type $G/G/s$ as described in Section 6.6. In particular, the queue discipline is FCFS, and the service time of the nth customer does not depend on the server on which it is served. Let both systems have the same constituent distribution, i.e. the same joint distribution of the sequence of interarrival and service times. Assume that both systems are empty at $t = 0$. The two queues differ only in the number of servers, which are s' and s'', respectively, where $s' < s''$.

Clearly, every reader expects that 'the waiting times' in Σ' are 'larger' than in Σ''. More precisely, the waiting time distribution functions of the nth customer W_n' and W_n'' should satisfy

$$W_n'' \leq_{st} W_n' \quad \text{for } n = 1, 2, \dots . \tag{4.3.19}$$

The authors expect that most of the readers would be able to find their own proof of (4.3.19). And probably most of them would not try to use the mapping method, but instead a variant of the coupling method. (Since there are no formulae for the waiting times in general $G/G/s$ queues, the functional method does not work in the context of this example.)

Since for Σ' and Σ'' the constituent distributions coincide, it is natural to consider a unique arrival process realization $(T_n(\omega), B_n(\omega))$, where $T_n(\omega)$ is the arrival instant of the nth customer and $B_n(\omega)$ its service time. Here the sample point ω is an element of some Ω being the sample space.

This process goes to the two service nodes with s' and s'' servers. It is then straightforward to show that the waiting times in Σ' are larger than in Σ''. Perhaps like this: Consider the workload vectors $\mathbf{W}'_n(\omega)$ and $\mathbf{W}''_n(\omega)$ of Σ' and Σ'', as explained on page 228. The ith component $W'_{i,n}(\omega)$ of $\mathbf{W}'_n(\omega)$ is the ith smallest sum of residual service time plus sum of service times of customers in the queue which will go to the same server, immediately before arrival of nth customer, $i = 1, \ldots, s'$. The first component $W'_{1,n}(\omega)$ of $\mathbf{W}'_n(\omega)$ is the waiting time $W'_n(\omega)$ of the nth customer in Σ'. The definition for Σ'' is analogous but now the vector $\mathbf{W}''_n(\omega)$ has s'' components. For fixed $B_n(\omega)$ and $T_n(\omega)$, $\mathbf{W}'_{n+1}(\omega)$ results from $\mathbf{W}'_n(\omega)$ by a mapping Φ'_n,

$$\mathbf{W}'_{n+1}(\omega) = \Phi'_n(\mathbf{W}'_n(\omega)),$$

which is increasing with respect to the usual order for elements of $\mathbb{R}^{s'}$. For $\mathbf{W}''_{n+1}(\omega)$ and Σ'' there is a mapping Φ''_n doing the same. Formula (6.6.1) gives details of Φ'_n and Φ''_n.

It is possible to apply Φ'_n and Φ''_n on the same s''-dimensional vector $\mathbf{W}$ with non-negative components W_i with $W_1 \leq \ldots \leq W_{s''}$ if in the case of Φ'_n the vector $[\mathbf{W}]_{s'}$ consisting of the first s' components of $\mathbf{W}$ is taken. Then it is easy to see that

$$\Phi'_n([\mathbf{W}]_{s'}) \geq [\Phi''_n(\mathbf{W})]_{s'} \tag{4.3.20}$$

i.e. componentwise comparability of two s'-dimensional vectors, where the righthand side denotes the vector consisting of the first s' components of $\Phi''_n(\mathbf{W})$.

It can be shown by induction that for all n

$$W'_{i,n}(\omega) \geq W''_{i,n}(\omega) \quad \text{for } i = 1, \ldots s'. \tag{4.3.21}$$

Clearly, (4.3.21) holds with '=' for the first s' customers, $n \leq s'$. Assume that (4.3.21) is true for some n. Then the monotonicity of Φ'_n ensures

$$\mathbf{W}'_{n+1}(\omega) = \Phi'_n(\mathbf{W}'_n(\omega)) \geq \Phi'_n(\mathbf{W}''_n(\omega))$$

and the comparability relation (4.3.20) yields

$$\Phi'_n([\mathbf{W}''_n(\omega)]_{s'}) \geq [\Phi''_n(\mathbf{W}''_n(\omega))]_{s'} = [\mathbf{W}''_{n+1}(\omega)]_{s'},$$

i.e. (4.3.21) for $n + 1$.

Thus the waiting times $W'_{1,n} = W'_n$ and $W''_{1,n} = W''_n$ in Σ' and Σ'' satisfy

$$W''_n(\omega) \leq W'_n(\omega), \tag{4.3.22}$$

an inequality which is true for all ω. (Note that it is really true for all ω, not 'only' almost surely.) Clearly, (4.3.22) implies (4.3.19) for the corresponding distribution functions. This is a 'decoupled' result for Σ' and Σ'', which were coupled by the assumption that both service nodes are confronted with the same arrival process.

In the given case the coupling is also of practical interest since it says that s'' servers instead of s' decrease the waiting times pathwise for a given arrival process. It is a typical advantage of the coupling method that such strong comparison is possible.

Of more interest are comparison theorems as in Chao and Scott (2000) which say that customer numbers in G/M/s queues are comparable with respect to $\leq_{st}$ if the total service effort is constant for the various values of s (see Section 6.8).

Example 4.3.15. (Comparison of two inhomogeneous Poisson processes.) Let (N_t') and (N_t'') be two Poisson processes on $[0, \infty)$ with mean (or integrated intensity) functions $\Lambda'(t)$ and $\Lambda''(t)$, where $\Lambda'(t) \leq \Lambda''(t)$ for all $t \geq 0$. (N_t' is the number of events in $[0, t]$ and $\Lambda'(t)$ the corresponding mean, etc.) Then clearly

$$P(N_t' \leq n) \geq P(N_t'' \leq n) \quad \text{for all } n \text{ and } t, \tag{4.3.23}$$

which is a simple consequence of the comparability property of Poisson random variables with respect to $\leq_{st}$ (see Table 1.2 on page 63).

Let, furthermore, T_m' be the random time of first arrival of (N_t') at the level m or the mth point of the point process, $T_m' = \inf\{t : N_t' \geq m\}$ etc. As expected, it is

$$P(T_m' \leq t) \geq P(T_m'' \leq t) \quad \text{for all } m \text{ and } t. \tag{4.3.24}$$

This follows from

$$P(T_m \leq t) = P(N_t \geq m), \tag{4.3.25}$$

which is true since the sample path of (N_t) is monotonically increasing.

An alternative proof of these results is possible with the coupling method. This immediately leads to the much deeper result of complete comparison of the processes and can easier be generalized to the comparison of more general processes like the Cox processes considered in Section 7.4.

Because $\Lambda'(t) \leq \Lambda''(t)$, we have $\Delta(t) = \Lambda''(t) - \Lambda'(t) \geq 0$ for all t. Thus one can consider two independent Poisson processes on the same probability space, having mean functions $\Lambda_1(t)$ and $\Delta(t)$. The sum of both processes is again a Poisson process, and its mean function is equal to $\Lambda''(t) = \Lambda'(t) + \Delta(t)$. Denote the new processes corresponding to $\Lambda'(t)$ and $\Lambda''(t)$ by $(\hat{N}_t')$ and $(\hat{N}_t'')$. These processes are now coupled but not independent, rather

$$\hat{N}_t'(\omega) \leq \hat{N}_t''(\omega) \quad \text{for all } \omega \text{ and } t. \tag{4.3.26}$$

In terms of Section 5.1, this implies that Poisson processes with comparable mean functions are 'completely comparable' with respect to $\leq_{st}$; see Section 5.4 for more comparison results for Poisson processes on $[0, \infty)$; and Section 7.4.2 considers Poisson point processes in more general spaces. The inequality (4.3.26) clearly yields

$$P(\hat{N}_t' \leq n) \geq P(\hat{N}_t'' \leq n) \quad \text{for all } n \text{ and } t$$

and, since $(N_t') =_{st} (\hat{N}_t')$ and $(N_t'') =_{st} (\hat{N}_t'')$, (4.3.23) follows. Furthermore, (4.3.26) says that the sample paths of $(\hat{N}_t')$ and $(\hat{N}_t'')$ are comparable; always that for $(\hat{N}_t'')$ is above that for $(\hat{N}_t')$. Thus it is trivial that the corresponding first arrival times $\hat{T}_m'$ and $\hat{T}_m''$ are comparable:

$$\hat{T}_m'(\omega) \geq \hat{T}_m''(\omega) \quad \text{for all } \omega, m \text{ and } t$$

and finally the decoupled inequality (4.3.24) holds, which follows since the distribution of T_m' is completely given by that of (N_t') and since (N_t') and $(\hat{N}_t')$ have the same distribution, and the same for T_m'' etc.

As the coupling proof yields the complete comparison of the processes, much deeper results than (4.3.23) and (4.3.24) can be derived. For example, it immediately implies that

$$P(N_t' \leq g(t) \text{ for all } t) \geq P(N_t'' \leq g(t) \text{ for all } t) \qquad (4.3.27)$$

holds for any function g, which is a stronger result than (4.3.23).

Example 4.3.16. (Comparison of compounds.) This example gives a proof of Theorem 4.3.5 by means of the coupling method. The problem is to show that for non-negative random variables

$$\sum_{i=1}^{N'} X_i' \leq_{st} \sum_{i=1}^{N''} X_i'' \qquad (4.3.28)$$

if

$$(N', X_1', X_2', \dots) \leq_{st} (N'', X_1'', X_2'', \dots).$$

Strassen's theorem (Theorem 2.6.1) ensures that there is a probability space and on it there are random variables $\hat{N}', \hat{X}_1', \hat{X}_2', \dots$ and $\hat{N}'', \hat{X}_1'', \hat{X}_2'', \dots$ with

$$\hat{N}'(\omega) \leq \hat{N}''(\omega)$$

and

$$\hat{X}_i'(\omega) \leq \hat{X}_i''(\omega) \quad \text{for all } \omega \text{ and } i,$$

$$(N', X_1', X_2', \dots) =_{st} (\hat{N}', \hat{X}_1', \hat{X}_2', \dots) \qquad (4.3.29)$$

and

$$(N'', X_1'', X_2'', \dots) =_{st} (\hat{N}'', \hat{X}_1'', \hat{X}_2'', \dots). \tag{4.3.30}$$

Then clearly the sums are comparable,

$$\sum_{i=1}^{\hat{N}'(\omega)} \hat{X}_i'(\omega) \le \sum_{i=1}^{\hat{N}''(\omega)} \hat{X}_i''(\omega) \quad \text{for all } \omega,$$

which implies

$$E\left(f\left(\sum_{i=1}^{\hat{N}'} \hat{X}_i'\right)\right) \le E\left(f\left(\sum_{i=1}^{\hat{N}''} \hat{X}_i''\right)\right)$$

for any increasing f. Because of (4.3.29) and (4.3.30) this implies (4.3.28).

These three examples show three possible ways towards couplings of stochastic processes useful in the comparison context:
(1) Construct processes which can be compared based on a common stochastic process by rules which enable comparison.
(2) Use theorems of probability theory to introduce inequalities between processes which are closely related to the processes that have to be compared.
(3) Use Strassen's theorem to ensure the existence of a common probability space and of corresponding comparable random variables which are then used to construct comparable processes.

Clearly, in the situations of Section 4.3.2 where the mapping method can be applied for $\le_{st}$, the coupling method yields the same results; the monotonicity properties of the mappings appearing are helpful also for coupling proofs.

The third example is close to *coupling of stochastic processes*. Let (X_t') and (X_t'') be stochastic processes on $(S, \mathcal{S})$, where S is a partially ordered space with order relation $\le$ and $\mathcal{S}$ a suitable σ-algebra. A coupling of (X_t') and (X_t'') is a stochastic process $(\hat{X}_t) = (\hat{X}_t', \hat{X}_t'')$ on $(S \times S, \mathcal{S} \otimes \mathcal{S})$ whose marginals have the same distribution as (X_t') and (X_t''), respectively. A coupling is called order preserving if $\hat{X}_0' \le \hat{X}_0''$ implies $P(\hat{X}_t' \le \hat{X}_t''$ for all $t) = 1$. For Markov processes there are criteria ensuring the existence of order preserving couplings (see Section 5.2). It is quite easy to prove coupling analogues of Theorems 4.3.8 and 4.3.10 for $\le_{st}$. Lindvall (1992), Last and Brandt (1995), Szekli (1995) and Thorisson (2000) describe systematically coupling constructions.

Note that all three examples yield inequalities with respect to stochastic order $\le_{st}$. The convexity part of Strassen's theorem (see Theorem 1.5.20) enables also comparison with respect to $\le_{icx}$, $\le_{cx}$ or $\le_{icv}$ (see Theorem 1.5.23 for a typical application). However, for other stochastic orders such as $\le_{lo}$, $\le_{uo}$, $\le_{sm}$, or $\le_{dcx}$ the coupling method does not help because they do not satisfy the conditions of Strassen's theorem.

The standpoint of the authors to the application of the mapping method and the coupling method is quite pragmatic: use always the simpler and more natural method and follow your taste. Perhaps, try to use the mapping method when it is possible because coupling is often based on Strassen's theorem, which is deep and complicated to prove. The mapping method is typically applicable when one is interested in a characteristic which depends only on the value of a stochastic process evaluated at a fixed point of time or in the limit as time goes to infinity. If, however, one is interested in a quantity that depends on the whole path of a process then typically the coupling method yields deeper results as demonstrated in (4.3.27).

As an impressive example for a proof based on coupling, here the usual proof of Harris' inequality is presented. Thorisson (2000) speaks about an 'i.i.d. coupling'.

Theorem 4.3.17. *Let X be a real-valued random variable and let f and g be two bounded increasing real functions. Then*

$$\text{Cov}(f(X), g(X)) = E(f(X)g(X)) - Ef(X)Eg(X) \geq 0 \qquad (4.3.31)$$

Proof. Let Y be a further random variable with $X =_{st} Y$ which is independent of X. Then

$$
\begin{aligned}
0 &\leq E[(f(X) - f(Y))(g(X) - g(Y))] \\
&= E(f(X)g(X)) + E(f(Y)g(Y)) - E(f(X)g(Y)) - E(f(Y)g(X)) \\
&= 2(E(f(X)g(X)) - Ef(X)Eg(X)) \\
&= 2\text{Cov}(f(X), g(X)).
\end{aligned}
$$

The first inequality is true since for any ω either $X(\omega) \leq Y(\omega)$ or $X(\omega) \geq Y(\omega)$ and hence the two factors on the right-hand side have the same sign. $\square$

This theorem says that a univariate random variable X is associated.

4.4   Extremal Problems

By the term *extremal problem* for a stochastic model Σ we mean a variational problem of the following type: for a given system quantity $c_\Sigma(F_1, F_2, \ldots)$ of Σ, determine its supremum or infimum when some (or all) of the F_k vary in specified classes $\mathbb{F}_k$ of constituent distributions while the others (if any) remain fixed, i.e., symbolically, determine for example

$$\sup\{c_\Sigma(F_1, F_2, \ldots) : F_k \in \mathbb{F}_k, k = k_1, \ldots, k_m\}.$$

If, for example, c_Σ denotes a mean and the classes $\mathbb{F}_k$ are characterized by fixed moments, the results of Harris (1962), Hartley and David (1954) and Hoeffding (1955) on the extrema of expected values of functions of random variables can be used. Note that 'supremum' and 'infimum' are to be understood in the sense of the order of real numbers if c_Σ is real, and otherwise in the sense of a suitable partial order for which C_Σ is a lattice.

An extremal problem can certainly be solved by means of partial orders when c_Σ is increasing in the F_k with respect to $\preceq_k$ and the $\mathbb{F}_k$ possess $\preceq_k$-extremal elements $\max\limits_{\mathbb{F}_k} F_k$ or $\min\limits_{\mathbb{F}_k} F_k$. Then for example

$$\max\{c_\Sigma(F_1, F_2, \dots) : F_1 \in \mathbb{F}_1\} = c_\Sigma(\max_{\mathbb{F}_1} F_1, F_2, \dots).$$

In Chapters 5, 6 and 7 some extremal problems are solved and some extremality properties are established in this way, making use of comparability properties and of extremal elements given in Section 1.10.

An important area of application of extremality statements is the solution of problems with incomplete information on distribution functions or distributions. By such problems we mean the problem of determining a system quantity c_Σ when some constituent distribution F_k is not known precisely but only that it belongs to some class $\mathbb{F}_k$. Some typical applications in risk theory are considered in De Vylder (1996).

We consider here the example of determining the mean waiting time m_W in a $GI/GI/1$ queueing system when the service time distribution function F_B is given, but for the inter-arrival times only the mean m_A and the lower and upper bound a and b are known. Letting Σ denote $GI/GI/1$, $c_\Sigma =$ mean stationary waiting time m_W, $F_1 =$ service time distribution function F_B, $F_2 =$ inter-arrival time distribution function F_A, $\mathbb{F}_2 = \{$distribution functions on $[a, b]$ with mean $m_A\} = \mathbb{M}_{m_A} \cap \mathbb{M}^{[a,b]}$, we then have a problem with incomplete information in the sense just described. Using the increasing concave order $\leq_{icv}$ as in Section 6.3 it is solved by giving the bounds on c_Σ which belong to constant inter-arrival times and the two-point distribution $\frac{b-m_A}{b-a}\delta_a + \frac{m_A-a}{b-a}\delta_b$ with mean m_A and masses at a and b (see Example 1.10.5). Of course, it may be difficult to determine m_W for the given F_B and these inter-arrival distributions.

5

Monotonicity and Comparability of Stochastic Processes

5.1 Introduction

In the following we consider stochastic processes $\mathbf{X} = (X_t : t \in T)$ with discrete time, $T = \{0, 1, 2, \dots\}$, or continuous time, $T = \mathbb{R}_+$, and with the state space $(S, \mathcal{S})$. Usually (X_n) is written in the discrete case. Typically S is the real line or a subset of it (real-valued processes) or S is a very general set, for example a Polish space with a closed partial order, and $\mathcal{S}$ is a σ-algebra of subsets of S, for example the Borel σ-algebra. Other stochastic processes with other index sets will be considered in Chapter 7, for example, spatial point processes, random fields and random sets.

We denote by $\preceq$ a partial order on the space $\mathbb{P}_S$ of all probability distributions on $\mathcal{S}$. It is a general partial order, but in most cases it will be some integral stochastic order. As concrete examples we shall consider here mainly $\leq_{st}$ and $\leq_{icx}$, but we emphasize the possibility of proving monotonicity and comparability properties also for other relations which may be specifically tailored to given problems.

Definition 5.1.1. A stochastic process $\mathbf{X} = (X_t : t \in T)$ is said to be *increasing (decreasing) with respect to $\preceq$* if

$$X_s \preceq (\succeq) X_t \quad \text{for all } s \text{ and } t \text{ in } T \text{ with } s < t. \tag{5.1.1}$$

If $\preceq$ is a multivariate stochastic order, then the process $\mathbf{X}$ is said to be *strongly increasing (decreasing) with respect to $\preceq$* if for every $m = 1, 2, \dots$ and all $\mathbf{s} = (s_1, \dots, s_m)$ and $\mathbf{t} = (t_1, \dots, t_m)$

$$\mathbf{X_s} \preceq (\succeq) \mathbf{X_t} \quad \text{for all } \mathbf{s} \text{ and } \mathbf{t} \text{ in } T^m \text{ with } \mathbf{s} \leq \mathbf{t}, \tag{5.1.2}$$

where $\mathbf{X_s} = (X_{s_1}, \dots, X_{s_m})$ etc.

Sometimes instead of 'increasing' the term *monotone* is used.
Note that in the case of discrete time (5.1.1) holds if and only if

$$X_n \preceq (\succeq) X_{n+1} \quad \text{for } n = 0, 1, 2, \ldots . \tag{5.1.3}$$

Examples of monotone processes will appear in the following sections and in Chapters 6 and 7. It is clear that for a suitable mapping Φ, inequality (5.1.3) implies

$$\Phi(X_n) \preceq (\succeq) \Phi(X_{n+1}) \quad \text{for all } n. \tag{5.1.4}$$

For example, if $\preceq$ is $\leq_{icx}$ and Φ is a mapping from S into $\mathbb{R}$, then (5.1.4) is true if Φ is convex and increasing. If (Z_n) is a stochastic process constructed from i.i.d. random variables Y_n and a process (X_n) via

$$Z_n = \varphi(X_n, Y_n)$$

then $\leq_{icx}$-monotonicity of (X_n) implies also $\leq_{icx}$-monotonicity of (Z_n) if φ is convex and increasing in its first variable. This may be interesting in the context of 'hidden' stochastic processes such as hidden Markov chains (see, for example, MacDonald and Zucchini (1997)).

Definition 5.1.2. Let (X_t) and (Y_t) be stochastic processes with the state space $(S, \mathcal{S})$. Then the process (X_t) is *smaller than* (Y_t) *with respect to* $\preceq$, symbolically

$$(X_t) \preceq (Y_t),$$

if for all t

$$X_t \preceq Y_t. \tag{5.1.5}$$

If $\preceq$ is a multivariate stochastic order, then the process (X_t) is said to be *strongly smaller than* (Y_t) *with respect to* $\preceq$ if for every $m = 1, 2, \ldots$ and all $(t_1, \ldots, t_m)$,

$$(X_{t_1}, \ldots, X_{t_m}) \preceq (Y_{t_1}, \ldots, Y_{t_m}), \tag{5.1.6}$$

symbolically,

$$\mathbf{X} \text{ str-}\preceq \mathbf{Y}. \tag{5.1.7}$$

The process $\mathbf{X} = (X_t)$ is *completely smaller than* $\mathbf{Y} = (Y_t)$ if

$$\mathbf{X} \preceq \mathbf{Y},$$

where $\mathbf{X}$ and $\mathbf{Y}$ are now considered as random elements on the spaces $S^{\mathbb{N}}$ or $S^{\mathbb{R}+}$ of process paths.

Strong comparability and monotonicity with respect to $\leq_{st}$ was first studied in the context of reliability theory; see Pledger and Proschan (1973) and Barlow and Proschan (1976). They showed bounds for sojourn times and inequalities such as (5.1.8).

Similarly as in the case of monotonicity, transformed processes of the form $(\Phi(X_t))$ and $(\Phi(Y_t))$ can be considered. If the mapping Φ has suitable properties, $(X_t) \preceq (Y_t)$ implies $(\Phi(X_t)) \preceq (\Phi(Y_t))$.

Section 5.2 considers methods for proving comparability and strong comparability for the class of Markov processes. Here for general stochastic processes the relationship between strong comparability and complete comparability is discussed.

With $\preceq$ denoting $\leq_{st}$ or $\leq_{icx}$ the following theorem shows that (5.1.7) implies the validity of

$$f(\mathbf{X}) \preceq f(\mathbf{Y}) \tag{5.1.8}$$

for certain monotone functionals f on $S^{\mathbb{R}+}$. An important example is

$$f(\mathbf{X}) = \inf\{t : X_t < a\} \text{ for some real } a.$$

If $X_0 > a$ a.s. then $f(\mathbf{X})$ can be interpreted as the time of first passage of the level a by (X_t) from above. Also *default times* (times until absorption) are of interest; see Shanthikumar (1988), Li and Shaked (1997) and Kijima (1998), who studied monotonicity properties of such times with respect to $\leq_{st}$ and $\leq_{hr}$.

Let $\mathcal{T} = \{\mathcal{T}_n\}$ denote sequences of non-negative real numbers $\mathcal{T}_n = \{t_{n0}, t_{n1}, \dots, t_{nk_n}\}$ for which

$$0 = t_{n0} < \dots < t_{nk_n}, \ \lim_{n\to\infty} t_{nk_n} = \infty \text{ and } \lim_{n\to\infty} \max_j (t_{n,j+1} - t_{nj}) = 0.$$

Further, given the continuous time process $\mathbf{X} = (X_t)$ and $\mathcal{T}$, let $\mathbf{X}^{(n)} = (X_t^{(n)})$ denote the piecewise constant stochastic process

$$X_t^{(n)} = \begin{cases} X_{t_{nj}} & \text{for} \quad t_{nj} \leq t < t_{n,j+1} \text{ and } j < k_n, \\ X_{t_{nk_n}} & \text{for} \quad t \geq t_{nk_n}. \end{cases}$$

The distribution functions of the random variables $f(\mathbf{X})$ and $f(\mathbf{X}^{(n)})$ are denoted by F and $F^{(n)}$ respectively, and, similarly, G and $G^{(n)}$ for $f(\mathbf{Y})$ and $f(\mathbf{Y}^{(n)})$.

Theorem 5.1.3. (Franken and Kirstein (1977)) *Let $\preceq$ denote $\leq_{st}$ or $\leq_{icx}$ and let $f \in \mathcal{R}_{\mathcal{F}}$. Then $\mathbf{X}$ str-$\preceq$ $\mathbf{Y}$ implies $f(\mathbf{X}) \preceq f(\mathbf{Y})$ if the sequences $\{F^{(n)}\}$ and $\{G^{(n)}\}$ converge weakly for arbitrary $\mathcal{T}$ to F and G and if, additionally, in the case of $\leq_{icx}$*

$$\lim_{n\to\infty} Ef(\mathbf{X}^{(n)}) = Ef(\mathbf{X})$$

and similarly for $\{f(\mathbf{Y}^{(n)})\}$, respectively.

Proof. The strong comparability of the processes ensures

$$(X_{t_{n1}}, \dots, X_{t_{nk_n}}) \preceq (Y_{t_{n1}}, \dots, Y_{t_{nk_n}})$$

for all n and every $\mathcal{T}_n$. For f as stated, Theorem 4.3.2 then yields $f(\mathbf{X}^{(n)}) \preceq f(\mathbf{X}^{(n)})$, which is the same as $F^{(n)} \preceq G^{(n)}$. The Theorems 1.2.14 and 1.5.9 giving conditions for the closure of $\preceq$ under weak convergence then imply $f(\mathbf{X}) \preceq f(\mathbf{Y})$. $\qquad\square$

In the case of real-valued processes with paths in $D[0, \infty)$ (the space of all real functions on $[0, \infty)$ that are right-continuous and have left-hand limits), the assumptions of the theorem are satisfied by all functionals on $D[0, \infty)$ which are continuous with respect to the Skorokhod metric, which makes $D[0, \infty)$ a Polish space. (See Billingsley (1999), section 16, for an exposition of the theory of $D[0, \infty) = D_\infty$.)

Comparison results of this type can also be obtained by the coupling method; see Lindvall (1992) and Brandt and Last (1994). There one constructs two processes $\hat{\mathbf{X}} = (\hat{X}_t)$ and $\hat{\mathbf{Y}} = (\hat{Y}_t)$ defined on a common probability space $(\Omega, \mathcal{A}, P)$ that have the same distribution as (X_t) and (Y_t) and that are 'pathwise comparable', i.e.

$$\hat{X}_t(\omega) \le \hat{Y}_t(\omega) \quad \text{for all } t \text{ and all } \omega.$$

For monotone f this implies

$$f(\hat{\mathbf{X}}) \le f(\hat{\mathbf{Y}}),$$

$$Ef(\hat{\mathbf{X}}) \le Ef(\hat{\mathbf{Y}}),$$

and finally

$$Ef(\mathbf{X}) \le Ef(\mathbf{Y}),$$

i.e. comparison of functionals of the processes (X_t) and (Y_t) is possible.

In the case of $\le_{st}$, strong comparability of (X_n) and (Y_n) implies complete comparability of (X_n) and (Y_n), as shown by Kamae et al. (1977) and thus these properties are equivalent.

Theorem 5.1.4. *Let $S_1, S_2, \dots$ be Polish spaces with closed partial orders and let S^∞ be the product space of all these spaces. Let $P, Q \in \mathbb{P}_{S^\infty}$ and let $P^{(n)}$ be the marginal distribution of the first n coordinates of P so that $P^{(n)} \in \mathbb{P}_{S_1 \times \dots \times S_n}$ and $Q^{(n)}$ the same for Q. If*

$$P^{(n)} \le_{st} Q^{(n)} \quad \text{for } n = 1, 2, \dots, \tag{5.1.9}$$

then

$$P \le_{st} Q. \tag{5.1.10}$$

Proof. Let $(z_1, z_2, \dots)$ be any fixed element of S^∞. For each n, by the Strassen Theorem 2.6.3 there are random n-vectors $(X_{n1}, \dots, X_{nn})$ and $(Y_{n1}, \dots, Y_{nn})$ in $S_1 \times \cdots \times S_n$ with distributions $P^{(n)}$ and $Q^{(n)}$ and such that $X_{nj} \leq Y_{nj}$ a.s. for $j = 1, \dots, n$. From these vectors we obtain random elements $\mathbf{X}_n$ and $\mathbf{Y}_n \in S^\infty$ by defining $X_{nj} = Y_{nj} = z_j$ for $j = n+1, n+2, \dots$. Let P_n and Q_n be the distributions of $\mathbf{X}_n$ and $\mathbf{Y}_n$ respectively. Since $\mathbf{X}_n \leq \mathbf{Y}_n$ a.s., $P_n \leq_{st} Q_n$.

For fixed j the j-th one-dimensional marginal of P_n is independent of n for $n \geq j$; hence the sequence of such marginals is tight, and by Tykhonov's theorem it follows that the sequences (P_n) and (Q_n) are tight. Let R_n be the joint distribution of $(\mathbf{X}_n, \mathbf{Y}_n)$ in $S^\infty \times S^\infty$. The sequence (R_n) is also tight and by Prohorov's theorem (Billingsley (1999), p. 59) it has a subsequence (R_{n_i}) which converges weakly to a probability measure $R \in \mathbb{P}_{S^\infty \times S^\infty}$. The marginals (chosen appropriately) of R are $P^{(n)}$ and $Q^{(n)}$ and P and Q. Also, each R_n has support in the closed set $H = \{(x, y) \in S^\infty \times S^\infty : x \leq y\}$. By theorem 2.1 of Billingsley (1999),

$$R(H) = \limsup_{i \to \infty} R_{n_i}(H) = 1,$$

so that R has support in H. Thus $P \leq_{st} Q$. $\square$

A similar statement holds for continuous time processes with paths in $D_S[0, \infty)$; see Kamae et al. (1977).

The next result can be considered as a counterpart to Theorem 5.1.4 for the increasing convex order. Notice, however, that it shows a weaker statement than $\mathbf{X} \leq_{icx} \mathbf{Y}$, since the continuity assumption in the theorem cannot be dropped, as is shown in Example 5.1.7.

For simplicity it is assumed that $S_1 = S_2 = \cdots = S$, where S is a separable Banach space with a closed partial order. Let W be a convex subset of S. For every subset A of $\{1, 2, \dots\}$, W^A is endowed with the product topology and the componentwise order.

Theorem 5.1.5. *(Bassan and Scarsini (1991)) Let $\mathbf{X} = (X_n)$ and $\mathbf{Y} = (Y_n)$ be two stochastic processes with values in W. If*

$$(X_1, \dots, X_n) \leq_{icx} (Y_1, \dots, Y_n) \quad \text{for all } n = 1, 2, \dots$$

then $Ef(\mathbf{X}) \leq Ef(\mathbf{Y})$ holds for all continuous increasing convex functions $f : W^\mathbb{N} \to \mathbb{R}$.

Proof. Let $f : W^\mathbb{N} \to \mathbb{R}$ be a continuous, increasing and convex function for which all expectations in the following exist. First it is assumed that f is bounded from below; later this assumption will be dropped.

Define the functions $f_n : W^\mathbb{N} \to \mathbb{R}$ and $\tilde{f}_n : W^n \to \mathbb{R}$ by

$$f_n(x_1, x_2, \dots) = \tilde{f}_n(x_1, \dots, x_n) = \inf_{\substack{y_k \in W \\ k > n}} f(x_1, \dots, x_n, y_{n+1}, \dots).$$

Clearly, $\tilde{f}_n$ is increasing and convex. Thus

$$E\tilde{f}_n(X_1,\ldots,X_n) \le E\tilde{f}_n(Y_1,\ldots,Y_n)$$

or, equivalently,

$$Ef_n(\mathbf{X}) \le Ef_n(\mathbf{Y}).$$

The sequence (f_n) is increasing and pointwise convergent to f. Namely, for any n and $\mathbf{x} \in W^{\mathbb{N}}$ let $s_{n+1}^{(n)}, s_{n+2}^{(n)},\ldots$ be a sequence such that for

$$\mathbf{x}^{(n)} = \left(x_1,\ldots,x_n, s_{n+1}^{(n)}, s_{n+2}^{(n)},\ldots\right)$$

it is

$$\left|f\left(\mathbf{x}^{(n)}\right) - f_n(\mathbf{x})\right| < 2^{-n}.$$

Then

$$|f(\mathbf{x}) - f_n(\mathbf{x})| \le \left|f(\mathbf{x}) - f\left(\mathbf{x}^{(n)}\right)\right| + \left|f\left(\mathbf{x}^{(n)}\right) - f_n(\mathbf{x})\right|,$$

the continuity of f and the convergence of $(\mathbf{x}^{(n)})$ to $\mathbf{x}$ yield $\lim_{n\to\infty} f_n(\mathbf{x}) = f(\mathbf{x})$.

The monotone convergence theorem ensures that

$$Ef(\mathbf{X}) \le Ef(\mathbf{Y}). \tag{5.1.11}$$

The general case follows by the usual argument of approximating an unbounded function f monotonically by $f_n = \max\{f, -n\}$. $\qquad\square$

Remark 5.1.6. Obviously, the proof just given can be adapted for the case of $\le_{st}$, i.e. it yields an alternative proof for Theorem 5.1.4 without coupling.

Example 5.1.7. (The continuity assumption in Theorem 5.1.5 is necessary.) The continuity assumption in Theorem 5.1.5 cannot be dropped, i.e. it is not true that

$$(X_1,\ldots,X_n) \le_{icx} (Y_1,\ldots,Y_n) \quad \text{for all } n = 1,2,\ldots$$

implies $\mathbf{X} \le_{icx} \mathbf{Y}$. As a counterexample consider the case of independent and identically distributed sequences (X_n) and (Y_n) with $P(X_n = 1) = 1$ and

$$P(Y_n = 2) = P(Y_n = 0) = \frac{1}{2}.$$

Then $X_1 \le_{icx} Y_1$, and thus it follows from Theorem 3.4.4 that

$$(X_1,\ldots,X_n) \le_{icx} (Y_1,\ldots,Y_n) \quad \text{for all } n = 1,2,\ldots.$$

However, for the increasing convex function $f(\mathbf{x}) = \liminf x_n$, which indeed is linear, but *not* continuous,

$$Ef(\mathbf{X}) = 1 > 0 = Ef(\mathbf{Y}).$$

This is an immediate consequence of the lemma of Borel–Cantelli.

Until now the case of stochastic processes (X_t) with time parameter set $T = [0, \infty)$ or $T = \{0, 1, 2, \dots\}$ was considered. In this case the Strassen Theorem 2.6.3 ensures that for any family (P_t) of distributions which is increasing with respect to $\leq_{st}$ there is a stochastic process $(\hat{X}_t)$ which is increasing almost surely and where $\hat{X}_t$ has the distribution P_t. It is interesting that if *both* T and the state space S are only partially ordered this may not be true. It is possible to construct examples where $t_1 \preceq t_2$ implies $P_{t_1} \leq_{st} P_{t_2}$ ('stochastic monotonicity') but there is not a system of S-valued random variables $(\hat{X}_t)$ such that $\hat{X}_{t_1} \preceq \hat{X}_{t_2}$ whenever $t_1 \preceq t_2$ and P_t is the distribution of $\hat{X}_t$ for $t \in T$ ('realizable monotonicity').

This problem is discussed in detail in Fill and Machida (2001). Here only the introductory example of that paper is given. Similar results can also be found in Kellerer (1997).

Example 5.1.8. (A stochastic process which is stochastically monotone but not realizably monotone.) Let $S = T$ be the four-element set $\{x, y, z, w\}$ with the partial order defined by the relations $x \preceq y$, $x \preceq z$, $y \preceq w$ and $z \preceq w$ (and $x \preceq w$ by transitivity); y and z are not comparable.

Define a system (P_t) of distributions on S by

$$P_x = \frac{1}{2}\delta_x + \frac{1}{2}\delta_y,$$

$$P_y = \frac{1}{2}\delta_x + \frac{1}{2}\delta_w,$$

$$P_z = \frac{1}{2}\delta_y + \frac{1}{2}\delta_z,$$

$$P_w = \frac{q}{2}\delta_y + \frac{q}{2}\delta_w.$$

Clearly, this system is stochastically monotone or increasing with respect to the stochastic order $\leq_{st}$ corresponding to $\preceq$. Suppose that there exists a system $(\hat{X}_t)$ which realizes the monotonicity of (P_t). This leads to a contradiction as follows.

Consider the event $\{\hat{X}_x = y\}$. We have

$$\frac{1}{2} = P(\hat{X}_x = y) = P(\hat{X}_x = y, \hat{X}_y = w, \hat{X}_z = y, \hat{X}_w = w).$$

Similarly, for $\{\hat{X}_z = z\}$

$$\frac{1}{2} = P(\hat{X}_z = z) = P(\hat{X}_x = x, \hat{X}_z = z, \hat{X}_w = w).$$

Since the two events on the right-hand side are disjoint, $P(\hat{X}_w = w) = 1$ follows, which contradicts the assumption $P(\hat{X}_w = w) = \frac{1}{2}$.

5.2 Comparability and Monotonicity of Markov Processes

5.2.1 Monotone and Comparable Operators

Already in Chapter 4 Markov chains were investigated with respect to monotonicity and comparability. This was done in the context of stochastic recurrences, exploiting monotonicity properties of the governing mappings φ. In contrast, in this section the role of transition kernels and operators is studied.

Let $\mathbf{T}$, $\mathbf{T}'$ and $\mathbf{T}''$ be operators on $\mathbb{P}_S$ with general state space $(S, \mathcal{S})$, i.e. mappings from $\mathbb{P}_S$ into itself, and let $\preceq$ be a partial order on $\mathbb{P}_S$. (It is easy to generalize this to the case of mappings from $\mathbb{P}_{S_1}$ to $\mathbb{P}_{S_2}$).

Definition 5.2.1. An operator $\mathbf{T}$ is said to be $\preceq$-*monotone* if $\mathbf{T}P' \preceq \mathbf{T}P''$ holds for all P' and $P'' \in \mathbb{P}_S$ with $P' \preceq P''$.

The operator $\mathbf{T}'$ is said to be *smaller than* $\mathbf{T}''$ if $\mathbf{T}'P \preceq \mathbf{T}''P$ for all $P \in \mathbb{P}_S$; in symbols, $\mathbf{T}' \preceq \mathbf{T}''$; an equivalent terminology is that $\mathbf{T}''$ *dominates* $\mathbf{T}'$.

It is clear that monotonicity of $\mathbf{T}, \mathbf{T}_1$ and $\mathbf{T}_2$ implies monotonicity of products (or compositions) $\mathbf{T}_1 \mathbf{T}_2$ and hence of all powers $\mathbf{T}^n$. Moreover, if $\preceq$ has property (MI), which holds for all integral stochastic orders (see Theorem 2.4.2), then all mixtures $\sum_{i=1}^k p_i \mathbf{T}_i$ of monotone operators are monotone.

For applications to discrete time Markov processes, the comparability of state distributions P_n' and P_n'' defined by

$$P_n' = \mathbf{T}'^n P_0' \quad \text{and} \quad P_n'' = \mathbf{T}''^n P_0'' \quad \text{for } n = 1, 2, \dots$$

is of interest for different initial distributions P_0' and P_0'' and operators $\mathbf{T}'$ and $\mathbf{T}''$.

Theorem 5.2.2. *Let* $\mathbf{T}'$ *and* $\mathbf{T}''$ *be operators on* $\mathbb{P}_S$ *and* $P_0', P_0'' \in \mathbb{P}_S$. *Then*

$$P_0' \preceq P_0'' \tag{5.2.1}$$

implies

$$P_n' \preceq P_n'' \quad \text{for all } n = 1, 2, \dots \tag{5.2.2}$$

if there is a $\preceq$-*monotone operator* $\mathbf{T}$ *on* $\mathbb{P}_S$ *for which*

$$\mathbf{T}' \preceq \mathbf{T} \preceq \mathbf{T}''. \tag{5.2.3}$$

Proof. Condition (5.2.3) implies $\mathbf{T}'P_0' \preceq \mathbf{T}P_0'$ and $\mathbf{T}P_0'' \preceq \mathbf{T}''P_0''$ by definition, while the $\preceq$-monotonicity of $\mathbf{T}$ together with (5.2.1) implies (5.2.2) for $n = 1$. A simple induction step based on $P_{n+1}' = \mathbf{T}'P_n' \preceq \mathbf{T}P_n' \preceq \mathbf{T}P_n'' \preceq \mathbf{T}''P_n'' = P_{n+1}''$ completes the proof. $\qquad\square$

Condition (5.2.3) first appeared in Gaede (1973). As in the proof of Theorem 5.2.19 below it can be shown that in the case of a stochastic order $\preceq$ which stands for either $\leq_{st}$ with arbitrary S or $\leq_{icx}$ with $S = \mathbb{R}$, there are simple conditions such that a $\preceq$-monotone operator $\mathbf{T}$ with

$$\mathbf{T}' \preceq \mathbf{T} \preceq \mathbf{T}''$$

can be constructed explicitly.

In other notation Theorem 5.2.2 says: If (X_n') and (X_n'') are homogeneous Markov chains with transition operators $\mathbf{T}'$ and $\mathbf{T}''$ for which a $\mathbf{T}$ exists such that (5.2.3) is true, then

$$X_0' \preceq X_0''$$

implies

$$X_n' \preceq X_n'' \quad \text{for all } n.$$

If the transition operator $\mathbf{T}$ of the Markov chain (X_n) is $\preceq$-monotone, then

$$X_0 \preceq X_1$$

implies

$$X_n \preceq X_{n+1} \quad \text{for all } n.$$

If the state space S has a minimal element a, then $X_1 \equiv a$ implies that the chain (X_n) is $\preceq$-increasing.

Homogeneous Markov chains (X_n) on an arbitrary state space $(S, \mathcal{S})$ can be described by their transition kernel $Q(x, B)$, namely

$$Q(x, B) = P(X_{n+1} \in B | X_n = x) \text{ for } x \in S \text{ and } B \in \mathcal{S}$$

or, in the case of real-valued processes, by their transition distribution function $F(y|x) = P(X_{n+1} \leq y | X_n = x)$.

In the following, conditions are given on the transition kernels ensuring monotonicity or comparability of the corresponding transition operators; for simplicity we will speak about monotone and comparable kernels.

From here it is assumed that $\preceq$ is an integral stochastic order with generator $\mathcal{F}$ and $\preceq$ is replaced by $\leq_{\mathcal{F}}$. Note that for partial orders which are not of that type, not all the following results are true. For example, Whitt (1986) gave a simple example showing that Theorem 5.2.11 may fail.

Theorem 5.2.3. *A transition operator $\mathbf{T}$ on $\mathbb{P}_S$ is $\leq_{\mathcal{F}}$-monotone if and only if for every $f \in \mathcal{F}$ the functional $f_\mathbf{T}$ defined by*

$$f_\mathbf{T}(x) = \int f(t) Q(x, \mathrm{d}t) \quad \text{for} \quad x \in S \tag{5.2.4}$$

is an element of the maximal generator $\mathcal{R}_{\mathcal{F}}$.

Proof. Take any P' and $P'' \in \mathbb{P}_S$ and any $f \in \mathcal{R}_{\mathcal{F}}$ and assume that the integrals $\int_S f(t)\mathbf{T}P'(\mathrm{d}t)$ and $\int f(t)\mathbf{T}P''(\mathrm{d}t)$ exist. Then if $f_{\mathbf{T}} \in \mathcal{R}_{\mathcal{F}}$,

$$
\int f(t)\mathbf{T}P'(\mathrm{d}t) = \int f(t) \int Q(y, \mathrm{d}t)P'(\mathrm{d}y) = \int f_{\mathbf{T}}(y)P'(\mathrm{d}y)
$$
$$
\leq \int f_{\mathbf{T}}(y)P''(\mathrm{d}y) = \int f(t)\mathbf{T}P''(\mathrm{d}t)
$$

implies $\mathbf{T}P' \leq_{\mathcal{F}} \mathbf{T}P''$. Conversely, it follows from these same relations that if $\mathbf{T}P' \leq_{\mathcal{F}} \mathbf{T}P''$ whenever $P' \leq_{\mathcal{F}} P''$, then $f_{\mathbf{T}} \in \mathcal{R}_{\mathcal{F}}$.   □

Corollary 5.2.4. *If S is a Polish space with a closed partial order then $\mathbf{T}$ is monotone with respect to $\leq_{st}$ if and only if*

$$
Q(x, C) \leq Q(y, C)
$$

for all x, $y \in S$ with $x \leq y$ and all increasing closed $C \subset S$.

Theorem 5.2.5. *The transition operators $\mathbf{T}'$ and $\mathbf{T}''$ satisfy $\mathbf{T}' \leq_{\mathcal{F}} \mathbf{T}''$ if and only if their transition kernels Q' and Q'' satisfy*

$$
Q'(x, \cdot) \leq_{\mathcal{F}} Q''(x, \cdot) \quad \textit{for all } x \in S. \tag{5.2.5}
$$

Proof. For any $P \in \mathbb{P}_S$, take any $f \in \mathcal{R}_{\mathcal{F}}$ for which the integrals $\int f(x)\mathbf{T}'P(\mathrm{d}x)$ and $\int f(x)\mathbf{T}''P(\mathrm{d}x)$ exist. The property (5.2.5) holds if and only if the functions $f_{\mathbf{T}'}$ and $f_{\mathbf{T}''}$ defined as in (5.2.4) satisfy

$$
f_{\mathbf{T}'}(x) \leq f_{\mathbf{T}''}(x) \quad \text{for all} \quad x \in S. \tag{5.2.6}
$$

Thus, writing

$$
\int f(x)\mathbf{T}'P(\mathrm{d}x) = \int f_{\mathbf{T}'}(x)P(\mathrm{d}x) \quad \text{etc.}
$$

and using (5.2.6) shows that then $\mathbf{T}' \leq_{\mathcal{F}} \mathbf{T}''$, while the converse is proved by taking P to be a Dirac distribution at x and deducing (5.2.6).   □

Corollary 5.2.6. *On a Polish space with a closed partial order $\mathbf{T}' \leq_{st} \mathbf{T}''$ is equivalent to*

$$
Q'(x, C) \leq Q''(x, C)
$$

for all $x \in S$ and all increasing closed $C \subset S$.

In the case of real-valued processes it is readily shown by using the class $\mathcal{F}$ corresponding to $\leq_{\mathcal{F}}$ that the transition distribution functions must satisfy the conditions below. These conditions appeared first in Daley (1968) for $\leq_{st}$

and in Stoyan (1972b) for $\leq_{icx}$ and $\leq_{icv}$. The condition for $\leq_{icx}$ is strongly related to the concept of *stochastic convexity* of the transition kernel as studied in chapter 6 of Shaked and Shanthikumar (1994). In order to give an example for a stochastic order which is not an integral order also the case of $\leq_{hr}$ is mentioned, following Kijima (1997), p. 131.

Monotonicity:

$\leq_{st}$: for all y, $F(y|x)$ is decreasing in x, which is the same as $X_1 \uparrow_{st} X_0$, i.e. the mapping φ given by (4.1.3) on page 150 is increasing in x;

$\leq_{icx}$: for all y, $\pi(y|x) = \int\limits_{y}^{\infty} \bar{F}(t|x)dt$ is increasing and convex in x;

$\leq_{hr}$: $[1 - F(y_1|x_1)][1 - F(y_2|x_2)] \geq [1 - F(y_2|x_1)][1 - F(y_1|x_2)]$

for all x_1 and x_2 with $x_1 \geq x_2$ and all y_1 and y_2 with $y_1 \geq y_2$,

or, in other words, $\bar{F}(\cdot|\cdot)$ is log-supermodular or TP_2.

In the case of a Markov chain with discrete state space $S = \{1, 2, \ldots, n\}$, monotonicity can be expressed in terms of the transition probabilities p_{ij} and the transition matrix $\mathbf{P} = ((p_{ij}))$. Then the corresponding transition operator is monotone with respect to $\leq_{st}$ if and only if the mapping φ given by (4.1.4) is increasing in i or if and only if

$$\mathbf{U}^{-1}\mathbf{P}\mathbf{U} \geq \mathbf{0},$$

where

$$\mathbf{U} = \begin{bmatrix} 1 & 0 & 0 & \cdots & 0 \\ 1 & 1 & 0 & \cdots & 0 \\ 1 & 1 & 1 & \cdots & 0 \\ \vdots & \vdots & \vdots & \ddots & \vdots \\ 1 & 1 & 1 & \cdots & 1 \end{bmatrix}$$

and

$$\mathbf{U}^{-1} = \begin{bmatrix} 1 & 0 & 0 & \cdots & 0 & 0 \\ -1 & 1 & 0 & \cdots & 0 & 0 \\ 0 & -1 & 1 & \cdots & 0 & 0 \\ \vdots & \vdots & \vdots & \ddots & \vdots & \vdots \\ 0 & 0 & 0 & \cdots & 1 & 0 \\ 0 & 0 & 0 & \cdots & -1 & 1 \end{bmatrix}.$$

See Keilson and Kester (1977) and Kijima (1997), p. 129, for a proof and further matrix-theoretic characterizations. Kijima (1998) shows that a transition matrix $\mathbf{P}$ is monotone with respect to $\leq_{hr}$ if $\mathbf{P}\mathbf{U} \in \mathrm{TP}_2$; a matrix

$\mathbf{A} = ((a_{ij}))$ is called 'totally positive of order 2', denoted by $\mathbf{A} \in \mathbf{TP}_2$, if

$$a_{ij} a_{kl} \geq a_{il} a_{kj} \quad \text{for } i < k \text{ and } j < l.$$

$\mathbf{P}$ is monotone with respect to $\leq_{rh}$ if $\mathbf{P} \mathbf{U}^\top \in \mathbf{TP}_2$ (see Bloch-Mercier (2001) for applications in reliability theory), and it is monotone with respect to $\leq_{lr}$ if $\mathbf{P}^n \in \mathbf{TP}_2$ for all n.

Monotonicity with respect to $\leq_{rand}$ is still an open problem. Alberti (2000) showed that in the case of two states any monotone $((p_{ij}))$ has either equal columns or is doubly stochastic, i.e. $p_{1j} + p_{2j} = 1$ for $j = 1$ and 2. For more than two states he conjectures that a similar criterion holds; he found as a sufficient condition that the row distributions $P_i = (p_{ij})$ (for fixed i) must be 'equivalent', i.e. $P_{i_2} = (p_{i_2 j})$ is obtained from P_{i_1} by a permutation of the $p_{i_1 j}$. If for any fixed i all p_{ij} are different then all P_i have to coincide, i.e. $((p_{ij}))$ has equal columns.

Comparability:

$$\leq_{st}: F_1(y|x) \geq F_2(y|x) \text{ for all } x \text{ and } y;$$
$$\leq_{icx}: \pi_1(y|x) \leq \pi_2(y|x) \text{ for all } x \text{ and } y.$$

Note that the condition

$$[1 - F_2(y_1|x)][1 - F_1(y_2|x)] \geq [1 - F_2(y_2|x)][1 - F_1(y_1|x)]$$

for all x and all y_1 and y_2 with $y_1 \geq y_2$ is *not* sufficient for the comparability of two transition operators with respect to $\leq_{hr}$, since $\leq_{hr}$ is not closed under mixture (see Example 1.3.6).

In the case of a discrete state space, comparability with respect to $\leq_{st}$ can be expressed in terms of the transition matrices $\mathbf{P}'$ and $\mathbf{P}''$:

$$\mathbf{P}' \mathbf{U} \leq \mathbf{P}'' \mathbf{U}.$$

Finally, there is an interesting relationship between monotonicity with respect to $\leq_{st}$ and association.

Definition 5.2.7. A transition operator $\mathbf{T}$ given by the kernel Q is said to be *associated* if $Q(x, \cdot)$ is associated for every $x \in S$, i.e.

$$(fg)_{\mathbf{T}}(x) \geq f_{\mathbf{T}}(x) g_{\mathbf{T}}(x)$$

for all $x \in S$ and all increasing $f, g : S \to \mathbb{R}$.

Theorem 5.2.8. *(Liggett (1985), Szekli (1995)) The following two statements are equivalent for a $\leq_{st}$-monotone transition operator $\mathbf{T}$ on a partially ordered state space $(S, \mathcal{S})$:*
(a) $\mathbf{T}P$ is associated whenever P is associated,
(b) $\mathbf{T}$ is associated.

Proof. (a) $\Rightarrow$ (b). For every x, the Dirac distribution δ_x is associated, so by (a) $\mathbf{T}\delta_x$ has the same property, i.e. for any increasing f and g

$$\int f(t)g(t)\mathbf{T}\delta_x(\mathrm{d}t) \geq \int f(t)\mathbf{T}\delta_x(\mathrm{d}t) \int g(t)\mathbf{T}\delta_x(\mathrm{d}t),$$

which is the same as (b).

(b) $\Rightarrow$ (a). Suppose P is associated and f and g are increasing. Since $\mathbf{T}$ is monotone, by Theorem 5.2.3 $f_\mathbf{T}$ and $g_\mathbf{T}$ are also increasing. Therefore

$$\int f_\mathbf{T}(x)g_\mathbf{T}(t)P(\mathrm{d}x) \geq \int f_\mathbf{T}(x)P(\mathrm{d}x) \int g_\mathbf{T}(x)P(\mathrm{d}x).$$

By (b)

$$\int (fg)_\mathbf{T}(x)P(\mathrm{d}x) \geq \int f_\mathbf{T}(x)g_\mathbf{T}(x)P(\mathrm{d}x).$$

Combining these two inequalities gives

$$\int f(x)g(x)\mathbf{T}P(\mathrm{d}x) \geq \int f(x)\mathbf{T}P(\mathrm{d}x) \int g(x)\mathbf{T}P(\mathrm{d}x),$$

i.e. association of $\mathbf{T}P$. □

For the more complicated continuous-time case, see Liggett (1985), p. 80.

Lindqvist (1988) showed that monotonicity with respect to $\leq_{st}$ and association of $\mathbf{T}$ are equivalent to association of the distribution $P \otimes \mathbf{T}$ on $S \times S$ given by

$$P \otimes \mathbf{T}(A_1 \times A_2) = \int \mathbf{1}_{A_1}(x)Q(x, A_2)P(\mathrm{d}x) \quad \text{for all } A_1, A_2 \in \mathcal{S}$$

for all $P \in \mathbb{P}_\mathcal{S}$. Furthermore, monotonicity with respect to $\leq_{st}$ and association of two operators $\mathbf{T}_1$ and $\mathbf{T}_2$ imply that also $\mathbf{T}_1\mathbf{T}_2$ is monotone and associated.

5.2.2 *Monotonicity and Comparability Conditions for Markov Processes*

Using the results of Section 5.2.1 now the following theorems are formulated, which concern the monotonicity and comparability of homogeneous Markov processes on a general partially ordered state space $(S, \mathcal{S})$ in discrete and continuous time, in terms of transition kernels and operators. These theorems are important tools for proving monotonicity and/or comparability properties of stochastic models. Clearly, analogous theorems can be formulated in the inhomogeneous case.

Theorem 5.2.9. (monotonicity, discrete time). *A homogeneous Markov chain (X_n) with transition kernel Q is increasing (decreasing) with respect to $\leq_{\mathcal{F}}$ if*

$$X_0 \leq_{\mathcal{F}} X_1 \ (X_1 \leq_{\mathcal{F}} X_0) \tag{5.2.7}$$

and if the operator corresponding to Q is $\leq_{\mathcal{F}}$-monotone.

Under these conditions, if the distributions P_n of the X_n converge suitably towards a stationary distribution P, and if for all n, either $P_n \leq_{\mathcal{F}} P_{n+1}$ or $P_{n+1} \leq_{\mathcal{F}} P_n$, then the distributions $P_0, P_1, P_2, \ldots$ are bounds (in the sense of the partial order $\leq_{\mathcal{F}}$) for P. For practical purposes, the quality of such a bound will depend on the choice of P_0.

Theorem 5.2.10. (monotonicity, continuous time). *A homogeneous Markov process (X_t) with the transition kernels Q_t, namely*

$$Q_t(x, B) = P(X_t \in B | X_0 = x),$$

is increasing (decreasing) with respect to $\leq_{\mathcal{F}}$ if there is a positive number ϑ such that

$$X_0 \leq_{\mathcal{F}} X_t \ (X_t \leq_{\mathcal{F}} X_0) \quad \text{for} \quad 0 \leq t \leq \vartheta \tag{5.2.8}$$

and if the operators corresponding to the Q_ts are $\leq_{\mathcal{F}}$-monotone for $0 < t < \vartheta$.

Theorem 5.2.11. (comparability, discrete time). *Two homogeneous Markov chains (X'_n) and (X''_n) with the transition kernels Q' and Q'' satisfy*

$$(X'_n) \leq_{\mathcal{F}} (X''_n) \tag{5.2.9}$$

if

$$X'_0 \leq_{\mathcal{F}} X''_0 \tag{5.2.10}$$

and if a transition kernel Q with $\leq_{\mathcal{F}}$-monotone transition operator exists such that

$$Q'(x, \cdot) \leq_{\mathcal{F}} Q(x, \cdot) \leq_{\mathcal{F}} Q''(x, \cdot) \quad \text{for all } x \in S. \tag{5.2.11}$$

An important special case of Theorem 5.2.11 is the comparison of two Markov chains with the same transition operator, but starting in two different initial states. For $\leq_{st}$ we get the following corollary.

Corollary 5.2.12. *Two homogeneous Markov chains (X'_n) and (X''_n) with the same $\leq_{st}$-monotone transition kernel Q satisfy*

$$(X'_n) \leq_{st} (X''_n) \tag{5.2.12}$$

if

$$X'_0 = x \quad \text{and } X''_0 = y \text{ almost surely} \tag{5.2.13}$$

for some $x \leq y$.

For *real-valued processes* (X'_n) and (X''_n) with the stationary distributions P' and P'' we have

$$P' \leq_{\mathcal{F}} P''$$

if there exist a $\leq_{\mathcal{F}}$-monotone Q satisfying (5.2.11) and initial distributions P'_0 and P''_0 with $P'_0 \leq_{\mathcal{F}} P''_0$, presuming that the sequences (P'_n) and (P''_n) converge weakly to P' and P'' if $\leq_{\mathcal{F}}$ is $\leq_{st}$, and, in the case that $\leq_{\mathcal{F}}$ is $\leq_{icx}$, additionally

$$\lim_{n\to\infty} EX'_n = EX' \quad \text{and} \quad \lim_{n\to\infty} EX''_n = EX''.$$

If $\leq_{\mathcal{F}}$ represents $\leq_{st}$ and S is a subset of the real line, then the condition (5.2.11) is equivalent to

$$x' \leq x'' \quad \text{implies} \quad Q'(x', \cdot) \leq_{st} Q''(x'', \cdot), \tag{5.2.14}$$

as used in Kalmykov (1962) and Daley (1968).

Perhaps a little more convenient is the equivalent condition

$$Q'(x, \cdot) \leq_{st} Q''(x + a, \cdot) \quad \text{for all} \quad x \quad \text{and all} \quad a \geq 0 \tag{5.2.15}$$

used by Kalmykov (1962). The equivalence is true since the set of all distribution functions endowed with $\leq_{st}$ is a complete lattice (see Theorem 1.10.3), and therefore Q can be chosen for example as the $\leq_{st}$-supremum of all $Q'(y, \cdot)$ with $y \leq x$, i.e. $Q(x, (t, \infty)) = \sup_{a \geq 0} Q'(x - a, (t, \infty))$ for all t in S. This Q is obviously a transition kernel of a monotone operator which satisfies (5.2.11). (The proof of Theorem 5.2.19 below shows a corresponding construction for the case of $\leq_{icx}$.)

Example 5.2.13. (Simple random walk with reflecting barriers.) Consider the simple random walk on $\mathbb{Z} = \{0, \pm 1, \pm 2, \dots\}$ with transition probabilities

$$p_{ii+1} = p \quad \text{and} \quad p_{ii-1} = 1 - p$$

for $0 < p < 1$. Let X_n be the position of the walking object at time n and let (Z_n) be a sequence of independent random variables with

$$P(Z_n = 1) = p \quad \text{and} \quad P(Z_n = -1) = 1 - p.$$

Then

$$X_{n+1} = X_n + Z_n \quad \text{for} \quad n = 0, 1, 2, \dots$$

Obviously, the Markov chain (X_n) is monotone with respect to $\leq_{st}$, $\leq_{icx}$ and many other stochastic orders. Also comparability is given:

$$X'_0 \leq_{\mathcal{F}} X''_0 \quad \text{and} \quad p' \leq p'' \quad \text{imply} \quad X'_n \leq_{\mathcal{F}} X''_n \quad \text{for all } n$$

for $\leq_{\mathcal{F}}$ being $\leq_{st}$, $\leq_{icx}$ and other relations.

Introduce now a reflecting barrier at 0. Then

$$p_{ii+1} = p \quad \text{for all} \quad i \geq 0, \quad p_{ii-1} = 1 - p \quad \text{for all} \quad i > 0 \quad \text{and} \quad p_{00} = 1 - p.$$

The corresponding Markov chain (X_n) satisfies

$$X_{n+1} = \max\left\{0, X_n + Z_n\right\} \quad \text{for} \quad n = 0, 1, 2, \ldots$$

This chain still has the monotonicity and comparability properties as in the unbounded case for the orders $\leq_{st}$ and $\leq_{icx}$.

Finally, introduce now also a reflecting barrier at m such that

$$p_{ii+1} = p \qquad \text{for} \qquad 0 \leq i < m \quad \text{and} \quad p_{mm} = p$$
$$p_{ii-1} = 1 - p \qquad \text{for} \qquad 0 < i \leq m \quad \text{and} \quad p_{00} = 1 - p.$$

Then the corresponding Markov chain (X_n) satisfies

$$X_{n+1} = \min\left\{m, \max\{0, X_n + Z_n\}\right\} \quad \text{for} \quad n = 0, 1, 2, \ldots$$

This chain has the monotonicity and comparability properties as above only for $\leq_{st}$.

If the transition probabilities $p_{ii+1} = \alpha_i$ depend on i ($p_{ii-1} = 1 - \alpha_i$), then the corresponding random walk is monotone with respect to $\leq_{st}$, if the sequence (α_i) is increasing in i. If $\alpha_i > \alpha_{i+1}$ for some i, however, then $Q(i, \cdot) \not\leq_{st} Q(i+1, \cdot)$ and the random walk is not monotone.

Theorem 5.2.14. (comparability, continuous time).   *Two homogeneous Markov processes (X'_t) and (X''_t) with the transition kernels Q'_t and Q''_t satisfy*

$$(X'_t) \leq_{\mathcal{F}} (X''_t) \tag{5.2.16}$$

if

$$X'_0 \leq_{\mathcal{F}} X''_0 \tag{5.2.17}$$

and if a family (Q_t) of $\leq_{\mathcal{F}}$-monotone transition kernels exists such that

$$Q'_t(x, \cdot) \leq_{\mathcal{F}} Q_t(x, \cdot) \leq_{\mathcal{F}} Q''_t(x, \cdot) \quad \text{for all} \quad x \quad \text{and all} \quad t > 0. \tag{5.2.18}$$

For the comparability of stationary distributions it is enough to have the same conditions as suffice in the case of discrete time chains. In the case of real-valued processes, the sufficient conditions on the transition distribution functions for monotonicity (5.2.11) and (5.2.18) to hold (and for them to hold for $\leq_{st}$ for a Polish state space S with a closed partial order) are also necessary when they hold for all initial conditions satisfying (5.2.10) or (5.2.17) respectively.

For the relation $\leq_{st}$ it is possible to prove strong and complete comparability, thereby extending Theorems 5.2.11 and 5.2.14 in an essential way.

Theorem 5.2.15. (Stoyan (1972a), Franken and Stoyan (1975)) *The homogeneous Markov processes (X_t') and (X_t'') are completely (and hence strongly) comparable with respect to $\leq_{st}$ if the conditions of Theorems 5.2.11 or 5.2.14 are true for $\leq_{st}$.*

Proof. (Discrete time). We show by induction that

$$(X_0', \ldots, X_n') \leq_{st} (X_0'', \ldots, X_n''). \tag{5.2.19}$$

Obviously, (5.2.19) is true for $n = 0$. Assume that (5.2.19) is true for some $n - 1 \geq 0$ and show the validity for n. Let f be an arbitrary increasing functional on S^n for which the following integrals exist. Then

$$\int f(x_0, \ldots, x_n) \mathcal{P}_n'(dx_0, \ldots, dx_n)$$

$$= \int \int f(x_0, \ldots, x_n) Q'(x_{n-1}, dx_n) \mathcal{P}_{n-1}'(dx_0, \ldots, dx_{n-1})$$

and the same for $\mathcal{P}_n''$, where $\mathcal{P}_k'$ and $\mathcal{P}_k''$ are the joint distributions of $(X_0', \ldots, X_k')$ and $(X_0'', \ldots, X_k'')$, respectively, for $k = 0, 1, \ldots$. Let f_n' and f_n'' denote the functionals

$$f_n'(x_0, \ldots, x_{n-1}) = \int f(x_0, \ldots, x_n) Q'(x_{n-1}, dx_n),$$

and analogously for f_n'', and set

$$f_n(x_0, \ldots, x_{n-1}) = \int f(x_0, \ldots, x_n) Q(x_{n-1}, dx_n),$$

where Q is the intermediate kernel satisfying the analogue of (5.2.11). Because of the assumption on Q, f_n is increasing, and because of (5.2.11) it is

$$f_n'(x_0, \ldots, x_{n-1}) \leq f_n(x_0, \ldots, x_{n-1}) \leq f_n''(x_0, \ldots, x_{n-1})$$

for all $(x_0, \ldots, x_{n-1})$. Then the induction hypothesis yields

$$\int f d\mathcal{P}_n' = \int f_n' d\mathcal{P}_{n-1}' \leq \int f_n d\mathcal{P}_{n-1}'$$

$$\leq \int f_n d\mathcal{P}_{n-1}'' \leq \int f_n'' d\mathcal{P}_{n-1}'' = \int f d\mathcal{P}_n'',$$

i.e. (5.2.19) is true also for n. The complete comparability follows from Theorem 5.1.4. $\square$

The proof of Theorem 3.3.7 shows that Theorem 5.2.15 can be extended to some non-Markov processes. But an analogous theorem for $\leq_{icx}$ is not true; see Stoyan (1972a).

It is also not true that a homogeneous Markov process (X_t) satisfying the conditions of Theorems 5.2.9 and 5.2.10 is strongly increasing with respect to $\leq_{st}$. (Theorem 4.2.6 in Stoyan (1983) is wrong. If it were correct, then in (5.2.34) it had to be $\leq_{uo}$ and not $\geq_{uo}$.) But, Theorem 5.2.15 can be used to show for such a process

$$(X_{t_1}, \ldots, X_{t_m}) \leq_{st} (X_{t_1+k}, \ldots, X_{t_m+k}) \tag{5.2.20}$$

for any positive k.

The same comparison results as above can be obtained also by means of variants of the coupling method, see Lindvall (1992) and Kamae et al. (1977). Comparison and monotonicity of transition operators ensure the existence of order preserving couplings. It is even possible to construct the coupling $(\hat{X}_n, \hat{Y}_n)$ so that it is a Markov chain on $S \times S$. López, Martinez, and Sanz (2000) show that this is also possible for continuous time Markov processes on partially ordered countable sets.

In the rest of this subsection comparability of Markov chains with respect to supermodular order will be considered. Assume that (X'_n) and (X''_n) are two Markov chains with transition kernels Q' and Q''. For supermodular order it is necessary that the two Markov chains have the same marginals, i.e. $X'_n =_{st} X''_n$ (see Theorem 3.9.5 a). Therefore it will be necessary to assume that both Markov chains have the same stationary (initial) distribution π. For the proof of Theorem 5.2.18 below the following two lemmata are needed.

Lemma 5.2.16. *If* $f : \mathbb{R}^{n+1} \to \mathbb{R}$ *is supermodular, then*

$$g(x_1, \ldots, x_n) = f(x_1, \ldots, x_n, x_n)$$

is also supermodular.

Proof. Since $\Delta_i^\varepsilon \Delta_j^\delta g = \Delta_i^\varepsilon \Delta_j^\delta f \geq 0$ for $1 \leq i < j \leq n - 1$, it is sufficient to show $\Delta_i^\varepsilon \Delta_n^\delta g \geq 0$. However, it is easy to see that

$$\Delta_i^\varepsilon \Delta_n^\delta g(\mathbf{x}) = \Delta_i^\varepsilon \Delta_n^\delta f(x_1, \ldots, x_n, x_n + \delta) + \Delta_i^\varepsilon \Delta_{n+1}^\delta f(x_1, \ldots, x_n, x_n) \geq 0.$$

$\square$

Lemma 5.2.17. *Assume that* $f : \mathbb{R}^{n+1} \to \mathbb{R}$ *is supermodular and that the transition kernel* Q *is* $\leq_{st}$-*monotone. Then the function*

$$g(x_1, \ldots, x_n) = \int f(x_1, \ldots, x_n, x_{n+1}) Q(x_n, dx_{n+1})$$

is also supermodular.

Proof. As in Lemma 5.2.16 it is sufficient to show $\Delta_i^\varepsilon \Delta_n^\delta g \geq 0$. Since $\Delta_i^\varepsilon f$ is increasing in all variables $x_j \neq x_i$, the $\leq_{st}$-monotonicity of Q implies

$$
\begin{aligned}
\Delta_i^\varepsilon \Delta_n^\delta g(\mathbf{x}) &= \int \Delta_i^\varepsilon f(x_1, \ldots, x_n + \delta, x_{n+1}) Q(x_n + \delta, \mathrm{d}x_{n+1}) \\
&\quad - \int \Delta_i^\varepsilon f(x_1, \ldots, x_n, x_{n+1}) Q(x_n, \mathrm{d}x_{n+1}) \\
&\geq \int \Delta_i^\varepsilon f(x_1, \ldots, x_n, x_{n+1}) Q(x_n + \delta, \mathrm{d}x_{n+1}) \\
&\quad - \int \Delta_i^\varepsilon f(x_1, \ldots, x_n, x_{n+1}) Q(x_n, \mathrm{d}x_{n+1}) \\
&\geq 0.
\end{aligned}
$$

$\square$

In Theorem 5.2.18 below the following assumption is needed: the transition kernel Q as well as the transition kernel Q_r of the time reversion, i.e. $Q_r(s, A) = P(X_1 \in A | X_2 = s)$ are stochastically monotone. The theorem is a generalization of lemma 1 in Bäuerle and Rolski (1998). They prove their result by a coupling method, where the finiteness of the state space is crucial, whereas the functional method used here works for arbitrary state spaces $S \subset \mathbb{R}$. For transition probability matrices with more special structures the result has already been obtained earlier by Bäuerle (1997b) and Szekli et al. (1994).

Theorem 5.2.18. *Let Q be a $\leq_{st}$-monotone transition kernel with stationary distribution π and assume that the transition kernel Q_r of the time reversed process is also $\leq_{st}$-monotone. Moreover, let $0 \leq c' \leq c'' \leq 1$ and assume that (X_n') and (X_n'') are the Markov chains with transition kernels $Q' = c'Q + (1-c')I$ and $Q'' = c''Q + (1-c'')I$, respectively, where $I(s, \cdot) = \delta_s$, $s \in S$, is the identity. Then (X_n') and (X_n'') both have the stationary distribution π and*

$$
(X_0', \ldots, X_n') \geq_{sm} (X_0'', \ldots, X_n'') \quad \text{for all } n.
$$

Proof. We proceed by induction on n. For $n = 0$ there is nothing to show. Hence let us assume

$$
(X_0', \ldots, X_n') \geq_{sm} (X_0'', \ldots, X_n'').
$$

and let $\mathcal{P}_n'$ and $\mathcal{P}_n''$ be the corresponding $(n+1)$-dimensional distributions. Assume that $f : \mathbb{R}^{n+2} \to \mathbb{R}$ is a supermodular function. According to Lemma 5.2.16 and 5.2.17 the functions

$$
g(x_0, \ldots, x_n) = f(x_0, \ldots, x_n, x_n) = \int f(x_0, \ldots, x_n, x_{n+1}) I(x_n, \mathrm{d}x_{n+1})
$$

and

$$h(x_0,\dots,x_n) = \int f(x_0,\dots,x_n,x_{n+1})Q(x_n,\mathrm{d}x_{n+1})$$

are then again supermodular. Therefore

$$
\begin{aligned}
Ef(X_0',\dots,X_{n+1}') &= c'\int h\,\mathrm{d}\mathcal{P}_n' + (1-c')\int g\,\mathrm{d}\mathcal{P}_n' \\
&\geq c'\int h\,\mathrm{d}\mathcal{P}_n'' + (1-c')\int g\,\mathrm{d}\mathcal{P}_n''.
\end{aligned}
$$

Since

$$Ef(X_0'',\dots,X_{n+1}'') = c''\int h\,\mathrm{d}\mathcal{P}_n'' + (1-c'')\int g\,\mathrm{d}\mathcal{P}_n''$$

it remains to show that

$$\int h\,\mathrm{d}\mathcal{P}_n'' - \int g\,\mathrm{d}\mathcal{P}_n'' \leq 0 \tag{5.2.21}$$

To this end let $Q_r'' = c''Q_r + (1-c'')I$ be the transition kernel of the time reversion of (X_n''). Then Q_r'' is $\leq_{st}$-increasing and hence from Lemma 5.2.17 it follows by induction that

$$f^*(x_n,x_{n+1}) = \int\cdots\int f(x_0,\dots,x_{n+1})Q_r''(x_1,\mathrm{d}x_0)\dots Q_r''(x_n,\mathrm{d}x_{n-1})$$

is supermodular. But according to Lorentz's inequality (which says that $\leq_{sm}$ has property (P5), see Theorem 3.9.8) it holds for any random variables $Z_1, Z_2 \sim \pi$ that

$$(Z_1,Z_2) \leq_{sm} (Z_1,Z_1).$$

Hence in particular $\pi \otimes Q_r \leq_{sm} \pi \otimes I$ and therefore

$$\int h\,\mathrm{d}\mathcal{P}_n'' - \int g\,\mathrm{d}\mathcal{P}_n'' = \int f^*\,\mathrm{d}(\pi \otimes Q_r) - \int f^*\,\mathrm{d}(\pi \otimes I) \leq 0.$$

$\square$

5.2.3 Homogeneous Markov Processes with Discrete State Space

It is often hard to verify the conditions of Theorems 5.2.10 and 5.2.14 because they relate to not merely one transition kernel or distribution function but families and because the transition kernels of Markov processes with continuous time are only seldom known. It is therefore desirable to have conditions on the infinitesimal operators of the processes which ensure monotonicity and comparability. For the case of a discrete state space $S = \{0,1,\dots\}$ such conditions, i.e. conditions on the transition rates, are known, as it will be shown below. Note that the case of diffusion processes is considered in Skorokhod (1965), Anderson (1972), Yamada (1973), O'Brien (1980), Ikeda and Watanabe (1981), Day (1983), Huang (1984), Rogers and Williams (1987), and Lindvall (1992), pp. 216–218. Such processes are monotone with respect to $\leq_{st}$ for suitable initial distributions; comparison with respect to $\leq_{st}$ is possible for comparable drift functions and fixed dispersion functions. Under some conditions also comparison with respect to $\leq_{icx}$ is possible; then the dispersion functions can be different.

For a homogeneous Markov process with the state space $S = \{0, 1, 2, \ldots\}$ let $p_{ij}(t)$ denote the transition probabilities, i.e.

$$p_{ij}(t) = P(X_{\tau+t} = j | X_\tau = i) \quad \text{for} \quad 0 \leq t, \ \tau < \infty \ \text{ and } \ i, j = 0, 1, \ldots$$

where $p_{ii}(0) = 1$ for all i and $p_{ij}(0) = 0$ for all $i \neq j$. The transition rates $q_{..}$ are defined by

$$q_{ii} = \lim_{t \to 0}(p_{ii}(t) - 1)/t$$

and

$$q_{ij} = \lim_{t \to 0} p_{ij}(t)/t.$$

Assume that the process is conservative, i.e.

$$-q_{ii} = \sum_{j \neq i} q_{ij} < \infty \quad \text{for all } i.$$

Then the transition probabilities $p_{ij}(t)$ satisfy the Kolmogorov backward equations; see Anderson (1991), Kijima (1997) or Yin and Zhang (1998), a system of linear differential equations for the $p_{ij}(t)$ in which the q_{ij} are coefficients. Assume furthermore that the solution of the system with the initial conditions $p_{ii}(0) = 1$ and $p_{ij}(0) = 0$ $(i \neq j)$ is unique. The following theorem gives sufficient conditions for the comparison of such Markov processes with respect to $\leq_{st}$ and $\leq_{icx}$.

Theorem 5.2.19. (Kirstein (1976)) *Let (X_t) and (Y_t) be homogeneous Markov process on the state space $S = \{0, 1, \ldots\}$ with transition rates $q_{ij}^{(x)}$ and $q_{ij}^{(y)}$ respectively and satisfying $X_0 \leq_{\mathcal{F}} Y_0$. It holds that*

$$(X_t) \leq_{\mathcal{F}} (Y_t) \tag{5.2.22}$$

and, additionally, for $\leq_{st}$

$$(X_t) \ str\text{-}\leq_{st} (Y_t) \tag{5.2.23}$$

as well as complete comparability, if the following conditions are satisfied.

(a) If $\leq_{\mathcal{F}}$ denotes $\leq_{st}$,

$$\sum_{k \geq \ell} q_{ik}^{(x)} \leq \sum_{k \geq \ell} q_{jk}^{(y)} \qquad (5.2.24)$$

for all $i \leq j$ and for all ℓ which satisfy $\ell \leq i$ or $\ell > j$.

(b) If $\leq_{\mathcal{F}}$ denotes $\leq_{icx}$, writing $\pi_z(k|i) = \sum_{m>k}(m-k)q_{im}^{(z)}$ for $z = x$ and y,

$$\pi_x(k|i) \leq \pi_y(k|j) \qquad (5.2.25)$$

for all $i \leq j \leq k$ and for all $i = j > k$ and

$$\pi_x(k|\ell) \leq \left[\frac{\ell - i}{j - i}\pi_y(k|j) + \frac{j - \ell}{j - i}\pi_y(k|i)\right] \qquad (5.2.26)$$

for all $k \leq i < \ell < j$ if $k < \ell$, and all $i < \ell < j \leq k$ if $k > \ell$ and

$$\sup_{i \in S}(-q_{ii}^{(z)}) < \infty \quad \text{for } z = x \text{ and } y. \qquad (5.2.27)$$

Proof. The proof of the continuous time result (5.2.16) is reduced to checking that two corresponding coupled embedded discrete time chains are similarly ordered. We start by recalling that any continuous time Markov chain (X_t) on S with uniformly bounded transition rates q_{ij} with $\sup_{i \in S}|q_{ii}| \leq \lambda < \infty$ has its matrix of transition probabilities $(p_{ij}(t))$ expressible as

$$\left(p_{ij}(t)\right) = e^{-\lambda t}\sum_{n=0}^{\infty}\left(\frac{(\lambda t)^n}{n!}\,(p_{ij,n})\right).$$

Here the $p_{ij,n}$ are the n-step transition probabilities of a discrete time Markov chain, (X_n^d) say, with transition probabilities

$$p_{ij,1} = p_{ij} = \begin{cases} q_{ij}/\lambda & \text{for } i \neq j, \\ 1 - |q_{ii}|/\lambda & \text{for } i = j. \end{cases}$$

$((X_t)$ is called a 'subordinated Markov process' with respect to (X_n^d), Kijima (1997).) The sample paths of (X_t) can be constructed by starting from sample paths of (X_n^d) for which X_0^d and X_0 have the same distribution, and identifying X_t with X_n^d for those t for which $T_n < t \leq T_{n+1}, T_0 \equiv 0$ and $\{T_n : n = 1, 2, \dots\}$ being the successive epochs of an independent Poisson process on $(0, \infty)$ with intensity λ. For (Y_t) an analogous construction is made and the corresponding characteristics obtain indices x and y.

The mixture property (MI) of $\leq_{\mathcal{F}}$ ensures that, if the discrete time chains (X_n^d) and (Y_n^d) satisfy $(X_n^d) \leq_{\mathcal{F}} (Y_n^d)$, then the Poisson mixtures of their distributions are comparable, i.e. $(X_t) \leq_{\mathcal{F}} (Y_t)$.

(a) We show now by means of Theorems 5.2.11 and relation (5.2.14) that $(X_n^d) \leq_{st} (Y_n^d)$. The corresponding transition probabilities satisfy

$$\sum_{k \geq \ell} p_{ik}^{(x)} \leq \sum_{k \geq \ell} p_{jk}^{(y)} \qquad (5.2.28)$$

for all $i \leq j$ and all ℓ since the expression on the left-hand side here equals

$$\sum_{k \geq l} p_{ik}^{(x)} = \begin{cases} \sum_{k \geq \ell} q_{ik}^{(x)}/\lambda & \text{for} \quad \ell > i, \\ 1 - \sum_{k \geq \ell} q_{ik}^{(x)}/\lambda & \text{for} \quad \ell \leq i, \end{cases}$$

and a similar relation holds for the right-hand side of (5.2.28).

Since λ can be taken arbitrarily large, the cases $i < \ell \leq j$ can be excluded.

(b) Also in the case of $\leq_{icx}$ Theorem 5.2.11 is used. A $\leq_{icx}$-monotone transition kernel p has to be found which satisfies $p^{(x)}(i, \cdot) \leq_{icx} p(i, \cdot) \leq_{icx} p^{(y)}(i, \cdot)$. This means that the corresponding integrated survival function $\pi(\cdot|i)$ must be smaller than all $\pi_y(\cdot|j)$ with $j > i$, and hence, according to the Theorems 1.10.2 and 1.10.3, has to be smaller than the convex hull of all these functions. Moreover the $\leq_{icx}$-monotonicity requires that $\pi(\cdot|\cdot)$ must be convex in the second argument, too. The $\pi(\cdot|\cdot)$ can be constructed if for all k and all $i \leq j$

$$E((X_1^d - k)_+|X_0^d = i) \leq E((Y_1^d - k)_+|Y_0^d = j)$$

or

$$\pi_x(k|i) \leq \pi_y(k|j)$$

and if for all $i < j$ and $i < l < j$

$$\pi_x(k|\ell) \leq \frac{\ell - i}{j - i}\pi_y(k|j) + \frac{j - \ell}{j - i}\pi_y(k|i).$$

It is then given by

$$\pi(k|\ell) = \inf_{m \geq \ell} \inf \left\{ \frac{s - m}{s - r}\pi_y(k|r) + \frac{m - r}{s - r}\pi_y(k|s) \right\}$$

where the inner infimum is taken over all r and s satisfying $r \leq m$ and $s > m$.

Expressing the $\pi_z(k|\ell)$ in terms of the intensity rates and omitting the cases which can be eliminated for large λ yields the two conditions (5.2.25) and (5.2.26).

In order to remove the constraint (5.2.27) (that the chain is 'uniformizable', Keilson and Kester (1977)) in the case of $\leq_{st}$, consider in place of the Markov process (X_t) the process (X_t^a) on state space $\{0, \ldots, a\}$ with $X_0^a = \min\{a, X_0\}$ and with transition rates (q_{ij}^a) defined by

$$
q_{ij}^a = \begin{cases}
q_{ij} & \text{for} \quad i, j < a, \\
\sum_{k \geq a} q_{ik} & \text{for} \quad i < a, \ j = a, \\
0 & \text{for} \quad i = a \geq j,
\end{cases}
$$

so that a is an absorbing state for (X_t^a). For every fixed finite t, the transition probabilities of the conservative processes satisfy $p_{ij}^a(t) \to p_{ij}(t)$ (for $a \to \infty$), and consequently, for $0 \leq t \leq T < \infty$ where T is any fixed finite quantity, (X_t^a) converges in distribution towards (X_t) for $a \to \infty$. Using a similar construction for (Y_t), the conditions (5.2.24) for $0 \leq i \leq j < a$ suffice to ensure that $(X_t^a) \leq_{st} (Y_t^a)$ because (5.2.27) holds for finite a. Letting $a \to \infty$ the conclusion still holds, but without requiring (5.2.27) to be satisfied.

The proof of (5.2.23) is similar to that of Theorem 5.2.15.   □

Note that if (5.2.16) holds for all comparable initial distributions, then the conditions (5.2.24) to (5.2.26) are necessarily satisfied; see Kirstein (1976).

For $q_{ij}^{(x)} = q_{ij}^{(y)}$ Theorem 5.2.19 yields monotonicity conditions for the operators $\mathbf{T}_t$ defined by the transition matrices $(p_{ij}(t))$. In particular, for $\leq_{st}$ the following theorem holds.

Theorem 5.2.20. *All transition operators $\mathbf{T}_t$ of a homogeneous Markov process on $\{0, 1, \ldots\}$ with transition rates q_{ij} are monotone with respect to $\leq_{st}$ if and only if*

$$
\sum_{k \geq \ell} q_{ik} \leq \sum_{k \geq \ell} q_{i+1,k} \tag{5.2.29}
$$

for all i and all $\ell \neq i+1$. This monotonicity is always given if the process has only two states.

For homogeneous birth-and-death processes, the transition rates are

$$
q_{ij} = \begin{cases}
\lambda_i & \text{for} & j = i+1, \\
\mu_i & \text{for} & j = i-1, \\
-(\lambda_i + \mu_i) & \text{for} & i = j, \\
0 & \text{otherwise.}
\end{cases}
$$

Assume $\mu_0 = 0$ and, if the state space is finite so that $S = \{0, \ldots, n\}$ say, $\lambda_n = 0$. The following result is an easy consequence of Theorems 5.2.20 and 5.2.19.

Theorem 5.2.21. *a) Every homogeneous birth-and-death process satisfies (5.2.29), so that its transition operator is $\leq_{st}$-monotone. If such a process starts with $X = 0$, it is increasing with respect to $\leq_{st}$.*

b) For homogeneous birth-and-death processes (X_t) and (Y_t) with the transition rates $\lambda_i^{(x)}, \lambda_i^{(y)}$ and $\mu_i^{(x)}, \mu_i^{(y)}$, $X_0 \leq_{st} Y_0$ implies $(X_t) \leq_{st} (Y_t)$ and (X_t) str-$\leq_{st} (Y_t)$ if

$$\mu_i^{(x)} \geq \mu_i^{(y)} \quad and \quad \lambda_i^{(x)} \leq \lambda_i^{(y)} \quad for \quad i = 0, 1, \ldots,$$

and the processes are then completely comparable.

Lindvall (1992) gives proofs of these theorems by means of the coupling method; see also van Doorn (1981). He also studies so-called *general birth-and-death processes*, which can be applied for the investigation of queueing networks with exponential service times and Poisson inputs; see also Szekli (1995), section 2.12.

Keilson and Kester (1977) and van Doorn (1981) give conditions for the monotonicity of birth-and-death processes in terms of the transition rates λ_i and μ_i and the initial distribution (p_i), $p_i = P(X_0 = i)$. As stated above, a birth-and-death process is $\leq_{st}$-monotone if $X_0 = 0$. But this is also true if

$$\mu_{i+1}p_{i+1} - \lambda_i p_i \leq 0 \quad \text{for all } i \tag{5.2.30}$$

and not all left-hand sides in (5.2.30) vanish. Notice that it is trivial that any pure birth (death) process is increasing (decreasing) with respect to $\leq_{st}$.

Theorem 5.2.19 leads also to monotonicity properties of the transition operators $\mathbf{T}_t$ of pure birth or death processes with respect to $\leq_{icx}$. This monotonicity is given for a pure birth process if the λ_i form a convex sequence, i.e. if

$$\lambda_\ell \leq \frac{\ell - i}{j - i}\lambda_j + \frac{j - \ell}{j - i}\lambda_i$$

for $i < \ell < j$. In the case of a pure death process, concavity of the sequence $\{\mu_i\}$ ensures monotonicity with respect to $\leq_{icx}$. Thus under these conditions two pure birth or death processes with the same transition mechanism can be compared with respect to $\leq_{icx}$.

For a birth-and-death process (X_t) the mean value function EX_t has interesting properties as shown by van Doorn (1981) and Lindvall (1992). It is convex for a pure birth process with $\lambda_0 \leq \lambda_1 \leq \ldots$ and for every initial distribution and concave and increasing for a birth-and-death process with $\lambda_0 \geq \lambda_1 \geq \ldots$ and $\mu_1 \leq \mu_2 \leq \ldots$ starting in 0. See also Kella and Sverchkov (1994) for the concavity of means of processes with stationary increments.

Relationships analogous to those of Theorem 5.2.21 are true also in the inhomogeneous case; see Massey (1987). Massey (1985) considered the particular case of the time-dependent queue $M/M/1$ and showed that

$$\lambda'(t) \leq \lambda''(t) \quad \text{and} \quad \mu'(t) \geq \mu''(t) \quad \text{for all } t$$

imply comparison of queue lengths with respect to $\leq_{st}$. Karpelevich and Rybko (2000) use this relation in the context of infinite dimensional systems. A forerunner of such work is Gaede (1965).

Glasserman and Yao (1994), section 5.4, prove monotonicity results for continuous-time Markov chains that describe generalized semi-Markov processes where all clock times are mutually independent and exponentially distributed.

Kijima (1998) characterizes monotonicity with respect to $\leq_{hr}$ and $\leq_{rh}$. He shows that a continuous-time Markov chain is a birth-and-death process iff the operators $\mathbf{T}_t$ are monotone with respect to both $\leq_{hr}$ and $\leq_{rh}$. Cover and Thomas (1991) contains some monotonicity properties with respect to entropy for Markov chains and other stochastic processes.

Lund, Meyn, and Tweedie[l] (1996) and Lund and Tweedie (1996) derive bounds for the exponential convergence rates of monotone Markov chains and processes (with discrete and continuous time).

5.2.4 *Monotonicity Properties of Second Order Characteristics of Markov Chains*

Since Daley (1968) it has been known that the covariances of (stationary) monotone real-valued Markov chains have monotonicity properties. This section will present some monotonicity properties for second order characteristics of Markov chains.

Let (X_n) be a $\leq_{st}$-monotone homogeneous Markov chain, i.e. the X_n are increasing with respect to $\leq_{st}$ and the transition operator is $\leq_{st}$-monotone. The state space is $(S, \mathcal{S})$, where S is equipped with the partial order $\leq$. Then

$$E(f(X_n)g(X_{n+k})) \leq E(f(X_m)g(X_{m+k})) \text{ for } n \leq m \text{ and } k = 1, 2, \ldots (5.2.31)$$

for any non-negative increasing f and g. This is an easy consequence of (5.2.20) since any mapping h given by $h(x, y) = f(x)g(y)$ is increasing for non-negative increasing f and g.

Inequality 5.2.31 is perhaps not very surprising since it fits to the general monotonicity behavior of the chain. A more interesting inequality is

$$E(f(X_n)g(X_{n+k})) \geq E(f(X_n)g(X_{n+k+1})) \tag{5.2.32}$$

for f and g as above, i.e. the means of the products are decreasing with increasing time lag. In the stationary case, (5.2.32) leads to decreasing covariances.

Consider now a real-valued Markov chain (X_n) which is *decreasing* with respect to the stochastic order $\leq_{st}$ and has monotone transition distribution functions $F(y|x) = P(X_{n+1} \leq y | X_n = x)$, i.e. $x_1 \leq x_2$ implies $F(y|x_1) \geq F(y|x_2)$. Assume (without loss of generality; see Section 4.1) that it is given in the form

$$X_{n+1} = \varphi(X_n, U_n) \tag{5.2.33}$$

where the U_n are independent and uniformly distributed on $(0,1)$ and φ is monotone in the first variable, i.e. $x_1 \leq x_2$ implies $\varphi(x_1, u) \leq \varphi(x_2, u)$ for all u.

Theorem 5.2.22. *(Bergmann and Stoyan (1978)) Under the conditions above, for any non-negative n and k*

$$(X_n, X_{n+k}) \geq_{uo} (X_n, X_{n+k+1}). \tag{5.2.34}$$

In particular, for all non-negative increasing functions f and g

$$E(f(X_n)g(X_{n+k})) \geq E(f(X_n)g(X_{n+k+1})). \tag{5.2.35}$$

Proof. The proof goes by induction on k. Let $C(F_n, F_{n+1})$ be the distribution function of (X_n, X_{n+1}), where C is the copula and F_n and F_{n+1} are the marginal distribution functions. According to the assumptions $F_n \geq_{st} F_{n+1}$. Therefore it follows from Theorem 3.3.8 and Theorem 3.8.2 that

$$C(F_n, F_{n+1}) \leq_{st} C(F_n, F_n) \leq_c C^+(F_n, F_n),$$

where as usual C^+ denotes the copula of the upper Fréchet bound. Since $\leq_{st}$ as well as $\leq_c$ imply $\leq_{uo}$ this implies $(X_n, X_{n+1}) \leq_{uo} (X_n, X_n)$, i.e. the assertion for $k = 0$. For the induction step from k to $k + 1$ notice that due to the monotonicity of φ it follows from Theorem 3.3.18 that

$$(X_n, X_{n+k}) \geq_{uo} (X_n, X_{n+k+1})$$

implies

$$(X_n, \varphi(X_{n+k}, u)) \geq_{uo} (X_n, \varphi(X_{n+k+1}, u))$$

for all u, and the property (MI) of $\leq_{uo}$ yields

$$(X_n, X_{n+k+1}) = (X_n, \varphi(X_{n+k}, U)) \geq_{uo} (X_n, \varphi(X_{n+k+1}, U)) = (X_n, X_{n+k+2}).$$

Equation (5.2.35) then follows from Theorem 3.3.16. $\qquad\square$

When the Markov chain is stationary, i.e. all X_n have the same marginal distribution, then the condition that the process is decreasing is automatically fulfilled, and $\leq_{uo}$ can be replaced by the stronger concordance or supermodular order, which are equivalent in this case (see Theorem 3.8.2). This yields the following corollary which essentially can be traced back to Daley (1968).

Corollary 5.2.23. *Let (X_n) be a stationary Markov chain with $\leq_{st}$-monotone transition kernel. Then it is for any non-negative n and k*

$$(X_n, X_{n+k}) \geq_{sm} (X_n, X_{n+k+1}) \quad for\ n = 0, 1, \ldots$$

and, in particular, for all increasing functions f

$$\mathrm{Cov}(f(X_0), f(X_n)) \geq \mathrm{Cov}(f(X_0), f(X_{n+1})) \quad for\ n = 0, 1, \ldots. \tag{5.2.36}$$

Note also that it is possible to show (5.2.35) for a real-valued Markov chain of the form (5.2.33) which is increasing with respect to $\leq_{icx}$ and has a transition operator which is monotone with respect to $\leq_{icx}$. Then the functions f and g have to be increasing, convex and non-negative, see Bergmann and Stoyan (1978).

More can be obtained for stationary reversible real-valued Markov chains (also for the case of continuous time). Recall that a Markov chain with initial distribution π and transition kernel Q is *reversible* if the 'detailed balance relation'

$$\pi(\mathrm{d}x)Q(x,\mathrm{d}y) = \pi(\mathrm{d}y)Q(y,\mathrm{d}x)$$

is satisfied; π is then invariant for Q. Hu and Joe (1995) proved that for such a chain association of (X_0, X_1) implies (5.2.36). Hu and Pan (2000) showed for a stationary chain (which is not necessarily reversible) that for $\mathbf{X_r} = (X_{r_1}, \dots, X_{r_m})$ and $\mathbf{X_s} = (X_{s_1}, \dots, X_{s_m})$

$$\mathbf{X_r} \geq_{sm} \mathbf{X_s} \tag{5.2.37}$$

holds for all $\mathbf{r}$ and $\mathbf{s}$ with

$$(r_1, r_2 - r_1, \dots, r_m - r_{m-1}) \leq (s_1, s_2 - s_1, \dots, s_m - s_{m-1}),$$

in the case of $\leq_{st}$-monotonicity of Q and of the transition kernel Q_r of the time reversion as introduced on p. 191, i.e. the process is decreasing in $\leq_{sm}$ with increasing lag (but not strongly decreasing). These authors proved also for stationary Markov chains (X'_n) and (X''_n) with the same stationary distribution

$$(X'_0, \dots, X'_m) \leq_{sm} (X''_0, \dots, X''_m) \quad \text{for all } m \geq 1 \tag{5.2.38}$$

if Q' and Q''_r (or Q'_r and Q'') are $\leq_{st}$-monotone and $(X'_0, X'_1) \leq_{sm} (X''_0, X''_1)$, which is an extension of Theorem 5.2.18.

In the context of Markov chain Monte Carlo the following problem has arisen. Consider a Markov chain (X_n) with state space $(S, \mathcal{S})$ and transition kernel Q which is reversible with respect to the distribution π, and let $f : S \to \mathbb{R}$ be an arbitrary function. Consider the sample average

$$\overline{X} = \frac{1}{n} \sum_{i=1}^{n} f(X_i).$$

Its variance is $\gamma_0 + \frac{2}{n} \sum_{i=1}^{n-1} (n-i)\gamma_i$, and for $n \to \infty$ the limit

$$v(f, Q) = \gamma_0 + 2 \sum_{i=1}^{\infty} \gamma_i$$

is obtained, where

$$\gamma_i = \mathrm{Cov}(f(X_0), f(X_i)) \quad \text{for } i = 0, 1, \dots.$$

Of course, it is assumed that Q and f are such that $v(f, Q)$ is finite.

Consider now two such transition kernels Q' (of (X'_n)) and Q'' (of (X''_n)) with the same π. The problem is to give sufficient conditions which ensure

$$v(f, Q') \le v(f, Q'') \tag{5.2.39}$$

for all f. Peskun (1973) found the condition 'Q' dominates Q'' off the diagonal', i.e. $Q'(x, A \setminus \{x\}) \ge Q''(x, A \setminus \{x\})$ for all x and all $A \in \mathcal{S}$. This condition means something like Q' mixes more than Q''. (Note the relationship to Theorem 5.2.18.) Thus use of Q' is more efficient than that of Q'' for determining $Ef(X_0)$ by Monte Carlo simulation.

A better condition (which is weaker than domination off the diagonal) is

$$\mathrm{Cov}(g(X'_1), g(X'_2)) \le \mathrm{Cov}(g(X''_1), g(X''_2))$$

for all measurable $g : S \to \mathbb{R}$ for which the covariance is finite. This is shown in Mira and Geyer (2000) by means of spectral theory of Hilbert space operators. Clearly, (5.2.38) implies

$$\gamma'_i \le \gamma''_i \quad \text{for } i = 0, 1, \dots.$$

In the case $S = \mathbb{R}^d$ and $\mathcal{S}$ being the corresponding Borel σ-algebra, Theorem 4.3.13 yields the association of a Markov chain of the form

$$Z_{n+1} = \varphi(Z_n, X_n) \quad \text{for } n = 0, 1, \dots$$

with i.i.d. X_n and independent Z_0. It implies

$$\mathrm{Cov}(f(Z_n), f(Z_{n+k})) \ge 0$$

for all n and k. For more information on association of Markov processes see Szekli (1995), section 3.2 and 3.7, which is partly based on Liggett (1985), and Lindqvist (1988).

5.2.5   Application of Monotone Markov Chains: Perfect Simulation

An important task in many fields of statistics and applied probability is the generation of samples from the stationary distribution of an ergodic Markov chain. Frequently, this Markov chain is only an auxiliary construction, which is built so that the stationary distribution is equal to some (partially) unknown distribution.

Two typical examples stem from statistical mechanics and Bayesian statistics.

(1) Consider point configurations x in a bounded area W. They are characterized by the energy $E(x; \theta)$ specified explicitly in the definition of the corresponding model, where θ is some model parameter. The probability density $f(x; \theta)$ of the patterns is given by

$$f(x; \theta) = \frac{\exp(-E(x; \theta))}{Z(\theta)}, \qquad (5.2.40)$$

where $Z(\theta)$ is the normalizing constant making f a density, called partition function. Very often $Z(\theta)$ cannot be given analytically but physicists and statisticians are interested in samples from $f(x; \theta)$.

(2) In Bayesian statistics one wants to use the posterior distribution of some parameter θ conditional on observed data D. Its density function is given by

$$f(\theta|D) = \frac{f(\theta)\, p(D|\theta)}{\int f(\theta)\, p(D|\theta)\, d\theta} \qquad (5.2.41)$$

where $f(\theta)$ is the prior density of θ and $p(D|\theta)$ the likelihood, giving the probability of observing D if θ is the true parameter. If the model is sufficiently complex (for example, if θ is high-dimensional) then it is practically very difficult to determine analytically the integral in the denominator of the right-hand side of (5.2.41).

In such cases Markov chain Monte Carlo (MCMC) is a helpful tool to obtain samples from distributions such as those given by the densities $f(x; \theta)$ and $f(\theta|D)$; see Gilks, Richardson, and Spiegelhalter (1996) and Møller (1999). A Markov chain is constructed whose stationary distribution coincides with the distribution of interest. For this construction there exist many possibilities described in the references; famous examples are the Gibbs sampler or the more general Metropolis–Hastings algorithm. Given the Markov chain, it is simulated for a long time and, after some burning-in time, samples are taken in the hope that they are good approximations for the stationary distribution. Unfortunately, it is difficult to choose an appropriate burning-in time and the statistician is not sure of having a long enough burning-in time. An important problem in this context is also the speed of convergence towards the stationary distribution, where monotonicity properties can be exploited; see Roberts and Tweedie (2000).

In this situation the methods of 'perfect' (or 'exact') simulation are helpful. An example is coupling from the past introduced by Propp and Wilson (1996). It can be described as follows, see Kendall and Møller (2000), Thönnes (2000) and Dimakos (2001).

Let (X_n) be the Markov chain to be simulated and let it be given in the form typical for simulation:

$$X_{n+1} = f(X_n, U_n) \quad \text{for} \quad n = 0, \pm 1, \pm 2, \dots \qquad (5.2.42)$$

where the U_n form an i.i.d. sequence of uniform random numbers. If we were able to start (X_n) at $n = -\infty$ then at $n = 0$ the chain would be in the stationary state. Of course, we are unable to simulate (X_n) for infinitely many steps. But it is possible to exploit simulations made in finite time intervals so that we obtain an X_0 as if it came from an infinite simulation.

We choose a time $-n_1$ and start the Markov chain from every possible state using the same realizations of $U_{-n_1}, U_{-n_1+1}, \ldots$. Therefore, paths starting from different states will coalesce when they meet. Assume now that, fortunately, all paths have coalesced into a single path by time $n = 0$. Then we can assume to have obtained a result which would be the same if we had simulated the chain since $n = -\infty$.

In the case of no coalescence we choose $n_2 > n_1$ and start simulation at time $-n_2$. We use new realizations of $U_{-n_2}, \ldots, U_{-n_1-1}$, but beginning with $-n_1$ we use the old realizations of the first step. If there is now complete coalescence we stop; otherwise we go further into the past and so on.

In practice it is difficult to observe all the sample paths corresponding to the different states of the Markov chain. Here it can be helpful to exploit monotonicity with respect to $\leq_{st}$; see also the discussion in the case of the Ising model in Section 7.4.1. Assume that the state space S of (X_n) is partially ordered with respect to $\preceq$. If
(a) the mapping f in (5.2.42) is monotone in the first variable, i.e.

$$x \preceq y \quad \text{implies } f(x, u) \preceq f(y, u) \quad \text{for all } u$$

and
(b) S has a minimal and a maximal element
the procedure above can be greatly simplified. Then the chain is started at $-n_1, -n_2, \ldots$ from the two extremal states only. If there is coalescence of the corresponding sample paths, then there would be coalescence also for all other starting states. This is an immediate consequence of Corollary 5.2.12.

Note that many Markov chains used in MCMC are monotone, perhaps with respect to sophisticated cleverly chosen partial orders.

Example 5.2.24. (Simulating a stationary simple random walk with reflecting barriers.) Consider the random walk with two barriers discussed in Example 5.2.13, which depends on the parameters p and m. Start it from the extremal states 0 and m and look for coalescence. It is not difficult to show that the starting point of simulation which leads to coalescence is almost surely finite, and that the expected number of needed simulations is finite, too.

The approach of Propp and Wilson described above has been generalized and extended; see, for example, Fill (1998), Thönnes (1999), Thorisson (2000), Møller and Schladitz (1999), Mira, Møller, and Roberts (2001), Dimakos (2001), and Kendall and Møller (2000). The last paper considers the case where there is only a minimal element but not a maximal element. It introduces a 'dominating process' which acts as a kind of stochastic maximum. Wilson (2000) gives an extension where it is not necessary to store realizations of the random variables.

5.2.6 Markov Decision Processes

In a Markov decision process (MDP) there is a decision maker who can control the evolution of a (discrete time) Markov process. She will do this in a manner to maximize some expected rewards, which of course are related to the evolution of the process. Here we will only consider the case of a stationary finite horizon model with a finite number N of stages and the discounted expected reward criterion. A good introduction to this topic is given in Puterman (1994). For a more advanced treatment the reader is referred to Hinderer (1970) or Bertsekas and Shreve (1978).

The model can be described briefly as follows: The Markov chain has some *state space* S; its transition probability measure, however, depends on the actions a chosen by the decision maker from the *action space* A. If the process is in state $s \in S$ then she can choose some action $a \in D(s)$, where $D(s)$ is the set of feasible decisions allowed in state s. The set $D \subset S \times A$ with

$$D = \{(s,a) : a \in D(s)\}$$

is called the *restriction set*.

Choosing the action a leads to a transition kernel $Q(s,a,\mathrm{d}s')$. Moreover, a reward $r(s,a,s')$ is obtained, depending on current state s, action a and next state s'. These rewards are discounted by a discount factor $\beta > 0$. The decision maker has to choose a policy $\pi = (f_0, \ldots, f_{N-1})$ consisting of decision functions $f_n : S \to A$ with $f_n(s) \in D(s)$ for all s. Here $f_n(s)$ is the action chosen at stage n, if the current state is s. Given such a policy π the expected total return is

$$V_{0\pi}(s_0) = E\left(\sum_{n=0}^{N-1} \beta^n r(X_n, f_n(X_n), X_{n+1}) + V_N(X_N) \right), \qquad (5.2.43)$$

where $V_N : S \to \mathbb{R}$ is the terminal reward paid at the end of the process, and the sequence $(X_0, X_1, \ldots, X_N)$ is an inhomogeneous Markov process with transition kernels $Q(s, f_n(s), \cdot)$ and fixed initial state $X_0 \equiv s_0$.

The decision maker faces the problem of maximizing $V_{0\pi}(s_0)$ over all feasible policies π. The corresponding maximal expected reward is denoted by

$$V_0(s_0) = \max_\pi V_{0\pi}(s_0).$$

It is well known that under some weak regularity conditions ensuring the existence of all mentioned quantities this problem can be solved via backward induction by solving the so called *Bellman equations*

$$V_n(s) = \max_{a \in D(s)} \left\{ \int r(s,a,s') + \beta V_{n+1}(s') Q(s,a,\mathrm{d}s') \right\} \quad \text{for } n = 0, \ldots, N-1.$$
$$(5.2.44)$$

The functions V_n are called *value functions*. The optimal policy $\pi^* = (f_0^*, \dots, f_{N-1}^*)$ is obtained by choosing for $f_n^*(s)$ the action which maximizes the right-hand expression in equation (5.2.44). For a proof of this result see Puterman (1994) or Bertsekas and Shreve (1978).

Typically, the effort for solving the Bellman equations (5.2.44) is extremely high. This is known as the *curse of dimensionality*. In this situation monotonicity properties are an important issue for reducing the computational effort. The following result about monotonicity of the value function is due to Hinderer (1984), theorem 1. The proof follows by induction.

Theorem 5.2.25. *Let the state space S be ordered and assume that*

(i) $D(\cdot)$ *is increasing, i.e. $s \le t$ implies $D(s) \subseteq D(t)$;*

(ii) *for each fixed $a \in A$ the transition kernel $Q(s,a,\cdot)$ is $\le_{st}$-monotone in s (on its domain);*

(iii) $r(\cdot, a, s')$ *is increasing for all a and all s';*

(iv) V_N *is increasing.*

Then V_n is increasing in s for all $n = 0, \dots, N$.

Similar results can be shown regarding convexity and concavity of the value function. The following results are obvious modifications of theorems 2 and 3 in Hinderer (1984).

Theorem 5.2.26. *Let the state space S and the actions space A be convex sets and assume that*

(i) D *is convex;*

(ii) $(s,a) \mapsto \int Q(s,a,\mathrm{d}s')f(s')$ *is concave for all concave functions f;*

(iii) $(s,a) \mapsto \int Q(s,a,\mathrm{d}s')r(s,a,s')$ *is concave;*

(iv) V_N *is concave.*

Then V_n is concave in s for all $n = 0, \dots, N$.

Theorem 5.2.27. *Let the state space S be some convex set and assume that*

(i) $D = S \times A$;

(ii) *for each fixed $a \in A$ the transition kernel $Q(s,a,\cdot)$ is $\le_{cx}$-monotone in s;*

(iii) $s \mapsto \int Q(s,a,\mathrm{d}s')r(s,a,s')$ *is convex for all a;*

(iv) V_N *is convex.*

Then V_n is convex in s for all $n = 0, \dots, N$.

Comparison theorems for MDPs were given in Müller (1997a). Suppose that there are given two Markov decision processes $\mathrm{MDP}(Q')$ and $\mathrm{MDP}(Q'')$ differing only with respect to their transition kernels, which are denoted by $Q'(s, a, \cdot)$ and $Q''(s, a, \cdot)$. All other data are the same for the two models. The value functions of the two models are denoted by $V_n^{Q'}$ and $V_n^{Q''}$, respectively. In this case the following general result for integral stochastic orders can be found in Müller (1997a), theorem 4.1.

Theorem 5.2.28. *If there is a generator $\mathcal{F}$ such that for all $(s, a) \in D$*

(i) $Q'(s, a, \cdot) \leq_{\mathcal{F}} Q''(s, a, \cdot)$ *for all $(s, a) \in D$;*

(ii) $r(s, a, \cdot) \in \mathcal{F}$ *for all $(s, a) \in D$;*

(iii) $V_n^{Q'}, V_n^{Q''} \in \mathcal{F}$ *for all n;*

then $V_n^{Q'}(s) \leq V_n^{Q''}(s)$ for all n and all s. Moreover, if π^* is an optimal policy for $MDP(Q')$ then

$$V_{0\pi^*}^{Q''}(s) \geq V_0^{Q'}(s) \quad \text{for all } s.$$

There are many examples where the assumptions of Theorem 5.2.28 are fulfilled with $\leq_{\mathcal{F}}$ being one of the orders $\leq_{st}, \leq_{cx}, \leq_{icx}$ or $\leq_{icv}$. For these orders the results in the Theorems 5.2.25, 5.2.26 and 5.2.27 can be used to verify assumption (iii) of Theorem 5.2.28. The following example is taken from Müller (1997a).

Example 5.2.29. (Optimal stopping.) Let $X_1, \ldots, X_N$ be i.i.d. random variables with distribution P, which can be observed sequentially at a cost c per observation. If the decision maker stops after the kth observation, she receives an immediate reward of $\max\{X_1, \ldots, X_k\}$. The problem is to find an optimal stopping rule. This is a familiar problem of optimal stopping that can be solved by backward induction; see, for example, Chow, Robbins, and Siegmund (1971). In the economic literature this problem occurs in the theory of search. Lippman and McCall (1976) call it the 'job search problem (with recall)'.

It is well known that the solution of this problem is given by the following Bellman equations.

$$V_n^P(s) = \max\left\{s, -c + \beta \int V_{n+1}^P(\max\{s, x\}) P(\mathrm{d}x)\right\}$$

with $V_N(s) = s$. Here $V_n^P(s)$ is the optimal expected reward, if n observations have been received, s is the best offer so far, and P is the distribution of X_i. The data of the underlying MDP are defined as follows: $A = \{0, 1\}$, where action 1 means 'stop' and action 0 means 'continue'. The reward function is $r(s, 0, s') = -c$ and $r(s, 1, s') = s$. The transition kernels are $Q(s, 1, \cdot) = \delta_\infty$

(where ∞ denotes the absorbing state) and

$$Q(s, 0, \cdot) = \int \delta_{\max\{s,x\}}(\cdot) P(\mathrm{d}x).$$

It easily follows from Theorem 5.2.25 and 5.2.27 that V_n is increasing and convex for all n, and therefore it follows from Theorem 5.2.28 that

$$P' \leq_{icx} P'' \quad \text{implies} \quad V_n^{P'}(s) \leq V_n^{P''}(s)$$

for all n and all s.

Stochastic orders can also be used to derive bounds for the value of the optimal stopping of a dependent sequence $X_1, \ldots, X_N$; see Müller (2001a) and Müller and Rüschendorf (2001).

Many optimal control problems for networks of queues can be formulated as MDPs. There structural properties of the value function and of optimal policies play an important role. Especially the concept of *multimodularity*, which is a generalization of supermodularity, is of interest in that field; see, for example, the seminal paper of Stidham and Weber (1993). Menich and Serfozo (1991) use stochastic orders to show structural properties of MDPs occurring in the context of routing and servicing in dependent parallel processing systems. Rieder and Zagst (1994) prove monotonicity results of partially observable MDPs with respect to convex order.

5.3   Monotonicity and Comparability of Non-Markov Processes

Of course, monotonicity properties can also be proved for other classes of processes and they are of interest in many situations.

Every *submartingale* is increasing with respect to $\leq_{icx}$ and every *supermartingale* is decreasing with respect to $\leq_{icv}$. This is an immediate consequence of Corollary 1.5.21. Conversely, if (P_n) is a sequence of distributions increasing with respect to $\leq_{icx}$ then there is a probability space and a submartingale which has the marginal distributions P_n. This can simply be achieved by choosing a Markovian submartingale with transition kernels obtained via Corollary 1.5.21. This result has first been stated by Strassen (1965), theorem 9. The extension to continuous time parameter is due to Kellerer (1972), see Theorem 2.6.9.

Chong (1975) studied for martingales (X_t) the problem of comparison of X_σ and X_τ with comparable σ and τ, where σ and τ are random variables with values in the parameter set of the process.

Processes which are increasing with respect to the randomness order $\leq_{rand}$ are studied in the context of dissipative systems of physics. Certain types of such systems are described by systems of ordinary differential equations

$$\frac{\mathrm{d}}{\mathrm{d}t}\mathbf{P}(t) = v(\mathbf{P}(t))$$

where $\mathbf{P}(t) = (P_1(t), \dots, P_n(t))$ gives the time-dependent state probabilities and v is a continuously differentiable vector field, see Alberti and Crell (1984, 1986).

The relation (5.1.4) gives the possibility of comparing non-Markov processes of a special structure. If (X_n) and (Y_n) are comparable Markov chains, then $(\Phi(X_n))$ and $(\Phi(Y_n))$ are comparable for suitable Φ, but if Φ is not a bijection then $(\Phi(X_n))$ and $(\Phi(Y_n))$ may be non-Markov. This can be easily extended to the case of mappings Φ and Ψ applied to (X_n) and (Y_n), respectively. See Section 4.3 for the properties Φ and Ψ have to satisfy.

As a more interesting example of comparison of a non-Markov stochastic process (X_n) with a Markov chain (Y_n) on the same state space S with transition kernel Q_Y and transition operator $\mathbf{T}_Y$, a result by Whitt (1986) is presented here. A similar theorem for continuous-time Markov processes can be found in Massey (1987). The construction by Whitt may remind some readers of the use of supplementary variables to construct Markov processes.

Assume that there is a discrete-time stochastic process (Z_n) on a state space S' such that $((X_n, Z_n))$ is a discrete-time Markov process with product state space $S \times S'$, transition kernel $Q_Z((x, x'), A)$ for $A \in S \otimes S'$ and transition operator $\mathbf{T}_Z$. Let π be the projection from $S \times S'$ onto S, thus $\pi(X_n, Z_n) = X_n$.

Let on $\mathbb{P}_S$ be given an integral stochastic order $\leq_{\mathcal{F}}$ determined by the generator $\mathcal{F}$. Let P_Z and P_Y be initial distributions for $((X_n, Z_n))$ and (Y_n).

Theorem 5.3.1. *Let the transition operator $\mathbf{T}_Y$ of (Y_n) be monotone with respect to $\leq_{\mathcal{F}}$, let*

$$Q_Z((x, x'), \pi^{-1}(\cdot)) \leq_{\mathcal{F}} Q_Y(x, \cdot) \quad \text{for all } (x, x') \in S \times S' \qquad (5.3.1)$$

and let

$$P_Z(\pi^{-1}(\cdot)) \leq_{\mathcal{F}} P_Y. \qquad (5.3.2)$$

Then

$$\mathbf{T}_Z^n P_Z(\pi^{-1}(\cdot)) \leq_{\mathcal{F}} \mathbf{T}_Y^n QZ \quad \text{for } n = 1, 2, \dots, \qquad (5.3.3)$$

i.e. $X_n \leq_{\mathcal{F}} Y_n$ for all n.

Proof. This is the second proof given in Whitt (1986), which is based on Theorem 5.2.11. The idea is to extend $\leq_{\mathcal{F}}$, $\mathcal{F}$ and $\mathbf{T}$ to $S \times S'$. The new partial order for distributions on $S \otimes S'$ is the integral stochastic order corresponding to $\hat{\mathcal{F}}$ being the set of all measurable real-valued functions $\hat{f}$ on $S \times S'$ such that for each fixed x' in S' the function $f_{x'}(x) = \hat{f}(x, x')$ is in $\mathcal{F}$. Corresponding to Q_Y, a new transition kernel $\hat{q}$ is constructed by

$$\hat{Q}_Y((x, x'), A \times A') = Q_Y(x, A)\mathbf{1}_{A'}(x')$$

for $(x, x') \in S \times S'$, $A \in S$ and $A' \in S'$. The corresponding initial distribution is $\hat{P}_Y$, $\hat{P}_Y(A \times A') = P_Y(A)$. It is easy to show that $\hat{Q}_Y$ is monotone with respect to the extended order relation $\leq_{\hat{\mathfrak{F}}}$, that (5.3.2) implies $P_Z \leq_{\hat{\mathfrak{F}}} \hat{P}_Y$, and that

$$Q_Z((x, x'), \cdot) \leq_{\hat{\mathfrak{F}}} \hat{Q}_Y((x, x'), \cdot).$$

Theorem 5.2.11 thus yields the result. □

As a general class of real-valued non-Markov processes now *semi-Markov processes* (or Markov renewal processes) are considered, following Sonderman (1980). In order to show what can happen and what one can expect, first an example is considered.

Example 5.3.2. (Alternating Renewal Process.) This process describes a system having two states: 1 – work and 0 – repair. The length of work times has the distribution function F and that of repair times G. All these times are independent. At time $t = 0$ the system is with probability p in state 1 and with $1 - p$ in state 0, and it starts with a fresh work or repair time. Denote by X_t the state of the system at time t.

An interesting comparison in this context would mean that for two such systems Σ' and Σ'' with distribution functions F', G' and F'', G'' the system with the longer work times and the shorter repair times is 'more often' in the 'greater' state 1. Remarkably, this is not true if the stochastic order $\leq_{st}$ is used. Assume that Σ' and Σ'' start in state 1 ($p' = p'' = 1$), i.e. $X'_0 = X''_0 = 1$, and that all work and repair times are deterministic, with work times being 2 and 4 and repair times being 2 and 1, respectively. Then clearly

$$F' \leq_{st} F'' \quad \text{and} \quad G'' \leq_{st} G',$$

but $X'_t = 1$ and $X''_t = 0$ for $4 \leq t < 5$ while $X''_t = 1$ for $0 \leq t < 4$. In other words,

$$X'_t \leq_{st} X''_t \quad \text{for } 0 \leq t < 4$$

and

$$X''_t \leq_{st} X'_t \quad \text{for } 4 \leq t < 5.$$

So it is clear that a stronger order of the distributions is necessary. As Sonderman (1980) discovered, the hazard rate order $\leq_{hr}$ is the right order in the given context.

Theorem 5.3.3. *If $p' \leq p''$, $F' \leq_{hr} F''$ and $G'' \leq_{hr} G'$, then there are two alternating renewal processes $(\hat{X}'_t)$ and $(\hat{X}''_t)$ on the same probability space having the same distribution as (X'_t) and (X''_t) and satisfying $P(\hat{X}'_t \leq \hat{X}''_t$ for all $t \geq 0) = 1$.*

This result is a particular case of a more general theorem for semi-Markov processes given below.

Remember that a semi-Markov process is a pure jump process (X_t) with state space $S = \{0, 1, \ldots, n\}$. It is specified by the pair (p, Q), where $p \equiv (p_i)$ is a probability vector representing the initial distribution (that of X_0) and $Q \equiv (Q_{ij}(t))$ is a matrix of subdistribution functions such that $\lim_{t\to\infty} \sum_j Q_{ij}(t) = 1$ for each i, where $Q_{ij}(t)$ denotes the probability that the next transition from i goes to state j and that it takes place before t time units since the last transition, which has led to state i.

Assume that there exist subprobability density functions $q_{ij}(t)$ such that

$$Q_{ij}(t) = \int_0^t q_{ij}(u)\mathrm{d}u \quad \text{for all } i, j \text{ and } t \geq 0$$

and use the notation

$$F_i(t) = \sum_{j=0}^n Q_{ij}(t).$$

Then the hazard rates of the Q_{ij} are defined as

$$r_{ij}(t) = q_{ij}(t)/[1 - F_i(t)] \quad \text{for all } i, j \text{ and } t \geq 0. \tag{5.3.4}$$

Theorem 5.3.4. *(Sonderman (1980)) If for two semi-Markov processes the initial distributions and hazard rates satisfy*

$$p' \leq_{st} p'',$$

$$\sum_{k=\ell}^n r'_{ik}(s) \leq \sum_{k=\ell}^n r''_{jk}(t) \quad \text{for all } s, t \text{ and } i \leq j < \ell \tag{5.3.5}$$

and

$$\sum_{k=0}^m r''_{jk}(s) \leq \sum_{k=0}^m r'_{ik}(t) \quad \text{for all } s, t \text{ and } m < i \leq j \tag{5.3.6}$$

then there exist two semi-Markov processes $(\hat{X}'_t)$ and $(\hat{X}''_t)$ on a common probability space with the same distributions as the processes (X'_t) and (X''_t) and satisfying

$$P(\hat{X}'_t \leq \hat{X}''_t \text{ for all } t \geq 0) = 1.$$

Sonderman's proof uses the construction with a subordinated Markov chain as in the proof of Theorem 5.2.19.

A still more general class of models are *generalized semi-Markov processes*. Glasserman and Yao (1994), section 3, prove comparison theorems for these processes, which describe typically the influence of model parameters or constituent distributions such as speeds or clock laws.

Theorem 5.3.4 can be seen also as a particular case of more general results given in Rolski and Szekli (1991) and Kwieciński and Szekli (1991), which were further developed in Last (1993) and still more generalized in Brandt and Last (1994). Here point process methods are systematically used, in particular the 'dynamical approach', i.e. martingale methods as described in Last and Brandt (1995). In the spirit of these papers, the proof of Theorem 5.2.19 as well as Sonderman's proof of Theorem 5.3.3 involve thinnings of Poisson processes. Brandt and Last (1994) compare *jump processes* with values in a partially ordered Polish space S with Borel σ-algebra $\mathcal{S}$. Such a process (X_t) is constructed as

$$X_t = X_0 \quad \text{for } 0 < T_1$$

and

$$X_t = m_n \quad \text{for } T_n \leq t < T_{n+1}$$

where $([T_n; m_n])$ is a marked point process with points T_n in $\mathbb{R}$ and marks m_n in S. The comparison result concerns path-wise comparison, i.e. comparison of distributions on the space $D_S[0, \infty)$ with respect to $\leq_{st}$, starting from the comparison of stochastic intensities. It is obtained by means of the coupling method.

5.4 Comparison of Point Processes

This section describes some concepts of comparison of point processes on the positive half line $[0, \infty)$; Section 7.3.2 below will consider the case of point processes in $\mathbb{R}^d$ and more general state spaces. Most of the comparability criteria presented here use special properties of one-dimensional point processes based on the natural linear order on the real axis. Many of the relations are closely related to the stochastic order $\leq_{st}$. The 'greater' process is always the process with 'more' or 'denser' points. The exposition is inspired by Szekli (1995), who also discusses association of point processes. We consider a point process sometimes as a random set (a locally finite point sequence consisting of the epoch-points) and sometimes as a random measure; in the latter case $N(A)$ is the random number of points in the set A, which is finite if A is bounded.

Let $\Pi = (T_n)$ be a simple point process on $[0, \infty)$, i.e. a strictly increasing sequence of event epochs T_n with $\lim_{n \to \infty} T_n = \infty$. The corresponding interpoint distances are $A_n = T_n - T_{n-1}$. The corresponding counting process (N_t) is defined by

$$N_t = \max\{n : T_n \leq t\},$$

so that the paths of (N_t) are right-continuous, where $T_0 = 0$ is used. This jump process is a particular case of the processes considered at the end of the preceding section. As there, hazard rates play an important role in the context of comparability. They are here the hazard rates $r_{n+1}(t; t_1, \dots, t_n)$ corresponding to the distributions of the A_{n+1} given that $T_1 = t_1, \dots, T_n = t_n$ for $n \geq 1$. The corresponding stochastic order $\leq_{hr}$ is defined as follows:

$$\Pi' \leq_{hr} \Pi'' \quad \text{if} \quad r'_{n+1}(t; t_1, \dots, t_n) \leq r''_{n+k+1}(t; s_1, \dots, s_{n+k})$$

for all k, $n \geq 0$, $\{t_1, \dots, t_n\} \subseteq \{s_1, \dots, s_{n+k}\}$ and $t > s_{n+k}$.

The following two stochastic orders are defined in the spirit of the coupling method:

$\Pi' \subseteq_{st} \Pi''$ if there are two point processes $\hat{\Pi}'$ and $\hat{\Pi}''$ on a common probability space with $\Pi' =_{st} \hat{\Pi}'$, $\Pi'' =_{st} \hat{\Pi}''$ and $\hat{\Pi}' \subseteq \hat{\Pi}''$ almost surely (where $\subseteq$ is the set-theoretic inclusion sign and the point processes are considered as the random sets consisting of the points T_n)

and

$\Pi' \leq_{st-D} \Pi''$ if there are two point processes $\hat{\Pi}'$ and $\hat{\Pi}''$ on a common probability space with $\Pi' =_{st} \hat{\Pi}'$, $\Pi'' =_{st} \hat{\Pi}''$ and with counting processes $(\hat{N}'_t)$ and $(\hat{N}''_t)$ satisfying $\hat{N}'_t \leq \hat{N}''_t$ for all t.

The relation $\subseteq_{st}$ is the same as $\leq_{st-\mathcal{N}}$ in Szekli (1995); the character $\subseteq_{st}$ is taken from Norberg (1992) and $\leq_{st-D}$ from Szekli (1995).

By means of the Strassen theorem, $\subseteq_{st}$ and $\leq_{st-D}$ can be reformulated as follows:

$$\Pi' \subseteq_{st} \Pi'' \quad \text{if} \quad Ef(\Pi') \leq Ef(\Pi'') \tag{5.4.1}$$

for every increasing (with respect to $\subseteq$) $f : \mathbb{S} \to \mathbb{R}$, where $\mathbb{S}$ is the set of all locally finite point sequences on $[0, \infty)$, and

$$\Pi' \leq_{st-D} \Pi'' \quad \text{if} \quad Eg(N') \leq Eg(N'') \tag{5.4.2}$$

for every increasing $g : D[0, \infty) \to \mathbb{R}$, where 'increasing' is defined with respect to the partial order on $D[0, \infty)$ given by $x \leq y$ if $x(t) \leq y(t)$ for all $t \geq 0$.

Results by Rolski and Szekli (1991) and Norberg (1992) (see also Theorem 7.4.6) ensure that $\subseteq_{st}$ is equivalent to

$$(N'(K_1)), \dots, N'(K_n)) \leq_{st} (N''(K_1)), \dots, N''(K_n))$$

for all n and all relative compact subsets $K_1, \dots, K_n$ of $[0, \infty)$, where $\leq_{st}$ denotes the usual stochastic order for random vectors.

The continuous time counterpart to Theorem 5.1.4 yields that $\leq_{st-D}$ is equivalent to

$$\left(N'_{t_1}, \dots, N'_{t_k} \right) \leq_{st} \left(N''_{t_1}, \dots, N''_{t_k} \right)$$

for all k and all $0 \leq t_1 < \cdots < t_k$.

For some applications stochastic orders are important which refer directly to the epochs T_n or to the inter-epoch distances A_n. One can compare the sequences (T_n):

$$(T_1', T_2', \dots) \leq_{st} (T_1'', T_2'', \dots)$$

or

$$(T_1', \dots, T_k') \leq_{st} (T_1'', \dots, T_k'') \quad \text{for } k = 1, 2, \dots .$$

By Theorem 5.1.4 both relations are equivalent. Instead of $\leq_{st}$ also other stochastic orders for random vectors can be used here, e.g. $\leq_{hr}$; see also below.

Let P_n be the joint distribution of T_1 and the differences $A_2, \dots, A_n$. (Remember that $T_0 = 0$ and $T_1 > 0$.) If originally a stationary point process is given, then P_n is the joint Palm distribution of the differences between the first n points of the process in $(0, \infty)$. (Note that the Palm distribution P_Π^0 is the point process distribution under the condition that in $T_0 = 0$ there is a point of the process.) Then comparison of the n-dimensional distributions with respect to integral stochastic orders yields interesting orders for point processes:

$$\Pi_1 \leq_{\mathcal{F}.P} \Pi_2 \quad \text{if } P_k^2 \leq_{\mathcal{F}} P_k^1 \quad \text{for all } k = 1, 2, \dots \tag{5.4.3}$$

for $\leq_{\mathcal{F}}$ being $\leq_{st}$, $\leq_{cx}$, $\leq_{icx}$, $\leq_{sm}$ or some other order.

The relation $\leq_{st.P}$ is the same as $<_{st-\infty}$ in Szekli (1995); the 'P' is related to 'Palm'. Note that $\leq_{cx.P}$ enables the comparison of stationary point processes of equal intensity.

Between the relations introduced above the following implications can be proved:

The relation $\leq_{hr}$ appeared already in the context of the Theorems 5.3.3 and 5.3.4 and the more general comparison of jump processes. Similarly as there it can be shown that

$$\Pi' \leq_{hr} \Pi'' \quad \text{implies } \Pi' \subseteq_{st} \Pi'';$$

see Szekli (1995). Theorem 4.3.2 yields that

$$\Pi' \subseteq_{st} \Pi'' \quad \text{implies } \Pi' \leq_{st-D} \Pi''$$

and

$$\Pi' \leq_{st.P} \Pi'' \quad \text{implies } \Pi' \leq_{st-D} \Pi'',$$

but the converse of the latter implication is not always true and also in general $\Pi' \subseteq_{st} \Pi''$ does not imply $\Pi' \leq_{st.P} \Pi''$. Criteria for $\subseteq_{st}$, $\leq_{st-D}$ and $\leq_{st.P}$ in terms of compensators are given in Szekli (1995).

Particular point process models can be found in Kwieciński and Szekli (1996), where monotonicity properties of so called *self-exciting* point processes are considered. They also discuss association properties of these processes, where association of point processes means association of the random vectors $(N(K_1), \dots , N(K_n))$, $(N_{t_1}, \dots , N_{t_n})$ or $(T_1, \dots , T_n)$.

It is possible to prove conditions which ensure the preservation of point process orders under such typical operations as superposition, thinning, shift and random change of time; see Deng (1985). Clearly, in the case of $\subseteq_{st}$ this is quite simple.

In the case of non-homogeneous *Poisson processes* Π' and Π'' with intensity functions $\lambda'(t)$ and $\lambda''(t)$ and integrated intensity functions $\Lambda'(t)$ and $\Lambda''(t)$ $(\Lambda(t) = \int\limits_{0}^{t} \lambda(x)\mathrm{d}x)$ all comparability properties are given if $\lambda'(t) \leq \lambda''(t)$ for all t. If only $\Lambda'(t) \leq \Lambda''(t)$ for all t, then still

$$(T_1', \dots , T_k') \leq_{st} (T_1'', \dots , T_k'') \quad \text{for all } k,$$

(see Example 4.3.15 in Section 4.3.3). Even the converse is true; see Shaked and Szekli (1995).

Belzunce, Lillo, Ruiz, and Shaked (2001) showed that $\lambda'(t) \leq \lambda''(t)$ implies

$$(T_1', \dots , T_k') \leq_{hr} (T_1'', \dots , T_k'') \quad \text{for all } k,$$

where $\leq_{hr}$ is the multivariate hazard rate order introduced in Definition 3.11.7.

Shaked and Szekli (1995) showed that

$$F' \leq_{disp} F'' \quad \text{implies} \quad \Pi'' \leq_{st.P} \Pi',$$

where $F(t) = 1 - \exp(-\Lambda(t))$. (Observe the relation to Theorem 1.7.7.)

In the case of *renewal processes*, i.e. if the interpoint distances are i.i.d., $\Pi' \leq_{st.P} \Pi''$, $\Pi' \leq_{st-D} \Pi''$ and $F'' \leq_{st} F'$ are equivalent, where F' and F'' are the interpoint distance distribution functions.

$F'' \leq_{hr} F'$ is not sufficient for $\Pi' \subset_{st} \Pi''$, but if the hazard rates fulfill the sharper condition $r'(t) \leq r''(s)$ for all $0 \leq s \leq t$, then $\Pi' \subset_{st} \Pi''$ holds; see Szekli (1995), p. 106.

It is clear that $F' \leq_{\mathcal{F}} F''$ is equivalent to $\Pi'' \leq_{\mathcal{F}.P} \Pi'$ whenever $\leq_{\mathcal{F}}$ has the property (IN), hence e.g. for $\leq_{st}$, $\leq_{cx}$, $\leq_{icx}$, $\leq_{sm}$ and many other orders.

Examples of stationary non-renewal point processes which are comparable with (pieces of) stationary Poisson processes (on $[0, \infty)$) with respect to $\leq_{cx.P}$ are Markov renewal processes and some doubly stochastic Poisson processes, including Markov modulated point processes.

In the case of a *Markov renewal process* (which is sometimes also called semi-Markov process) the point process is constructed by means of a stationary irreducible finite Markov chain (X_n). If $X_n = i$, then the difference A_n

between the $(n-1)$th and nth point has the distribution function F_i; given the values of (X_n), the differences $A_1, A_2, \ldots$ are independent. Two Markov renewal processes with the same transition probabilities p_{ij} of the controlling Markov chain and with interpoint distance distribution functions F_i' and F_i'' are $\leq_{cx.P}$-comparable if $F_i' \leq_{cx} F_i''$ for all i; see Rolski (1984). If the transition probabilities are different, comparison with respect to $\leq_{cx.P}$ is obviously possible if the transition probabilities satisfy the conditions of Theorem 5.2.11 for $\leq_{st}$ in reverse order (the process with $'$ is the bigger) and if

$$F_i' \leq_{cx} F_{i+1}' \quad \text{and} \quad F_i'' \leq_{cx} F_{i+1}'' \quad \text{for all } i$$

and

$$F_i'' \leq_{cx} F_i' \quad \text{for all } i.$$

A doubly stochastic Poisson process is defined here starting from a non-negative stochastic process (λ_t) satisfying $\int_0^t \lambda_s \mathrm{d}s < \infty$ for all $t \geq 0$. Given a sample of this process, the Poisson process with this function as intensity function is considered. Point sequences generated by this two-step random mechanism are called Cox or doubly stochastic Poisson processes directed by (λ_t). If (λ_t) is a stationary process, then the Cox process is stationary as well. A particular case is a Markov modulated point process where the process (λ_t) is a stationary continuous-time Markov chain with finite state space. The process (λ_t) defines a random measure Λ satisfying $\Lambda([a, b]) = \int_a^b \lambda_s \mathrm{d}s$ for all $0 \leq a \leq b$. Clearly, we may also introduce a Cox process directed by a general random measure Λ such that $\Lambda([0, t])$ has a finite mean for all $t \geq 0$.

Theorem 5.4.1. *Let (λ_t) be a stationary process with finite and positive mean λ and satisfying $\int_0^\infty \lambda_s \mathrm{d}s = \infty$. Then the stationary Poisson process Π_λ with intensity λ is $\leq_{cx.P}$-smaller than the Cox process C directed by the random intensity function (λ_t).*

Proof. We present a here an abridged and simplified version of the original proof given in Rolski (1984) (theorem 3.5). It is convenient to assume that the stationary process (λ_t) is defined for all $t \in \mathbb{R}$ and to consider point processes on $\mathbb{R}$ rather than on $[0, \infty)$. If Π is any point process on $\mathbb{R}$ with a locally finite intensity measure, then we let P_Π^x denote the Palm distribution, i.e. the conditional distribution of Π given that x is a point of Π (see Daley and Vere-Jones (1988) for more details). A point process Π^x with distribution P_Π^x has almost surely a point at x. Removing this point, we get another point process whose distribution is the so-called *reduced Palm distribution* $P_\Pi^{x!}$ of Π (at x).

Assume now that Π is a Poisson process. Slivnyak's theorem (see, for instance, example 12.1(b) in Daley and Vere-Jones (1988)) says that then $P_\Pi^{x!}$ coincides with the original distribution of Π (for almost all x with respect to the intensity measure of Π). Therefore, it easily follows that the reduced Palm distribution $P_C^{0!}$ of C is the distribution of a Cox process C_0, say, directed by a random measure Λ_0 that is distributed according to the Palm distribution Π_Λ^0 of Λ. The latter Palm distribution is defined exactly in the same way as for point processes (see again Daley and Vere-Jones (1988)). Letting for $t \geq 0$

$$\Lambda_0^{-1}(t) = \inf\{s \geq 0 : \Lambda_0((0, s]) \geq t\},$$

it is a straightforward consequence of the so-called refined Campbell theorem for Λ_0 that

$$E\Lambda_0^{-1}(t) = t/\lambda,$$

for all $t \geq 0$; see Geman and Horowitz (1973).

Let $T_1 < T_2 < \ldots$ denote the points of Π_λ in $(0, \infty)$ and assume (without restricting generality) that this sequence is independent of Λ_0. We may then further assume (see Grandell (1976), ch. 1) that the points of C_0 in $(0, \infty)$ are given as $\Lambda_0^{-1}(\lambda T_i)$. Taking any convex function $f : \mathbb{R}^n \to \mathbb{R}$ and letting $\mathbf{T}_n$ denote the random vector $(T_1, \ldots, T_n)$ we obtain from the n-dimensional conditional version of Jensen's inequality (or Strassen's theorem as stated in Theorem 2.6.6):

$$\begin{aligned}
E(f(\Lambda_0^{-1}(\lambda T_1), &\Lambda_0^{-1}(\lambda T_2) - \Lambda_0^{-1}(\lambda T_1), \ldots, \Lambda_0^{-1}(\lambda T_n) - \Lambda_0^{-1}(\lambda T_{n-1}))) \\
&= E(E[f(\Lambda_0^{-1}(\lambda T_1), \ldots, \Lambda_0^{-1}(\lambda T_n) - \Lambda_0^{-1}(\lambda T_{n-1}))|\mathbf{T}_n]) \\
&\geq E(f(E[\Lambda_0^{-1}(\lambda T_1)|\mathbf{T}_n], \ldots, E[\Lambda_0^{-1}(\lambda T_n) - \Lambda_0^{-1}(\lambda T_{n-1})|\mathbf{T}_n])) \\
&= Ef(T_1, \ldots, T_n - T_{n-1}).
\end{aligned}$$

The theorem is thus proved.   □

Kulik and Szekli (2001) use $\leq_{cx.P}$, $\leq_{sm.P}$, $\leq_{dcx.P}$, and $\leq_{plcx.P}$ in the context of stationary point processes with long range count dependence and apply them for a series of interesting models.

6

Monotonicity Properties and Bounds for Queueing Systems

6.1 Basic Facts for $GI/GI/1$ and $G/G/1$

$GI/GI/1$ or $GI/GI/1/\infty$ denotes the following single-server queueing system. Customers arrive for service at the epochs T_n given by

$$T_0 = 0, \quad T_{n+1} = A_1 + \cdots + A_{n+1} = T_n + A_{n+1} \quad \text{for } n = 0, 1, \ldots.$$

The customer arriving at T_n is called the nth customer, the time A_n between the $(n-1)$th and nth arrival is the nth interarrival time. The nth customer requires service for a time B_n. (A_n) and (B_n) are independent sequences of random variables which are i.i.d., and $X_n = B_n - A_n$ denotes the difference between these two quantities. Their distribution functions and parameters are:

$$F_A(t) = P(A_n \le t), \quad m_A = EA_n, \quad \sigma_A^2 = \text{Var}(A_n),$$

$$F_B(t) = P(B_n \le t), \quad m_B = EB_n, \quad \sigma_B^2 = \text{Var}(B_n),$$

$$F_X(t) = P(X_n \le t),$$

$$\varrho = m_B/m_A = \lambda m_B = \quad \text{(relative) traffic intensity.}$$

Customers are served in order of arrival (FCFS, 'first-come first-served' discipline), and there are infinitely many waiting places in the system.

If the interarrival times may be interdependent but the service times B_n are still i.i.d. and independent of the interarrival times, the model is denoted by $G/GI/1$; if the service times may also be interdependent the symbol $G/G/1$ is used.

The model $GI/GI/1$, and to a lesser extent $G/G/1$, has been extensively studied in queueing theory. We recommend the books by Wolff (1989) for a broad survey of the subject, Cohen (1982) for *GI/GI/1* in particular, Baccelli and Brémaud (1994) and Brandt et al. (1990), who discuss *G/G/1* at some length. Below those results are recalled which are needed for the discussion of this book.

The waiting time W_n of the nth customer satisfies the recursion formula

$$W_{n+1} = \max\{0, W_n + X_n\} \equiv (W_n + X_n)_+ \quad \text{for } n = 0, 1, \ldots. \quad (6.1.1)$$

Assume W_0 to be a random variable independent of all A_n and B_n and having the distribution function $F_{W,0}$. Then corresponding to (6.1.1) the distribution function $F_{W,n}$ of W_n satisfies

$$F_{W,n+1}(t) = \int_{[0,\infty)} F_X(t - x) F_{W,n}(dx) = \int F_{W,n}(t - x) F_X(dx) \quad \text{for } t \geq 0. \quad (6.1.2)$$

Lindley (1952) proved that in the case $\varrho < 1$, and this assumption is made in the following, for any initial distribution function $F_{W,0}$ the distribution functions $F_{W,n}$ converge weakly towards a unique distribution function F_W, the stationary waiting time distribution function, which is the unique distribution function solution of the integral equation

$$F_W(t) = \int_{[0,\infty)} F_X(t - x) F_W(dx) = \int F_W(t - x) F_X(dx) \quad \text{for } t \geq 0. \quad (6.1.3)$$

If $m_A < \infty$, then the nth moment of F_W is finite if and only if the $(n + 1)$th moment of F_B is finite; see Kiefer and Wolfowitz (1955). Denote the first moment of F_W, the mean stationary waiting time, by

$$m_W = \int_{[0,\infty)} t F_W(dt),$$

and the quantity $1 - F_W(0)$ by p_W. If the function

$$w(t) = \int_{[0,\infty)} F_X(t - x) F_W(dx)$$

is continuous at 0, p_W equals the probability that a customer has to wait in the stationary state.

The Laplace–Stieltjes transform ℓ_W of F_W satisfies

$$\ell_W(\theta) = \int_{[0,\infty)} e^{-\theta t} F_W(dt) = \exp\left(\sum_{n=1}^{\infty} n^{-1} \int_{[0,\infty)} (1 - e^{-\theta x}) x F_X^{*n}(dx)\right) \quad (6.1.4)$$

where F_X^{*n} is the n-fold convolution power of F_X, and hence

$$m_W = \sum_{n=1}^{\infty} n^{-1} \int_{[0,\infty)} x F_X^{*n}(dx). \qquad (6.1.5)$$

If $W_0 = 0$, then the mean $m_{W,n}$ of the waiting time W_n is given by

$$m_{W,n} = EW_n = \sum_{k=1}^{n} k^{-1} \int_{[0,\infty)} x F_X^{*k}(dx). \qquad (6.1.6)$$

The rth order cumulant s_r of the stationary waiting time is expressible as

$$s_r = \sum_{k=1}^{\infty} k^{-1} \int_{[0,\infty)} x^r F_X^{*k}(dx); \qquad (6.1.7)$$

here $s_1 = m_W$ and $s_2 = \sigma_W^2$, the variance of the stationary waiting time.

Given the stationary waiting time distribution function F_W the stationary distribution of the number of waiting customers can be determined. Let N_t be the number of customers at time t in the system, let P_k^0 be the probability that an arriving customer in a stationary system finds at least k customers already waiting, and let

$$P_k = \lim_{t \to \infty} P(N_W(t) \geq k) = \lim_{t \to \infty} P(N_t \geq k+1) \quad \text{for } k = 1, 2, \ldots,$$

where $N_W(t) = (N_t - 1)_+$ denotes the number of customers *waiting* at time t. Then

$$P_k^0 = \int_{[0,\infty)} F_A^{*k}(t) F_W(dt), \qquad (6.1.8)$$

and

$$P_k = \int_{[0,\infty)} (F_A^{*(k-1)} * F_{A,e})(t) F_W(dt) \quad \text{for } k = 1, 2, \ldots, \qquad (6.1.9)$$

where

$$F_{A,e}(t) = \lambda \int_0^t [1 - F_A(u)] du \quad \text{with } \lambda = 1/m_A.$$

In the case of dependent service and interarrival times, i.e. for *G/G/1*, the following is true. If the sequence of interarrival and service times $(\{A_n, B_n\})$ for $n = 0, \pm 1, \pm 2, \ldots$ is stationary and ergodic, and if

$$\varrho \equiv m_B/m_A = \lambda m_B = EB_0/EA_0 < 1$$

then there exists a stationary sequence (W_n) of waiting times satisfying

$$W_{n+1} = (W_n + B_n - A_n)_+ \quad \text{for } n = 0, \pm 1, \pm 2, \ldots . \tag{6.1.10}$$

This has been proved by Loynes (1962); see also Brandt et al. (1990) and Baccelli and Brémaud (1994).

Finally

$$P(N_t \geq 1) = \varrho \tag{6.1.11}$$

and

$$L_W = \lambda m_W \tag{6.1.12}$$

('Little's formula'), where L_W denotes the mean queue length; see Baccelli and Brémaud (1994).

6.2 Monotonicity Properties of *GI/GI/1* and *G/G/1* Queues

The transition operator of the homogeneous Markov chain (W_n) of waiting times given by (6.1.1) is monotone with respect to $\leq_{st}$ and $\leq_{icx}$, as can be shown easily using Theorems 4.3.10 and 5.2.9. Consequently, the following is true.

Theorem 6.2.1. *If*

$$F_{W,0} \preceq F_{W,1}, \tag{6.2.1}$$

$\preceq$ *denoting one of $\leq_{st}$ and $\leq_{icx}$ then for $n = 0, 1, \ldots$*

$$F_{W,n} \preceq F_{W,n+1} \tag{6.2.2}$$

and if the stationary waiting time distribution function F_W exists, then

$$F_{W,n} \preceq F_W \quad \text{for } n = 0, 1, \ldots \tag{6.2.3}$$

where $m_W < \infty$ is assumed in the case of $\leq_{icx}$.

In $G/G/1$ queues with ergodic $(\{A_n, B_n\})$ and $W_0 = 0$ we have

$$W_n \leq_{st} W_{n+1} \quad \text{for } n = 0, 1, \ldots ; \tag{6.2.4}$$

see Stoyan (1983), p. 80.

Similar statements for the *queue lengths* are true in general for $\leq_{st}$ (see Theorem 6.7.2); they also hold for $\leq_{icx}$ in the particular cases *M/GI/1* and *GI/M/1* of exponentially distributed interarrival and service times, respectively.

Let N_n^* be the number of customers in an $M/GI/1$ queueing system immediately after the departure of the nth customer. Then

$$N_{n+1}^* = (N_n^* - 1 + Y_n)_+ \quad \text{for } n = 0, 1, \ldots \tag{6.2.5}$$

where Y_n denotes the number of customers arriving during the service of the nth customer. Because of the properties of the Poisson process the Y_n are i.i.d., and (N_n^*) is a Markov chain with the same monotonicity properties as (W_n).

Let N_n^0 be the number of customers in a $GI/M/1$ system immediately before the arrival of the nth customer. Then

$$N_{n+1}^0 = (N_n^0 + 1 - Z_n)_+ \quad \text{for } n = 0, 1, \ldots \tag{6.2.6}$$

where Z_n denotes the number of customers which can be served during the nth interarrival time. Because of the properties of the exponential distribution, the Z_n are i.i.d., and so an analogue of Theorem 6.2.1 exists for (N_n^0) with respect to $\leq_{st}$ and $\leq_{icx}$.

By Theorem 4.3.13 the process (W_n) is *associated* if W_0 is independent of the service and interarrival times. This implies that in the stationary case the covariances $\mathrm{Cov}(W_n, W_{n+m})$ are positive for all n and m. Also Corollary 5.2.23 can be applied and yields that these covariances are decreasing in m.

6.3   Comparison Properties of $GI/GI/1$ and $G/G/1$

Because of the simple form of the recursion formula (6.1.1)

$$W_{n+1} = (W_n + X_n)_+ \quad \text{for } n = 0, 1, \ldots,$$

it is particularly easy to use the mapping method to study the influence of the joint distribution of the differences $(X_1, X_2, \ldots)$ on the distribution functions $F_{W,n}$ of waiting times W_n and the stationary waiting time distribution function F_W. Simply the (max,+) recursion idea is applied. The mapping $\varphi : \mathbb{R}^2 \to \mathbb{R}$ defined by

$$\varphi(x, z) = (x + z)_+,$$

is increasing, convex and supermodular as a function of (x, z). Theorem 4.3.7 therefore shows that W_{n+1} is an increasing, convex and supermodular function of W_0 and $X_0, \ldots, X_n$,

$$W_{n+1} = \Psi_n(W_0, X_0, \ldots, X_n) \quad \text{for } n = 0, 1, \ldots. \tag{6.3.1}$$

Therefore the following general result for arbitrary $G/G/1$ queues holds.

Theorem 6.3.1. *If*

$$(W_0', X_0', \ldots, X_n') \preceq (W_0'', X_0'', \ldots, X_n'') \quad \text{for } n = 0, 1, \ldots,$$

then

$$W'_n \preccurlyeq W''_n \quad for \ n = 0, 1, \dots$$

when $(\preceq, \preccurlyeq)$ *is one of the pairs* $(\leq_{st}, \leq_{st})$, $(\leq_{icx}, \leq_{icx})$ *or* $(\leq_{ism}, \leq_{icx})$.
Especially, in all these cases $EW'_n \leq EW''_n$ *holds for* $n = 0, 1, \dots$.

For the case of $GI/GI/1$ queues this yields the result below.

Theorem 6.3.2. *Consider two GI/GI/1 queues with the difference distribution functions* F'_X *and* F''_X *and the waiting time distribution functions* $F'_{W,n}$ *and* $F''_{W,n}$ *of the nth customer, respectively. Then if* $\preceq$ *denotes* $\leq_{st}$ *or* $\leq_{icx}$

$$F'_X \preceq F''_X \tag{6.3.2}$$

and

$$F'_{W,0} \preceq F''_{W,0} \tag{6.3.3}$$

imply that for $n = 0, 1, \dots,$

$$F'_{W,n} \preceq F''_{W,n}. \tag{6.3.4}$$

If the corresponding stationary waiting time distribution functions F'_W *and* F''_W *exist, and, additionally when* $\preceq$ *denotes* $\leq_{icx}$ *they have finite means* m'_W *and* m''_W, *then (6.3.2) is sufficient to ensure that*

$$F'_W \preceq F''_W \tag{6.3.5}$$

and

$$m'_W \leq m''_W. \tag{6.3.6}$$

Proof. (of the relations for the stationary characteristics.) For $W'_0 = W''_0 = 0$, it is $F'_{W,0} \leq_{st} F'_{W,1}$ and $F''_{W,0} \leq_{st} F''_{W,1}$; hence $F'_{W,n} \leq_{st} F'_{W,n+1}$ and $F''_{W,n} \leq_{st} F''_{W,n+1}$ and, since F'_W and F''_W exist, $(F'_{W,n})$ and $(F''_{W,n})$ converge weakly to F'_W and F''_W as $n \to \infty$. Again, from the monotone convergence of $(F'_{W,n})$ and $(F''_{W,n})$, the corresponding means converge monotonically. Since $F'_{W,0} = F''_{W,0}$ for these processes, it follows (6.3.4), and hence the limits satisfy (6.3.5) and (6.3.6). $\qquad \square$

Remark 6.3.3. (1) Sufficient conditions for (6.3.2) for $\leq_{icx}$ are

$$F''_A \leq_{icv} F'_A \text{ and } F'_B \leq_{icx} F''_B. \tag{6.3.7}$$

Also, recall from (1.4.7) that, in the case of equal means of F'_A and F''_A, $F'_A \leq_{icx} F''_A$ and $F''_A \leq_{icv} F'_A$ are equivalent.

Another sufficient condition is $(A'_n, B'_n) \leq_{sm} (A''_n, B''_n)$ for two queues where the pairs (interarrival time, service time) are independent but where there are correlations between the service time of a customer and the interarrival time for the following customer, see Müller (2000).

(2) Theorem 6.3.2 yields the intuitively clear statement that 'lengthening' service or 'shortening' interarrival times (or both) leads to 'longer' waiting times. The extremality statement, already known in the 1960s (Rogozin (1966) and Rossberg (1968)), that in the case of constant service or interarrival times the mean stationary waiting times are shorter than for all distribution functions with the same means, can be embedded in the comparison statement of Theorem 6.3.2 for $\leq_{icx}$. To see this, recall from Section 1.10 that constant service times are $\leq_{icx}$-minimal and constant interarrival times are $\leq_{icv}$-maximal in the sets of random variables with fixed means.

A simpler embedding of this extremality result into a general statement is:

> It is generally believed that the more regular (in some appropriate sense) each of these processes is, the better any of the usual performance measures will be. (Wolff (1977a))

Note, however, that the term 'any' has to be used with care; for example, the probability p_W that a customer has to wait does not necessarily follow this pattern; see Atkinson (2000).

The comparability of all moments of order greater than one of the stationary waiting times follows from Theorem 6.3.2, when (6.3.2) is true for $\leq_{st}$ or $\leq_{icx}$. The following theorem shows that the cumulants are comparable even under weaker conditions (see Definition 1.6.2 for $\leq_{r-icx}$).

Theorem 6.3.4. *Let s'_n and s''_n be the nth order cumulants of the stationary waiting times in two GI/GI/1 systems, and for some $r \geq 1$ suppose that $F'_X \leq_{r-icx} F''_X$. Then*

$$s'_n \leq s''_n \text{ for all } n \geq r. \tag{6.3.8}$$

Proof. Because of the convolution property (C) of $\leq_{r-icx}$,

$$F'^{*k}_X \leq_{r-icx} F''^{*k}_X \quad \text{for all } k = 1, 2, \ldots,$$

and so for all $n \geq r$,

$$\int_{[0,\infty)} x^n F'^{*k}_X(\mathrm{d}x) \leq \int_{[0,\infty)} x^n F''^{*k}_X(\mathrm{d}x).$$

(6.3.8) then follows from (6.1.7). $\qquad\qquad\qquad\qquad\qquad\qquad\qquad\square$

Theorem 6.3.4 yields extremal properties of cumulants. In the set of all distribution functions of interarrival (or service) times with fixed means and variances, two-point distribution functions yield extremal values of the cumulants for $n \geq 2$, in particular, of the variance σ^2_W of the stationary waiting time; see Bergmann, Daley, Rolski, and Stoyan (1979). When more than two moments are known then the results of Denuit, Lefevre, and Shaked (2000) can be used to determine the extremal distributions, and thus bounds for cumulants. Bergmann (1979) proved monotonicity properties of covariances of waiting times with respect to $\leq_{st}$.

Statements about queue lengths that are analogous to (6.3.4) and (6.3.5) can be proved for $\leq_{st}$ (see Theorem 6.7.2). In the stationary case, the proof can be accomplished by using the formulae (6.1.8) and (6.1.9). For the $M/GI/1$ system, use of the recursion formulae (6.2.5) and (6.2.6) enables the case $\leq_{icx}$ to be treated also; Rolski (1976) shows how this extension is possible for queues with group arrivals or service. Finally, the formulae for $W(t)$ for these systems enable a proof of the sharper results below.

Theorem 6.3.5. (a) For two $M/GI/1$ queues with arrival rates λ' and λ'' suppose that $\lambda' \leq \lambda''$, $F_B' \leq_{cx} F_B''$, and set $\varrho' = \lambda'm_B$ and $\varrho'' = \lambda''m_B$ with $\varrho'' < 1$. Then the corresponding stationary waiting times satisfy

$$W' \leq_{st} W''. \qquad (6.3.9)$$

(b) For two $GI/M/1$ queues with mean service times $1/\mu'$ and $1/\mu''$ suppose that $\mu' \geq \mu''$, $F_A'' \leq_{Lt} F_A'$, $\varrho' \equiv 1/\mu'm_A' < 1$ and $\varrho'' \equiv 1/\mu''m_A'' < 1$. Then the corresponding stationary waiting times satisfy (6.3.9).

Proof. It is well-known that in an $M/GI/1$ queue

$$W =_{st} \sum_{n=1}^{N} X_n,$$

where $X_1, X_2, \ldots$ are i.i.d. with distribution function

$$F_e(t) = \frac{1}{m_B} \int_0^t [1 - F_B(x)]\mathrm{d}x$$

and N is geometrically distributed with parameter ϱ. Since F_B' and F_B'' have the same means, $F_B' \leq_{cx} F_B''$ is equivalent to $F_B' \leq_{icx} F_B''$ and hence to

$$\frac{1}{m_B} \int_0^t [1 - F_B'(x)]\mathrm{d}x \geq \frac{1}{m_B} \int_0^t [1 - F_B''(x)]\mathrm{d}x \quad \text{for } t \geq 0$$

(see Theorem 1.5.7) and therefore $F_e' \leq_{st} F_e''$. Thus (6.3.9) follows from Theorem 4.3.5.

For $GI/M/1$ queues the formula

$$F_W(t) = 1 - \delta \exp(-\mu(1 - \delta)t) \quad \text{for } t \geq 0$$

is used, where δ is the unique root in $0 < z < 1$ of the equation

$$z = \int_{[0,\infty)} \exp(-\mu(1 - z)t)F_A(\mathrm{d}t). \qquad (6.3.10)$$

The inequalities $F_A'' \leq_{Lt} F_A'$ and $\mu' \geq \mu''$ yield

$$\int_{[0,\infty)} \exp(-\mu''(1-z)t)F_A''(dt) \geq \int_{[0,\infty)} \exp(-\mu'(1-z)t)F_A'(dt)$$

for all real $0 < z < 1$, and so the corresponding roots of (6.3.10) satisfy $\delta' \leq \delta''$. □

Daley and Rolski (1984) strengthened these results. They showed that (6.3.9) is already true if the mean service times are comparable.

6.4 Bounds Obtained from Comparison Properties of $GI/GI/1$

For a given $GI/GI/1$ queue bounds for the stationary waiting time distribution function F_W can be obtained by using systems Σ' and Σ'' for which

$$F_X' \preceq F_X \preceq F_X''$$

where $\preceq$ denotes $\leq_{icx}$ or $\leq_{st}$ and appealing to Theorems 6.3.2 and 6.3.5. The relation $\leq_{icx}$ is of special interest, because it can lead to bounds from systems with the same traffic intensity ϱ. In particular, practically useful bounds for m_W are obtained by using the formulae for $M/GI/1$ and $GI/M/1$.

(a) If

$$F_A \leq_{icx} \text{Exp}(1/m_A), \tag{6.4.1}$$

then

$$m_W \leq \frac{EB_1^2}{2m_A(1-\varrho)} = \frac{\sigma_B^2 + m_B^2}{2m_A(1-\varrho)}. \tag{6.4.2}$$

Notice that (6.4.1) holds if and only if F_A is HNBUE, see Theorem 1.8.7.

(b) If F_B is HNBUE then

$$m_W \leq \delta m_B/(1-\delta) \tag{6.4.3}$$

where δ is the unique root in $0 < \delta < 1$ of the equation

$$\delta = \int_{[0,\infty)} \exp\left(-(1-\delta)t/m_B\right) F_A(dt).$$

In many cases this bound is better than (6.4.2).

A general lower bound for m_W is obtained by using the extremal property of constant interarrival times.

Theorem 6.4.1. *For any GI/GI/1 system with $\varrho < 1$,*

$$m_W \geq \frac{\sigma_B^2}{2m_A(1-\varrho)} - \frac{m_B}{2}, \qquad (6.4.4)$$

and if $b \equiv \sup\{t : F_B(t) = 0\} > 0$, then

$$m_W \geq \frac{\sigma_B^2}{2m_A(1-\varrho)} - \frac{m_B - b}{2}. \qquad (6.4.5)$$

Proof. When the interarrival distribution function is IFR, the bound

$$m_W \geq \frac{\sigma_A^2 + \sigma_B^2}{2m_A(1-\varrho)} - \frac{1}{2}m_A(\varrho + \sigma_A^2/m_A^2)$$

holds (see Stoyan (1983), p. 90), and in particular, since the distribution function of a constant (m_A) is IFR, this bound applies to $D/GI/1$ queues. By Theorem 6.3.2 m_W is bounded below by the mean waiting time in $D/GI/1$ with constant interarrival time m_A and service time distribution function F_B, which in turn is bounded as in (6.4.4).

The bound (6.4.5) is proved by applying (6.4.4) to a $D/GI/1$ queue with service time distribution function $F_B(t+b)$ and interarrival times $m_A - b$. (For this queue the X_n have the same distribution function as in the $D/GI/1$ queue above and thus the waiting time distribution function is not changed.) □

6.5 Bounds in the Case of Non-renewal Input

Methods similar to those used for independent interarrival and service times yield bounds in some cases where the interarrival and service times may be stochastically dependent.

Assume that the sequence $(\{A_n, B_n\})$ is ergodic and that the mean stationary waiting time is finite. If the service times are mutually independent and independent of the interarrival times, then as analogues of (6.4.4) and (6.4.5)

$$m_W \geq \frac{\sigma_B^2}{2m_A(1-\varrho)} - \frac{m_B}{2} \qquad (6.5.1)$$

and

$$m_W \geq \frac{\sigma_B^2}{2m_A(1-\varrho)} - \frac{m_B - b}{2} \qquad (6.5.2)$$

for any b for which $F_B(b) = 0$. The proof of these formulae is the same as in the case of a renewal input, for it uses the fact that m_W is greater than the mean stationary waiting time in $D/GI/1$.

If the arrival point process is smaller (larger) with respect to $\leq_{cx.P}$ than a Poisson process with the same intensity (as in the setting of Theorem 5.4.1) then, analogously to (6.4.2),

$$m_W \leq (\geq) \frac{\sigma_B^2 + m_B^2}{2m_A(1 - \varrho)}. \tag{6.5.3}$$

Such bounds are considered in Rolski (1981, 1984, 1986) for the cases of arrival processes which are Markov renewal processes or doubly stochastic or Cox processes; Theorem 5.4.1 is the base for the latter case.

For $G/GI/1$ queues also the *workload* or virtual waiting time has been studied in various papers. Note first that for two such queues with the same arrival process but with $\leq_{icx}$-comparable (i.i.d.) service times the workloads W_t are $\leq_{icx}$-comparable for any t. This is a simple consequence of Theorem 6.3.1.

More interesting is the comparison of two systems with the same service time distribution function and different arrival processes. The general tendency is that more variable arrival processes lead to longer waiting times. Ross (1978) conjectured that in any $G/GI/1$ queue with an arrival process with average intensity λ the mean waiting times are longer than in a corresponding $M/GI/1$ queue with the same intensity. Rolski (1981) proved this conjecture when the arrival process is a doubly stochastic Poisson process. Rolski (1986) gives an upper bound for the stationary workload in this model in terms of mixtures of the same characteristics in a family of $M/GI/1$ queues with varying intensities of the arrival process.

For the case of two Markov renewal processes with $\leq_{icx}$-comparable F_i (see Section 5.4) Rolski (1984) showed that the stationary workload distribution functions are also $\leq_{icx}$-comparable.

Chang, Chao, and Pinedo (1991) and Bäuerle and Rolski (1998) considered $\leq_{icx}$-comparability of the stationary workload distributions of queues with Markov modulated arrival processes. They showed that faster changing environmental processes lead to smaller workload characteristics. Similar results were also given in Szekli et al. (1994) and Bäuerle (1997b). The proofs of the results in these papers are typically based on Theorem 6.3.1 in combination with Theorem 5.2.18.

Li and Xu (2000) studied systems of s parallel single-server queues with correlated input vectors (in particular, s-dimensional interarrival time vectors). They proved comparison theorems for corresponding s-dimensional performance characteristics with respect to $\leq_{uo}$, $\leq_{lo}$ and $\leq_{sm}$. They also compared such systems with systems having input vectors with independent components, using the dependence properties of Section 3.10.

6.6   Basic Facts for the Multi-server System *GI/GI/s*

The queue *GI/GI/s* is a generalization of the model *GI/GI/1* and differs from this by the existence of s servers instead of one server. On their arrival, customers are assigned with equal probability to any empty server if such exist, or else, if all servers are busy, the customers wait in a common queue from which they go to service in order of their arrival (FCFS queue discipline) as soon as any server becomes free. The characteristics of the model are denoted as in the case of the model *GI/GI/1* in Section 6.1. *G/G/s* denotes an analogous system without independence assumptions.

GI/GI/s queues are, of course, much more complicated than *GI/GI/1* queues. For example, in studying waiting times, s-dimensional vectors must be considered, and consequently there is no such (relatively) simple integral equation for the waiting time distribution function as that of Lindley in (6.1.3). For further information on results for *GI/GI/s*, see Wolff (1989) and the original papers of Kiefer and Wolfowitz (1955, 1956). The case of systems *G/G/s* is treated in Baccelli and Brémaud (1994) and Brandt et al. (1990).

Following Kiefer and Wolfowitz (1955) the waiting time process in a *GI/GI/s* system can be studied as an s-dimensional vector process $(\mathbf{W}^*(t)) = (W_1^*(t), \dots, W_s^*(t))$ in which the kth component equals the workload of the kth server, i.e. the sum of the residual service time of the customer (if any) currently in service at time t with the kth server plus the service times of those customers (if any) waiting in the queue and subsequently served by the kth server, while $W_k^*(t) = 0$ if this sum is empty. Rather than $\mathbf{W}^*(t)$, it proves to be more convenient to rearrange its components in ascending order to yield the vector

$$\mathbf{W}(t) = (W_1(t), \dots, W_s(t))$$

where $0 \leq W_1(t) \leq \dots \leq W_s(t)$. Furthermore, $\mathbf{W}_n$ denotes the vector $\mathbf{W}(T_n-)$ at the epoch T_n of arrival of the nth customer,

$$\mathbf{W}_n = (W_1(T_n-), \dots, W_s(T_n-)) = (W_{1,n}, \dots, W_{s,n}),$$

yielding $W_{1,n}$ as the waiting time of the nth customer. Corresponding to (6.1.1) then

$$\mathbf{W}_{n+1} = (\mathbf{R}(\mathbf{W}_n + B_n \mathbf{e}_1 - A_n \mathbf{1}))_+ \quad \text{for } n = 1, 2, \dots \qquad (6.6.1)$$

where $\mathbf{e}_1$ and $\mathbf{1}$ denote the s-vectors $(1, 0, \dots, 0)$ and $(1, 1, \dots, 1)$ respectively, and for any s-vector $\mathbf{x}$, $(\mathbf{x})_+$ denotes $(\max\{0, x_1\}, \dots, \max\{0, x_s\})$ and $\mathbf{R}(\mathbf{x})$ consists of the components of $\mathbf{x}$ rearranged in ascending order. Finally, let $F_{W,n}$ denote the distribution function of $W_{1,n}$ and $F_{\mathbf{W},n}$ that of $\mathbf{W}_n$.

Kiefer and Wolfowitz (1955) showed that when

$$sm_A > m_B, \quad \text{i.e. } \varrho/s < 1 \quad \text{where } \varrho = m_B/m_A, \tag{6.6.2}$$

then there exists a uniquely determined stationary distribution function $\mathbf{F_W}$ for the Markov chain $(\mathbf{W}_n)$, and for any $\mathbf{W}_0$,

$$F_{\mathbf{W},n} \text{ converges weakly to } F_{\mathbf{W}} \quad \text{for } n \to \infty. \tag{6.6.3}$$

Since $F_{W,n}(t) = F_{\mathbf{W},n}(t, \infty, \dots, \infty)$, it follows from (6.6.3) that

$$F_{W,n} \text{ converges weakly to } F_W \quad \text{for } n \to \infty. \tag{6.6.4}$$

Kiefer and Wolfowitz (1956) also showed that the stationary mean waiting time m_W is finite if $EB_n^2 < \infty$, i.e. if the service time distribution function has a finite second moment.

In the case of constant service times, i.e. of systems of the type $GI/D/s$, it is enough to consider single-server queues of the type $GI/D/1$ for the study of waiting times, because when the customers are assigned to the servers *cyclically*, i.e. the $(js + k)$th customer goes to the kth server for all j irrespective of the state of the system, the waiting times are the same as under the FCFS discipline. With the cyclic discipline, $GI/GI/s$ works like s staggered $GI/GI/1$ systems with the interarrival distribution function A^{*s}.

Finally, if the stationary waiting time distribution function F_W is known, the stationary queue length distribution function can be determined via (6.1.8) and (6.1.9); if $F_{\mathbf{W}}$ is known the distribution function of the number of customers in the system can be determined as well.

6.7 Monotonicity Properties of $GI/GI/s$ Queues

In general the queue $GI/GI/s$ has similar monotonicity properties as $GI/GI/1$, but only with respect to $\leq_{st}$. For $\leq_{icx}$ similar statements are true only in particular cases, for example, $GI/D/s$ systems. This is a consequence of the mapping

$$\varphi : \mathbb{R}_+^{s+2} \to \mathbb{R}_+^s$$

given by

$$\varphi(\mathbf{W}, A, B) = (\mathbf{R}(\mathbf{W} + B\mathbf{e}_1 - A\mathbf{1}))_+, \tag{6.7.1}$$

being monotone in $\mathbf{W}$, A and B (increasing, decreasing, and increasing respectively), but not convex. Theorem 4.3.10 yields

Theorem 6.7.1. *If*

$$\mathbf{W}_0 \leq_{st} \mathbf{W}_1, \tag{6.7.2}$$

then for $n = 0, 1, \ldots,$

$$\mathbf{W}_n \leq_{st} \mathbf{W}_{n+1} \tag{6.7.3}$$

and, if the stationary waiting time distribution function $F_\mathbf{W}$ exists,

$$F_{\mathbf{W},n} \leq_{st} F_\mathbf{W}. \tag{6.7.4}$$

If $\mathbf{W}_0 \geq_{st} \mathbf{W}_1$, then the stationary waiting time distribution function $F_\mathbf{W}$ exists, and (6.7.3) and (6.7.4) hold with the reverse inequality.

The last part of the theorem is a consequence of $(\mathbf{W}_n)$ being decreasing in distribution and being almost surely bounded below by zero.

The sequence $(\mathbf{W}_n)$ is *associated* if $\mathbf{W}_0 = (0, \ldots, 0)$ or if the sequence is stationary. This is easily obtained as a particular case of Theorem 4.3.13. Even more is true: the components of the vectors $\mathbf{W}_n$ are associated. In the stationary case the association of the $\mathbf{W}_n$ implies that the waiting times are positively correlated.

Also for the *number of customers* in the system there exists a monotonicity statement, which can be proved by using a particular Markov chain as follows. Let N_n denote the number of customers in the system at T_n-, let $Z_{1,n}, \ldots, Z_{s,n}$ denote the residual service times of customers being in service at T_n-, specified in increasing order so that $Z_{1,n} \leq \ldots \leq Z_{s,n}$ and $Z_{1,n} = 0$, $\ldots, Z_{s-N_n,n} = 0$ if $N_n \leq s$, and when $N_n > s$, let $Y_{1,n}, \ldots, Y_{(N_n-s),n}$ denote the service times of the $N_n - s$ customers waiting at $T_n - 0$, specified in order of their arrival; if $N_n \leq s$ there are no such Y random variables. Then the sequence (X_n) defined by

$$X_n = \{N_n, Z_{1,n}, \ldots, Z_{s,n}, Y_{1,n}, \ldots, Y_{(N_n-s),n}\}$$

is a homogeneous Markov chain on state space $(S, \mathcal{S})$ where S is the set of all finite sequences $\boldsymbol{x} = (x_0, \ldots, x_j)$ where $j = \max\{s, x_0\}, x_0 \in \{0, 1, \ldots\}, x_i \geq 0$ (for $i = 1, \ldots, j$), and $\mathcal{S}$ the σ-algebra generated by the sets M of the form

$$M = \begin{cases} (i, \{0\}^{s-i} \times C) & \text{for } i = 0, \ldots, s-1, \\ (i, C) & \text{for } i = s, s+1, \ldots, \end{cases}$$

where C is any Borel subset of $\mathbb{R}_+^i$. X_{n+1} is then given by

$$X_{n+1} = \varphi(X_n, A_{n+1}, B_n)$$

where the mapping $\varphi : S \times \mathbb{R}_+ \times \mathbb{R}_+ \to S$ is measurable and has the monotonicity property that for $\mathbf{x}_k \in S$, a_k and $b_k > 0$ for $k = 1$ and 2

$$\mathbf{x}_1 \leq_S \mathbf{x}_2, \ a_1 \geq a_2, \ b_1 \leq b_2 \text{ implies } \varphi(\mathbf{x}_1; a_1, b_1) \leq_S \varphi(\mathbf{x}_2; a_2, b_2)$$

when as partial ordering $\mathbf{x} \leq_S \mathbf{y}$ on S is used which is satisfied if (i) $x_0 \leq y_0$, and (ii) for each $i \in I_x \equiv \{1, \ldots, \max\{s, x_0\}\}$ there exists $j(i) \in I_y \equiv$

$\{1, \ldots, \max\{s, y_0\}\}$ such that $x_i \leq y_{j(i)}$ and if $i_1 \neq i_2$, $j(i_1) \neq j(i_2)$, and also if $s < i_1 < i_2$, then $s < j(i_1) < j(i_2)$ while if $i \leq s$ then $j(i) \leq s$.

The lengthy proof is elementary and omitted here, being given in Stoyan (1973). This monotonicity of φ means that (X_n) satisfies the assumptions of Theorem 4.3.10 for $\leq_{st}$ (generated by the partial ordering $\leq_S$ on S) on the set of distributions on S, and hence the following theorem holds.

Theorem 6.7.2. *Let P_n be the distribution of X_n for $n = 0, 1, \ldots$. If*

$$P_0 \leq_{st} P_1, \tag{6.7.5}$$

then for all $n = 0, 1, \ldots$,

$$P_n \leq_{st} P_{n+1} \tag{6.7.6}$$

and, if a stationary distribution P exists, then

$$P_n \leq_{st} P. \tag{6.7.7}$$

If $P_0 \geq_{st} P_1$, then a stationary distribution P exists, and (6.7.6) and (6.7.7) hold with the reverse inequality.

For example, (6.7.5) is satisfied when $X_0 = (0, \mathbf{0})$. Of course, inequalities for the number of customers follow from (6.7.6) and (6.7.7).

6.8 Comparability Properties of $GI/GI/s$

In general, just as for monotonicity, comparability properties of $GI/GI/s$ can be proved only with respect to $\leq_{st}$. Examples can be given to show that a relation analogous to (6.8.1) with $\leq_{st}$ replaced by $\leq_{icx}$ is not true in general; see Wolff (1977a,b) and Whitt (1980b).

Theorem 6.8.1. *Consider two $GI/GI/s$ queues with the interarrival and service time distribution functions F'_A and F''_A and F'_B and F''_B, and let $F'_{\mathbf{W},n}$ and $F''_{\mathbf{W},n}$ be the corresponding distribution functions of the waiting time vectors at T_n-. Then*

$$F'_{\mathbf{W},0} \leq_{st} F''_{\mathbf{W},0}, F''_A \leq_{st} F'_A, \text{ and } F'_B \leq_{st} F''_B \text{ imply } F'_{\mathbf{W},n} \leq_{st} F''_{\mathbf{W},n} \tag{6.8.1}$$

for $n = 1, 2, \ldots$, while for the stationary distribution functions $F'_{\mathbf{W}}$ and $F''_{\mathbf{W}}$ (assuming they exist),

$$F''_A \leq_{st} F'_A \text{ and } F'_B \leq_{st} F''_B \text{ imply } F'_{\mathbf{W}} \leq_{st} F''_{\mathbf{W}}. \tag{6.8.2}$$

Analogous inequalities are true for the waiting time distribution functions F'_W and F''_W. Theorem 6.8.1 is an easy consequence of Theorem 4.3.9 observing that the mapping φ defined by (6.7.1) is increasing in $\mathbf{W}_n$ and B_n and decreasing in A_n.

The distributions of the number of customers in the system or in the queue at arrival instants have analogous monotonicity properties with respect to $\leq_{st}$.

Similarly, just as for $G/G/1$, comparability properties can be proved with respect to $\leq_{st}$ for $G/G/s$ (i.e. for dependent interarrival and service times).

Because of the relation between $GI/D/s$ and $GI/D/1$ noted at the end of Section 6.6, the following result for stationary waiting time distributions in $GI/D/s$ can be obtained.

Theorem 6.8.2. *Consider two $GI/D/s$ queues with distribution functions of interarrival times F'_A and F''_A and service times m'_B and m''_B. Then*

$$F''_A \leq_{st} F'_A \quad and \quad m'_B \leq m''_B \quad imply \quad F'_W \leq_{st} F''_W, \qquad (6.8.3)$$

$$F''_A \leq_{icv} F'_A \quad and \quad m'_B \leq m''_B \quad imply \quad F'_W \leq_{icx} F''_W. \qquad (6.8.4)$$

Using a proof analogous to that of Theorem 6.3.5(b) Daley and Rolski (1984) showed the following result.

Theorem 6.8.3. *For two $GI/M/s$ queues with mean service times $1/\mu'$ and $1/\mu''$ assume that $\mu' \geq \mu''$, $F''_A \leq_{Lt} F'_A$, and $\varrho' \equiv (s\mu'm'_A)^{-1} < 1$ and $\varrho'' \equiv (s\mu''m''_A)^{-1} < 1$. Then the corresponding stationary waiting time distribution functions satisfy*

$$F'_W \leq_{st} F''_W.$$

Perhaps surprisingly, comparability properties with respect to $\leq_{st}$ cannot be proved in general for such quantities as the number of customers in the system at (deterministic) time t or for the virtual waiting time; see Jacobs and Schach (1972). However, if the input of the two systems is the same, the following theorem is true.

Theorem 6.8.4. *Consider two initially empty $G/GI/s$ queues with the same arrival process and service time distribution functions F'_B and F''_B and denote by W'_t, W''_t and N'_t, N''_t the virtual waiting times and the numbers of customers at time t, respectively. Then*

$$F'_B \leq_{st} F''_B \qquad (6.8.5)$$

implies that for all $t \geq 0$,

$$W'_t \leq_{st} W''_t \qquad (6.8.6)$$

and

$$N'_t \leq_{st} N''_t. \qquad (6.8.7)$$

Proof. One can use a simple coupling argument and assume that both systems have the same sequence (A_n) of interarrival times. For given $t > 0$, suppose that

$$T_n = \sum_{i=1}^{n} A_i \leq t < \sum_{i=1}^{n+1} A_i = T_{n+1}.$$

By Theorem 6.8.1 the waiting times of the customers arriving at T_n satisfy $W_n' \leq_{st} W_n''$. Thus (6.8.6) follows from the relation $W_t = (W_n - (t - T_n))_+$, (where n is such that $T_n \leq t < T_{n+1}$). Relation (6.8.7) is proved similarly using Theorem 6.7.2. $\square$

An interesting theme in the study of multiserver queues is the comparison of systems with the same arrival process (i.e. the same T_n) but different numbers of servers where the total 'service effort' is constant. To be definite, consider two $G/M/s$ queues Σ' and Σ'' with the same arrival process, server numbers s' and s'' with $s' < s''$ and service rates μ' and μ'', satisfying $s'\mu' = s''\mu'' = C$.

Let X_t' and X_t'' be the numbers of customers in Σ' and Σ'' at time t, where $X_0' = X_0'' = 0$. Chao and Scott (2000) showed (among other results) the inequality

$$(X_t') \leq_{st} (X_t'').   \tag{6.8.8}$$

Its proof can be based on the following theorem, which is obtained by the coupling method and which can be used also for the proof of comparison results for systems with bounded waiting space.

Theorem 6.8.5. *Let Σ' and Σ'' be two $G/M/1$ queues being empty at $t = 0$ with the same arrival process and state dependent service rates μ_i' and μ_i'' respectively when there are i customers in the systems, which are bounded above by some finite Δ. If*

$$\mu_i'' \leq \mu_i' \leq \mu_{i+1}''   \quad for\ i = 1, 2, \ldots   \tag{6.8.9}$$

then

$$(X_t') \leq_{st} (X_t'') \leq_{st} (X_t' + 1).$$

Proof. The result is obtained by constructing two processes $(\hat{X}_t')$ and $(\hat{X}_t'')$ on a common probability space such that for every $t \geq 0$,

$$X_t' =_{st} \hat{X}_t' \quad and \quad X_t'' =_{st} \hat{X}_t''$$

and

$$\hat{X}_t' \leq \hat{X}_t'' \leq \hat{X}_t' + 1   \quad for\ all\ t.   \tag{6.8.10}$$

All the potential service completions can be generated by a Poisson process with rate Δ. Since the arrival process is independent of the service process it can be assumed that it is deterministic. Consider the superposition of both processes. Sort the points of the superposition in increasing order $0 = \tau_0 \le \tau_1 \le \cdots$. Since the states of the two queues do not change between τ_n and τ_{n+1}, it suffices to show that

$$\hat{X}_n' \le \hat{X}_n'' \le \hat{X}_n' + 1, \qquad \text{for } n = 0, 1, \dots, \tag{6.8.11}$$

where $\hat{X}_n' = \hat{X}_{\tau_n}'$ etc.

We prove (6.8.11) by induction. It is true for $n = 0$ as the queues are empty in the beginning. Assume that the result has been established for n and proceed to prove it for $n + 1$.

If τ_{n+1} is an arrival epoch then (6.8.11) is clearly satisfied for $n + 1$ because $\hat{X}_{n+1}' = \hat{X}_n' + 1$ and $\hat{X}_{n+1}'' = \hat{X}_n'' + 1$. Suppose now that τ_{n+1} is a potential service completion epoch.

Take a uniform random variable U_{n+1} independent of $\hat{X}_n'$ and $\hat{X}_n''$. At time τ_{n+1} the states of the two queueing systems are

$$\hat{X}_{n+1}' = \hat{X}_n' - 1_{[U_{n+1} \le \mu'_{\hat{X}_n'}/\Delta]} \tag{6.8.12}$$

and analogously for $\hat{X}_{n+1}''$.

Prove first $\hat{X}_{n+1}' \le \hat{X}_{n+1}''$. If $\hat{X}_n' < \hat{X}_n''$ then $\hat{X}_{n+1}' \le \hat{X}_{n+1}''$ is always satisfied by (6.8.12). If $\hat{X}_n' = \hat{X}_n''$ then it follows from (6.8.9) that $\mu'_{\hat{X}_n'} \ge \mu''_{\hat{X}_n''}$, and (6.8.12) yields $\hat{X}_{n+1}' \le \hat{X}_{n+1}''$.

Show next $\hat{X}_{n+1}'' \le \hat{X}_{n+1}' + 1$. If $\hat{X}_n'' < \hat{X}_n' + 1$ then it follows from (6.8.12) that $\hat{X}_{n+1}'' \le \hat{X}_{n+1}' + 1$ is always satisfied. If $\hat{X}_n'' = \hat{X}_n' + 1$, then again by assumption (6.8.9)

$$\mu''_{\hat{X}_n''} \ge \mu'_{\hat{X}_n'},$$

thus

$$1_{[U_{n+1} \le \mu''_{\hat{X}_n''}/\Delta]} \ge 1_{[U_{n+1} \le \mu'_{\hat{X}_n'}/\Delta]}.$$

Hence it follows from (6.8.12) that $\hat{X}_{n+1}'' \le \hat{X}_{n+1}' + 1$ is satisfied.

Thus it has been shown that (6.8.11) holds for all n and consequently (6.8.10) is satisfied for all $t \ge 0$. This completes the proof of the theorem. $\square$

Remark 6.8.6. a) The initial condition can be relaxed from being initially empty to

$$X_0' \le X_0'' \le X_0' + 1.$$

b) To see that (6.8.8) follows from Theorem 6.8.5, note that a $G/M/s$ queue can be considered as a $G/M/1$ queue with state dependent service rate

$\mu_i = \min\{i, s\}\frac{C}{s} = C\min\{1, \frac{i}{s}\}$. For two such queues with s' and $s'' = s' + 1$ servers, respectively, condition (6.8.9) is

$$\min\{1, \frac{i}{s'+1}\} \leq \min\{1, \frac{i}{s'}\} \leq \min\{1, \frac{i+1}{s'+1}\},$$

which obviously holds. The case of arbitrary $s' < s''$ follows by transitivity.

Chao and Scott (2000) also show that for the queue lengths the inequality in (6.8.8) is reversed. Moreover they prove that under additional conditions comparison is also possible in the case of nonexponential service times. If the service times B_n in a $G/GI/s$ queue have an IFR distribution then the number of customers in the queue is stochastically smaller than in a $G/GI/1$ queue with the same T_n but service times B_n/s.

6.9   Remarks on other Queueing Systems

In the sections above mainly classical results for classical systems are given. The reader should note that there is a vast literature on monotonicity and comparability results and bounds for many other systems of queueing theory, including, in particular, queueing networks, systems with infinitely many servers and loss systems. References are Stoyan (1983), Szekli (1995), Baccelli and Brémaud (1994), and Shaked and Shanthikumar (1994).

Glasserman and Yao (1994) present a general theory for the comparison of event epochs with respect to $\leq_{icx}$ using (max,+) recursions. These authors prove also comparison theorems for so-called *generalized semi-Markov process* with respect to $\leq_{st}$. Many systems of queueing, reliability and related theories can be described by such processes. Their results

> subsume many earlier results for specific models. Most comparison results, particularly in the queueing literature, proceed by induction to establish a sample-path ordering from which stochastic ordering follows. The conditions developed [here] isolate the essential features embedded in many such examples; once the key conditions are identified, it becomes possible to establish monotonicity results in a unified way, without recourse to case-by-case analysis. That the essential condition for monotonicity should also lead to many other interesting results is not surprising. (p. 70 in Glasserman and Yao (1994)).

In a series of papers *queueing disciplines* are compared. This is closely related to scheduling problems. We mention here papers which compare characteristics of $G/GI/s$ queues under the FCFS discipline with their analogues under any other discipline which allocates each arrival to a server independently of the service time of it and subsequent arrivals. An example of such a discipline is the cyclical assignment rule sending arrival $sj + i$ to server i.

Between the characteristics (those corresponding to FCFS are marked with an $'F'$) inequalities of different natures are true, where it is always assumed that the systems are empty at $t = 0$.

Pathwise comparison is true under the coupling assumption that B_n is the service time of the nth customer to *enter* service and not that of the nth customer arriving. Then

$$J_n^F \leq J_n \quad \text{for } n = 1, 2, \ldots$$

and

$$C_t^F \leq C_z \quad \text{for } t \geq 0,$$

where J_n is the nth ordered departure epoch from the queue and C_t the cumulative work performed by epoch t; see Wolff (1987).

Stochastic comparison is possible for the work in system V_t at time t,

$$V_t^F \leq_{st} V_t,$$

and the same for the work in queue; see Wolff (1987).

Only weaker comparison is possible for the work-load vectors $\mathbf{V}_n$. The ith component of $\mathbf{V}_n$ denotes the work load the ith server is facing immediately after the arrival of the nth customer, $i = 1, \ldots, s$. Daley (1987) (see also Foss (1980), who was the first to use ideas from majorization in the given context, and Foss and Chernova (2001)) showed

$$f(\mathbf{V}_n^F) \leq_{st} f(\mathbf{V}_n) \quad \text{for } n = 1, 2, \ldots$$

for all mappings $f : \mathbb{R}^s \to \mathbb{R}$ which are increasing, symmetric and convex. Examples are $f(x_1, \ldots, x_s) = \sum_{i=1}^{s} x_i$ and $f(x_1, \ldots, x_s) = \max x_i$.

Finally, the stationary waiting times and the stationary residual waiting times are $\leq_{icx}$-smaller under FCFS discipline, as shown by Wolff (1977b).

7

Applications to Various Stochastic Models

7.1 Monotonicity Properties and Bounds for the Renewal Function

Consider a system in which an item begins working at time $t = 0$ and fails at $t = X_1$, being replaced immediately by a new one which fails at $t = X_1 + X_2$, being in turn replaced by a new one, and so on. Assume that $X, X_1, X_2, \ldots$ are i.i.d. non-negative random variables with the distribution function F having $F(0) < 1$ and finite mean $m = EX = \lambda^{-1}$. Let $T_0 = 0$, $T_n = X_1 + \ldots + X_n$ and let N_t denote the number of renewals in $(0, t]$, i.e. $N_t = \sup\{n > 0 : T_n \leq t\}$, so that

$$P(N_t \geq i) = F^{*i}(t) \quad \text{for } i = 1, 2, \ldots . \tag{7.1.1}$$

The mean of N_t is denoted by $H(t)$ and called the *renewal function*, being the unique solution of the renewal equation

$$H(t) = F(t) + \int_0^t H(t - x) F(\mathrm{d}x), \tag{7.1.2}$$

namely,

$$H(t) = \sum_{n=1}^{\infty} F^{*n}(t); \tag{7.1.3}$$

see, for example, Barlow and Proschan (1975), p. 166. (Note that some authors define the renewal function as the mean number of renewals in $[0, t]$ which increases the function by 1.) Further,

$$t/m = \lambda t = \sum_{n=1}^{\infty} \left(F_e * F^{*(n-1)} \right) (t) \tag{7.1.4}$$

where F_e is the so called *equilibrium distribution* given by

$$F_e(t) = \lambda \int_0^t \bar{F}(x)\mathrm{d}x,$$

and $F^{*0}(t)$ is the distribution function of the degenerated random variable concentrated in 0.

For the renewal function H as well as for N_t some comparability properties hold. Since $\leq_{st}$ has the convolution property (C), it follows from (7.1.1) that, with the numbers of renewals N_t' and N_t'' corresponding to the distribution functions F' and F'',

$$F' \leq_{st} F'' \text{ implies } N_t'' \leq_{st} N_t' \tag{7.1.5}$$

and hence

$$EN_t'' = H''(t) \leq H'(t) = EN_t' \quad \text{for } t \geq 0. \tag{7.1.6}$$

Additionally, the renewal function has the following monotonicity property: If (h_n) is a sequence of functions defined by specifying h_1 and the recurrence relation

$$h_{n+1}(t) = F(t) + \int_0^t h_n(t-x)F(\mathrm{d}x) \quad \text{for } t \geq 0, \tag{7.1.7}$$

$h_1(t)$ being bounded on any finite interval and such that the integral at (7.1.7) with $n = 1$ is well-defined, then

$$h_1(t) \leq [\geq] h_2(t) \quad \text{for } t \geq 0 \tag{7.1.8}$$

implies that for all $t \geq 0$

$$h_n(t) \leq [\geq] h_{n+1}(t) \uparrow [\downarrow] H(t) \quad \text{for } n \to \infty. \tag{7.1.9}$$

The monotonicity of (h_n) follows easily from (7.1.7) and (7.1.8), while the convergence (and hence, boundedness) is a consequence of

$$\lim_{n \to \infty} \int_0^t h_1(t-x)F^{*n}(\mathrm{d}x) = 0.$$

Marshall (1973) used the idea behind (7.1.9) to establish the following.

Theorem 7.1.1. *Let A be the set of all $t \geq 0$ with $\bar{F}(t) > 0$, and define*

$$b_L = \inf_{t \in A} \frac{(F(t) - F_e(t))}{\bar{F}(t)}, \quad b_U = \sup_{t \in A} \frac{(F(t) - F_e(t))}{\bar{F}(t)}. \tag{7.1.10}$$

Then

$$\lambda t + b_L \leq H(t) \leq \lambda t + b_U \quad \text{for } t \geq 0. \tag{7.1.11}$$

Proof. It follows from (7.1.10) that $b_L \bar{F}(t) \leq F(t) - F_e(t) \leq b_U \bar{F}(t)$. Convolution with F^{*n} and summation over $n = 0, 1, \ldots$ yields (7.1.11). $\square$

Since the NBUE property is equivalent to $F_e \leq_{st} F$ (see Theorem 1.8.6), the following result is obtained.

Corollary 7.1.2. *If F is NBUE [NWUE], then*

$$H(t) \leq [\geq] \lambda t \quad for \ t \geq 0. \tag{7.1.12}$$

It is possible to improve the bound (7.1.11) by iteration as in (7.1.7) starting with $h_1(t) = \lambda t + b_L$, and similarly for the upper bound $\lambda t + b_U$. However, this upper bound need not necessarily to be finite nor as good as Lorden's bound in (7.1.14) below.

Since $(F(t) - F_e(t)) / \bar{F}(t) = -1 + \bar{F}_e(t)/\bar{F}(t) \geq -1$, Theorem 7.1.1 implies the well known bound

$$H(t) \geq \lambda t - 1 \quad \text{for } t \geq 0 \tag{7.1.13}$$

which can also be established by the sub-additivity of the function $H^{(0)}(t) = H(t) + 1$. Further, it follows from the theorem and the definition that when F is NBUE [NWUE], $b_L \leq 0$ $[b_U \geq 0]$.

A general upper bound was found by Lorden (1970):

$$H(t) \leq \lambda t + \lambda^2 \sigma^2 \quad \text{for } t \geq 0, \tag{7.1.14}$$

where σ^2 is the variance corresponding to the distribution function F; see Carlsson and Nerman (1986) for an elegant proof.

7.2   Reliability Applications

7.2.1   Coherent Systems

Let Σ be a reliability system consisting of n components, each of which can be in one of the two states 0 (down) and 1 (up). Let the corresponding indicator state variables $X_1, \ldots, X_n$ determine the state Z of the system uniquely, so that there exists a $\{0, 1\}$-valued system function Φ say, such that $Z = \Phi(X_1, \ldots, X_n)$. Such systems frequently occur in reliability. For a series system it is

$$\Phi(X_1, \ldots, X_n) = \prod_{i=1}^{n} X_i,$$

while for a parallel system

$$\Phi(X_1, \ldots, X_n) = 1 - \prod_{i=1}^{n}(1 - X_i).$$

A system with system function Φ is said to be *coherent*, if it has the following properties:

   (a) $\Phi(0,\dots,0) = 0$;

   (b) $\Phi(1,\dots,1) = 1$;

   (c) if $x_i \leq y_i$ for all $i = 1,\dots,n$ then $\Phi(x_1,\dots,x_n) \leq \Phi(y_1,\dots,y_n)$.

Clearly, any series or parallel system is coherent.

Now suppose that the component state variables are random and time-dependent and can be described by the stochastic processes $(X_{1,t}),\dots,(X_{n,t})$, so that the system state Z_t at time t is given by

$$Z_t = \Phi(X_{1,t},\dots,X_{n,t}).$$

Theorem 4.3.3 can be applied to coherent systems by virtue of monotonicity property (c) above, leading immediately to the following result.

Theorem 7.2.1. *Let Σ' and Σ'' be coherent systems with the same system function Φ and with component state processes $(X'_{i,t})$ and $(X''_{i,t})$ for $i = 1,\dots,n$ and system state processes (Z'_t) and (Z''_t). Then*

$$\left(X'_{1,t},\dots,X'_{n,t}\right) \leq_{st} \left(X''_{1,t},\dots,X''_{n,t}\right) \tag{7.2.1}$$

implies

$$(Z'_t) \leq_{st} (Z''_t). \tag{7.2.2}$$

Two corollaries follow for coherent systems in which all the components are new at $t = 0$ and none of them is replaced when it fails.

Corollary 7.2.2. *Let the components have independent lifetimes with distribution functions F'_i and F''_i for $i = 1,\dots,n$ and let G' and G'' be the corresponding lifetime distribution functions of the system. Then*

$$F'_i \leq_{st} F''_i \quad for \ i = 1,\dots,n \tag{7.2.3}$$

implies

$$G' \leq_{st} G''. \tag{7.2.4}$$

Proof. The relations (7.2.3) and (7.2.4) follow from Theorem 7.2.1 since one can write $\bar{F}_i(t) = P(X_{i,t} = 1)$ and $\bar{G}(t) = P(Z_t = 1)$. $\qquad\square$

Note that the assumption of independent components can be replaced by the weaker assumption that the joint distribution of the lifetimes of the components of the two systems have a common copula (see Theorem 3.3.8).

Corollary 7.2.3. *If in a coherent system Σ with independent lifetimes the lifetime distribution functions have IFR with means $m_1,\dots,m_n$ then the lifetime distribution function G of the system satisfies*

$$G(t) \leq G_{\exp}(t) \ for \ t < \min\{m_1,\dots,m_n\} \tag{7.2.5}$$

where $G_{\exp}$ denotes the lifetime distribution function of a system with the same system function as Σ and exponentially distributed lifetimes with the means $m_1, \ldots, m_n$.

Proof. An IFR distribution function with mean m satisfies $F(t) \le 1 - e^{-t/m}$ for $0 \le t < m$; see Barlow and Proschan (1975), p. 112. Thus (7.2.5) follows from Theorem 7.2.1 as in the previous corollary. $\qquad\square$

In some particular cases, comparison of system lifetime distribution functions is possible with respect to $\le_{icx}$ or $\le_{icv}$. This is the case if the system lifetime L can be expressed as a function of component lifetimes $L_1, \ldots, L_n$ which is constructed by addition and maximization (minimization) operations; this is a further application of (max, +) recursions. The (max, +) case is given for systems with components in parallel (max) and in reserve (+), while the (min, +) case appears for some systems with components in series and in reserve.

However, more is wanted in reliability theory. An important problem is comparison of the hazard rates $r'(t)$ and $r''(t)$ of two reliability systems Σ' and Σ''. In the notation of Section 1.3, the aim is to have $G' \le_{hr} G''$. It is natural to conjecture that

$$F_i' \le_{hr} F_i'' \quad \text{for } i = 1, \ldots, n \qquad (7.2.6)$$

is sufficient. This is indeed the case for series systems with independent components, since there the system rate $r(t)$ is equal to the sum of component rates $r_i(t)$. For more general systems the hazard rate $r(t)$ is given by

$$r(t) = \frac{\mathrm{d}}{\mathrm{d}t}(-\ln \bar{G}(t))$$

$$= \sum_{i=1}^{n} r_i(t) \frac{p_i \frac{\partial \Phi(\mathbf{p})}{\partial p_i}}{\Phi(\mathbf{p})} \Bigg|_{\mathbf{p} = (\bar{F}_1(t), \ldots, \bar{F}_n(t))},$$

where $\bar{G}(t) = 1 - G(t)$ is the system survival function.

In some cases this expression helps to prove the aimed comparison. An important example is a k-out-of-n system, which is up if and only if at least k of the n components are up. But already here it is difficult to show that the function

$$\sum_{i=1}^{n} \frac{p_i \frac{\partial \Phi(\mathbf{p})}{\partial p_i}}{\Phi(\mathbf{p})}$$

is decreasing in $\mathbf{p} = (p_1, \ldots, p_n)$; see Esary and Proschan (1963), and there are examples of coherent systems which show that (7.2.6) is not sufficient for $G' \le_{hr} G''$; see Boland, El-Neweihi, and Proschan (1994) and the references therein.

Clearly, the lifetime of a k-out-of-n system and the $(n-k)$th order statistic $L_{(n-k:n)}$ of the lifetimes $L_1, \ldots, L_n$ coincide. Therefore, results for order statistics are important in the reliability context. Boland et al. (1994) proved the following 'monotonicity' properties for the $L_{(i:n)}$: It is

$$L_{(i:n)} \leq_{hr} L_{(i+1:n)} \quad \text{if } i \leq j \text{ and } m - i \geq n - j$$

for independent but not necessarily identically distributed L_i. For the case of i.i.d. lifetimes $L_1, L_2, \ldots$ Raqab and Amin (1996) showed the stronger result that

$$L_{(i:m)} \leq_{lr} L_{(j:n)} \quad \text{if } i \leq j \text{ and } m - i \geq n - j.$$

If $L_1', \ldots, L_m'$ and $L_1'', \ldots, L_n''$ are two sequences of i.i.d. random variables, then for any m and n the following comparability results can be shown:

$$L_1' \leq_{st} L_1'' \quad \text{implies} \quad L_{(i:m)}' \leq_{st} L_{(j:n)}'',$$

$$L_1' \leq_{lr} L_1'' \quad \text{implies} \quad L_{(i:m)}' \leq_{lr} L_{(j:n)}'',$$

$$L_1' \leq_{hr} L_1'' \quad \text{implies} \quad L_{(i:m)}' \leq_{hr} L_{(j:n)}''$$

if $i \leq j$ and $m - i \geq n - j$. The first relation is an easy consequence of Corollary 1.2.18. The second relation was shown in Lillo, Nanda, and Shaked (2001), and the third one in Nanda and Shaked (2002).

Proschan and Sethuraman (1976) found a very interesting application of majorization. Their result shows that systems with more heterogeneous components are more reliable in the sense of having longer lifetimes. Let $L_1, \ldots, L_n$ be a sequence of independent lifetimes, and let $L = (L_{(1:n)}, \ldots, L_{(n:n)})$ be the corresponding vector of order statistics. Assume that the corresponding hazard rates $r_1, \ldots, r_n$ are proportional to a given (hazard rate) function r, i.e. there exist positive constants λ_i with $r_i(t) = \lambda_i r(t)$ for $i = 1, \ldots, n$. Now consider two such sequences with the same r.

If $(\lambda_1', \ldots, \lambda_n') \leq_M (\lambda_1'', \ldots, \lambda_n'')$ then $L' \leq_{st} L''$. The proof is based on a coupling argument. A generalization of this result to imperfect repair models is given in Hu (1996).

7.2.2   Comparison of Maintenance Policies

Typically, reliability theory has also to do with replacements and repairs. Often failed units are replaced by new ones or older by new (or newer) ones. There are several maintenance policies for single units; see Barlow and Proschan (1975). Here three such policies are briefly discussed and compared, as an application of comparison methods for counting processes as introduced in Section 5.4. See also Szekli (1995), section 2.11, Last and Szekli (1998) and the references therein. For all policies, it is assumed that the time for repair or replacement can be neglected.

Renewal replacement
Failed units are replaced by new ones. All lifetimes are i.i.d. with distribution function F, which is assumed to be continuous with $F(t) < 1$ for all t. Let N_t be the number of replacements in $(0, t]$. Clearly, (N_t) is a renewal process.

Minimal repair
If the unit fails at age x it is repaired to its condition just prior to failure so that the lifetime distribution function of the repaired unit is $1 - \bar{F}(t+x)/\bar{F}(x)$. Equivalently, the failure rate of the repaired unit is $r_x(t) = r(t + x)$, where $r(t)$ is the failure rate corresponding to F. Let N_t^m be the number of failures (or repairs) in $(0, t]$. It is easy to see that (N_t^m) is an inhomogeneous Poisson process with intensity function $r(t)$ and mean $E(N_t^m) = -\ln \bar{F}(t)$.

Block replacement
Let $S = (s_n)$ be a sequence of real numbers with $0 < s_1 < s_2 < \ldots$ and $s_n \to \infty$. (An important particular case is $s_n = nh$ for $h > 0$.) The s_n are instances of planned replacements by a new unit; in each s_n there begins an independent replacement process (for example, a renewal process or a minimal repair process). The corresponding counting process $N_t(S)$ counts the non-planned repairs.

Block, Langberg, and Savits (1990) gave a survey of various stochastic comparisons, namely comparisons of the same maintenance policy for units having different lifetime distributions, comparisons of different policies for the same unit and comparisons of block replacement strategies with $S_1 \subseteq S_2$. Here only the first case is considered.

The following theorem gives comparison relations for counting processes as consequences of comparison relations for the lifetime distribution functions. $(N_{F,t})$ denotes the renewal process corresponding to distribution function F etc.

Theorem 7.2.4. *Let F and G be continuous lifetime distribution functions satisfying $F(t) < 1$ and $G(t) < 1$ for all t. If $F \leq_{st} G$ then*

$$(N_{G,t}) \ \text{str-}\leq_{st} (N_{F,t}), \tag{7.2.7}$$

$$(N_{G,t}^m) \ \text{str-}\leq_{st} (N_{F,t}^m), \tag{7.2.8}$$

$$(N_{G,t}(S)) \ \text{str-}\leq_{st} (N_{F,t}(S)) \quad \text{for all } S \tag{7.2.9}$$

and

$$(N_{G,t}^m(S)) \ \text{str-}\leq_{st} (N_{F,t}^m(S)) \quad \text{for all } S. \tag{7.2.10}$$

Proof. Relation (7.2.7) is a consequence of the comparison properties of renewal processes in Section 5.4. Relation (7.2.8) follows from the comparison property of Poisson processes in (4.3.26), since the processes compared are non-homogeneous Poisson processes with comparable mean functions.

The proof of (7.2.9) and (7.2.10) is based on (7.2.7) and (7.2.8), respectively. For (7.2.9) it is as follows.

For any k, let $(d'_{n,k})$ and $(d''_{n,k})$ be infinite sequences of positive numbers satisfying $d'_{n,k} \leq d''_{n,k}$ for all k and n. Form two point sequences (t'_n) and (t''_n) on $(0, \infty)$ by the following rule (given for (t'_n)):
(t'_n) is the union of subsequences in $[s_k, s_{k+1})$ for $k = 0, 1, \dots$ given by $s_k + d'_{1,k}$, $s_k + d'_{1,k} + d'_{2,k}$, $\dots$ for $k = 0, 1, \dots$.

Let $n'_{t_n} = \max\{n : t'_n \leq t\}$. Obviously, $d'_{n,t} \leq d''_{n,t}$ implies $n'_t \geq n''_t$ for all t. Thus the mapping $((d_{n,1}), (d_{n,2}), \dots) \mapsto n_t$ is decreasing, and the mapping or coupling method yield the complete comparability of the processes $(N_{G,t}(S))$ and $(N_{F,t}(S))$ with respect to $\leq_{st}$. $\square$

7.3   PERT and Scheduling Problems

The PERT (Project Evaluation and Review Technique) model describes a complex project consisting of project activities of random durations X_i, $i = 1, \dots, m$. The order in which the activities have to be connected is given by a directed and acyclic network of n nodes and m arcs. The network has a single initial node, a single terminal node, and multiple arcs are possible. With the ith arc a_i the duration X_i is connected, and all X_i are assumed to be independent.

An important problem in the analysis of such networks is to obtain distributional information on the project duration Z. It is clear that Z is a function of the X_i, namely

$$Z = \max_{\pi_j \in \Pi} \sum_{a_i \in \pi_j} X_i, \tag{7.3.1}$$

where $\Pi = \{\pi_1, \pi_2, \dots\}$ is the (finite) set of network paths π_j from initial to terminal node. The length of path π_j is the sum $\sum_{a_i \in \pi_j} X_i$. The problem is that these sums are not independent. Indeed, they are positively correlated. It is an easy consequence of Theorem 3.10.5 that they are associated. (We note in passing that simpler expressions for Z are possible in which each X_i appears once.)

The determination of the distribution of Z is a difficult problem; see the discussion in Kamburowski (1992). Therefore Monte Carlo simulations, approximations or bounding techniques are used to obtain solutions. For example, a simple approximation for EZ is the classical PERT 'solution'

$$EZ \approx \max_{\pi_j \in \Pi} \sum_{a_i \in \pi_j} EX_i.$$

Since the right hand side of (7.3.1) is a convex function of the X_i, it is an easy consequence of Jensen's inequality that this is a lower bound.

Further approximations and bounds are based on the increasing convex order $\leq_{icx}$. To the best of our knowledge Hanisch and Stoyan (1978) were the first who observed that for two networks of the same structure but with different durations X_i' and X_i''

$$X_i' \leq_{icx} X_i'' \quad \text{implies} \quad Z' \leq_{icx} Z''.$$

Approximations of the distribution functions of Z can be obtained by bounding iteratively the distribution functions of partial sums and partial maxima by distribution functions from some suitable class. While Hanisch and Stoyan (1978) worked with uniform distributions, Kamburowski (1989) recommended the use of normal distributions. In this case it is used that sums of independent normally distributed random variables are normally distributed again, and for maxima of such variables the following recursively determined bounds can be used:

$$L_n \leq_{icx} \max\{Y_1, \dots, Y_n\} \leq_{icx} U_n$$

for $Y_i \sim \mathcal{N}(\mu_i, \sigma_i^2)$, $\sigma_i \leq \sigma_{i+1}$, with

$$L_j \sim \mathcal{N}\left(E \max\{L_{j-1}, Y_j\}, \left(\sum_{i=1}^{j} \frac{1}{\sigma_i^2}\right)^{-1}\right),$$

$$U_j \sim \mathcal{N}\left(E \max\{U_{j-1}, Y_j\}, \sigma_j^2\right)$$

for $j = 2, \dots, n$, where $L_1 = U_1 = Y_1$.

A straightforward calculation yields that for two independent normally distributed random variables $Y_1 \sim \mathcal{N}(\mu_1, \sigma_1^2)$ and $Y_2 \sim \mathcal{N}(\mu_2, \sigma_2^2)$ with $\sigma_1 < \sigma_2$ it is

$$E \max\{Y_1, Y_2\} = \mu_1 \Phi(\alpha) + \mu_2 \Phi(-\alpha) + a\varphi(\alpha)$$

where Φ and φ denote as usual distribution function and probability density of standard normal distribution, $a^2 = \sigma_1^2 + \sigma_2^2$ and $\alpha = (\mu_1 - \mu_2)/a$.

In Kamburowski (1992) approximations by discrete distributions such as in Section 1.11 are used, while in Kamburowski (1989) bounds for EZ are derived in the case of incomplete information in the sense of Section 4.4.

The PERT problem can be considered as a particular case of more general *scheduling problems*, for which there is a systematic theory; see Pinedo (1995). In the abbreviating code used there it is of type $P_\infty|prec|C_{\max}$. In scheduling theory terms, the description of PERT is as follows: There are m jobs subject to precedence constraints (the order of activities is prescribed) and an unlimited number of resources (or machines) in parallel. The character '$C_{\max}$'

is related to the objective of PERT of determining the 'makespan', which is the completion time of the last activity.

As a typical scheduling problem now the problem $1|\ |\sum C_i$ is considered. There are n jobs with independent processing times X_i which can be processed in arbitrary order on a single machine (therefore the '1'; the empty space between the two bars shows that there are no processing restrictions and constraints). The aim is to minimize the flowtime T, where $T = \sum_{i=1}^{n} C_i$ (therefore '$\sum C_i$') and the C_i are the completion times of the jobs. (Note that one assumes that all jobs appear at $t = 0$ at the machine; one is processed while the others have to wait. Thus the C_i contain both the X_i and waiting times.)

The jobs are numbered according to SEPT ($=$ shortest-expected-processing-time-first schedule), i.e. $EX_i \leq EX_{i+1}$ for $i = 1, \ldots, n-1$.

Let T_π be the flowtime corresponding to any schedule π (which gives the order of processing the jobs), while T_{LEPT} ($=$ largest-expected-processing-time-first schedule) is the flowtime when the job with largest processing time is processed first and so on.

Theorem 7.3.1. *(Chang and Yao (1993), Shanthikumar and Yao (1991))*

(i) If $X_i \leq_{lr} X_{i+1}$ for all i, then $T_{SEPT} \leq_{st} T_\pi \leq_{st} T_{LEPT}$.

(ii) If $X_i \leq_{hr} X_{i+1}$ for all i, then $T_{SEPT} \leq_{icx} T_\pi \leq_{icx} T_{LEPT}$.

(iii) $ET_{SEPT} \leq ET_\pi \leq ET_{LEPT}$.

Proof. The proof uses an interchanging method which is typical for scheduling theory. It is presented for $T_{SEPT} \leq T_\pi$; the proof of the other inequality is similar.

Let π be such that job i follows job j though $i < j$; after job i still $\ell - 1$ jobs follow. Let ϱ be the schedule which agrees with π except that i and j are interchanged; ϱ is 'closer' to SEPT than π. Then

$$T_\pi = (\ell + 1)X_j + \ell X_i + Y = X_j + \ell(X_i + X_j) + Y = g(X_j, X_i) + Y$$

and

$$T_\varrho = (\ell + 1)X_i + \ell X_j + Y = X_i + \ell(X_i + X_j) + Y = g(X_i, X_j) + Y.$$

Here Y is the contribution to flowtime resulting from the processing times of all other jobs; clearly Y and the g-terms are independent. Therefore (i) and (ii) are easy consequences of Theorem 1.9.4 and the convolution property of $\leq_{st}$ and $\leq_{icx}$, see Theorem 1.2.17 and 1.5.5, respectively.

The mean value inequality corresponding to (iii) is clear. $\square$

7.4 Comparison of Random Sets and Point Processes

7.4.1 Comparison of Random Closed Sets

Random closed sets play an important role in many fields of applied probability; see Jeulin (1997), Stoyan (1998) and Stoyan, Kendall, and Mecke (1995). Some papers also discuss or use comparison properties of random closed sets, and it is the aim of this section to present some of the ideas in this field. Here not the most general case is considered; the sets are subsets of a locally compact Polish space S (a complete and separable metric space); $S = \mathbb{R}^d$ is an important example. Note that a simple point process (more precisely: the support of the process, the set of point locations) in such a space is a particular case; therefore the results can be used also in the point process context, see Section 7.4.2.

Let $\mathbb{F}$ and $\mathbb{K}$ be the system of all closed and compact subsets of S, respectively. Matheron's σ-algebra $\mathcal{F}$ is the σ-algebra of subsets of $\mathbb{F}$ which is generated by the system of sets $\mathbb{F}_K = \{F \in \mathbb{F} : F \cap K \neq \emptyset\}$ for $K \in \mathbb{K}$. A *random closed set* Ξ is an $(\mathcal{A}, \mathcal{F})$-measurable mapping of a probability space $(\Omega, \mathcal{A}, P)$ into $(\mathbb{F}, \mathcal{F})$.

The functional T given by

$$T(K) = P(\Xi \cap K \neq \emptyset) \quad \text{for } K \in \mathbb{K}$$

is called the *capacity functional*. The Choquet theorem says that the capacity functional T determines uniquely the distribution of a random closed set on $\mathcal{F}$. In some sense it plays a role similar to that of the distribution function of a real-valued random variable. Unfortunately, it is not sufficient for comparisons of random closed sets.

Perhaps the most natural comparison relation for random closed sets is given in the following definition.

Definition 7.4.1. The random closed set Ξ'' is *stochastically larger than* Ξ', if there are a probability space $(\Omega, \mathcal{A}, P)$ and two random closed sets $\hat{\Xi}'$ and $\hat{\Xi}''$ defined on this space with $\Xi' =_{st} \hat{\Xi}'$, $\Xi'' =_{st} \hat{\Xi}''$ and $\hat{\Xi}' \subseteq \hat{\Xi}''$ almost surely. This is denoted by $\Xi' \subseteq_{st} \Xi''$.

It is clear that $\Xi' \subseteq_{st} \Xi''$ implies

$$T'(K) \leq T''(K) \quad \text{for all } K \in \mathbb{K}, \tag{7.4.1}$$

where $T'(K) = P(\Xi' \cap K \neq \emptyset)$ etc. The converse is not true, as the following example shows.

Example 7.4.2. ($\Xi' \subseteq_{st} \Xi''$ does not imply (7.4.1), Norberg (1992).) Let $S = \{a, b\}$ be equipped with the discrete topology. Let the distribution of Ξ' and Ξ'' be given by

$$P(\Xi' = \emptyset) = \frac{2}{3} \quad \text{and} \quad P(\Xi' = \{a, b\}) = \frac{1}{3}$$

and

$$P(\Xi'' = \{a\}) = P(\Xi'' = \{b\}) = \frac{3}{8} \quad \text{and } P(\Xi'' = \{a,b\}) = \frac{1}{4}.$$

It is easy to see that

$$P(\Xi' \cap K \neq \emptyset) \leq P(\Xi'' \cap K \neq \emptyset)$$

for all $K \subseteq S$. But it is $P(\Xi' = S) > P(\Xi'' = S)$, which contradicts $\Xi' \subseteq_{st} \Xi''$.

Thus, more is needed to ensure $\Xi' \subseteq_{st} \Xi''$. Norberg (1992) proved a necessary and sufficient condition, which in the case considered here is given in the following result.

Theorem 7.4.3. $\Xi' \subseteq_{st} \Xi''$ *if and only if*

$$P\left(\bigcap_{i=1}^{n}\{\Xi' \cap K_i \neq \emptyset\}\right) \leq P\left(\bigcap_{i=1}^{n}\{\Xi'' \cap K_i \neq \emptyset\}\right) \qquad (7.4.2)$$

for all n and all compact subsets $K_1, \ldots, K_n$ of S.

Norberg (1992) gives two proofs of the sufficiency of (7.4.2). One uses Strassen's theorem, while the other does not refer to that theorem. (Norberg even shows that Theorem 7.4.3 is equivalent to a particular case of Strassen's theorem.) The proof based on Strassen's theorem (Theorem 2.6.3) shows that (7.4.2) implies $Ef(\Xi') \leq Ef(\Xi'')$ for any bounded mapping $f : \mathbb{F} \to \mathbb{R}$ which is increasing (with respect to the ordering $\subseteq$ of set inclusion) and continuous (with respect to the topology in $\mathbb{F}$). The Polish space and the partial order in the application of Strassen's theorem thus are $\mathbb{F}$ and $\subseteq$.

The proof of $\Xi' \subseteq_{st} \Xi''$ for concrete random closed sets is frequently based on the construction of the sets. We give two simple examples.

Example 7.4.4. (Comparison of Boolean models.) The Boolean model is a random closed set in $\mathbb{R}^d$ constructed as follows. There are a homogeneous Poisson process with intensity λ, the points (called 'germs') of which are denoted as $x_1, x_2, \ldots$; and an infinite sequence of i.i.d. random compact sets $\Xi_0, \Xi_1, \ldots$ (called 'grains'). The set Ξ_0 is called the 'typical grain'. Every Ξ_i is shifted from the origin o to x_i; the result is $\Xi_i + x_i$ for $i = 1, 2, \ldots$. The union of all $\Xi_i + x_i$ is a random closed set Ξ, which is called the *Boolean model*; see Matheron (1975) and Stoyan et al. (1995). It is clear that two such models Ξ' and Ξ'' are comparable with respect to $\subseteq_{st}$ if the intensities λ' and λ'' satisfy $\lambda' \leq \lambda''$ and the typical grains satisfy $\Xi'_0 \subseteq_{st} \Xi''_0$.

Proof. This is simple by the coupling method:

Consider on some probability space a Boolean model $\hat{\Xi}_\delta$ with Poisson process intensity $\lambda'' - \lambda'$ and typical grain $\hat{\Xi}''_0$, $\hat{\Xi}''_0 =_{st} \Xi''_0$. (If $\lambda'' = \lambda'$ then $\hat{\Xi}_\delta = \emptyset$.) Furthermore, consider on the same probability space an independent

Poisson process of intensity λ' with points $\hat{x}_1, \hat{x}_2, \ldots$ and two sequences of independent random compact sets $\hat{\Xi}'_i$ and $\hat{\Xi}''_i$ independent of $\hat{\Xi}_\delta$ and the Poisson process with the same distribution as Ξ'_0 and Ξ''_0 and $\hat{\Xi}'_i \subseteq \hat{\Xi}''_i$ almost surely. (Because of the assumption $\Xi'_0 \subseteq_{st} \Xi''_0$ such sequences exist.) Then construct the sets $\hat{\Xi}' = \bigcup \hat{\Xi}'_i + \hat{x}_i$ and $\hat{\Xi}'' = \bigcup \hat{\Xi}''_i + \hat{x}_i$. Clearly $\hat{\Xi}' \subseteq \hat{\Xi}''$ and $\hat{\Xi}' =_{st} \Xi'$ and $\hat{\Xi}'' \cup \hat{\Xi}_\delta =_{st} \Xi''$. Thus $\Xi' \subseteq_{st} \Xi''$.

Example 7.4.5. (Comparison of the random sequential absorption model and the dead leaves model, Stoyan and Schlather (2000).) Now two random closed sets of different structure are compared, which are both of independent mathematical and physical interest, where the random sequential absorption model (RSA model, see Evans (1993)) is the more complicated one. Even the fundamental parameter p 'volume fraction' in this model can only be determined via simulation, while for the dead leaves model there is a simple explicit formula for p. (The volume fraction of a stationary or homogeneous random closed set Ξ is the mean volume occupied by the set in the unit cube. It is equal to the probability that the origin o is an element of Ξ.) Both models describe infinite systems of hard (non-overlapping) spheres of fixed radius r in $\mathbb{R}^d$.

The homogeneous RSA model is defined by means of the limit of a spatial birth process. Such a process is a family of homogeneous point processes in the space indexed by time t, or a time continuous Markov process whose states are point patterns. At isolated instances of time further points are born, i.e. added to the existing point pattern. Such 'births' are controlled by a birth-rate β, a positive function $\beta_\varphi(x)$ with

$$\int_W \beta_\varphi(x)\mathrm{d}x < \infty$$

for any bounded set W and any point configuration φ. The probability that a birth occurs in the set W in the infinitesimal time interval $[t, t + \mathrm{d}t]$, given that the process has configuration φ at time t, equals

$$\mathrm{d}t \int_W \beta_\varphi(x)\mathrm{d}x + o(t).$$

The set theoretic union of all spheres of radius r centered at the points $x \in \varphi$ is denoted by s_φ. Then the birth rate of the RSA birth process can be expressed as

$$\beta_\varphi(x) = \left(1 - \mathbf{1}_{s_\varphi \oplus b(o,r)}(x)\right),$$

where $A \oplus b(o, r)$ is the union all spheres of radius r centered at points in A. The birth rate simply vanishes in all the points x at which a sphere of radius r would intersect some other sphere of radius r with center in φ.

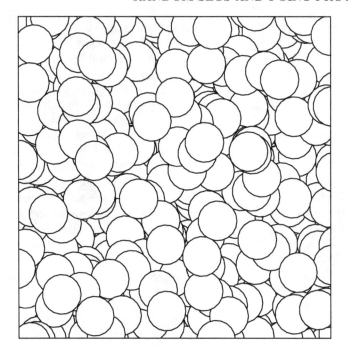

Figure 7.1 Simulated sample of the dead leaves model. The intact disks form the random closed set I discussed in the text.

The start configuration at $t = 0$ is the empty set, while the RSA model is the union of all spheres of radius r centered at the points of the limit of the birth process at $t \to \infty$, the so-called jamming limit. It is a homogeneous and isotropic random closed set R in $\mathbb{R}^d$. In the case $d = 1$ the model is known as Renyi's car parking model.

The volume fraction p of the RSA model is 0.747 ($d = 1$), 0.547 ($d = 2$) and 0.382 ($d = 3$); the latter two values are obtained by simulation.

The dead leaves model can be obtained by placing spheres of radius r randomly and uniformly in space during the time interval $(-\infty, 0]$ so that subsequent spheres may (partially) cover already existing spheres and infinitely many spheres are placed in any subset W of the space with positive volume. At time $t = 0$ the system is in its time-stationary state and yields a tessellation of the space as shown in Figure 7.1 for the planar case. (In the planar case, the spheres are disks and can be interpreted as dead leaves; the union of all the disks is then an infinite layer of dead leaves.)

A sphere is called 'intact' if it is not (partially) covered by any other sphere. The system of all intact spheres is denoted by I. The volume fraction of this random closed set is $p = 2^{-d}$.

The following shows that there is a coupling of both sets which enables us to prove that $I \subseteq_{st} R$.

Consider a homogeneous Poisson process of intensity one on $\mathbb{R}^d \times [0, \infty)$. Its points (x, t) can be interpreted as potential centers and instants of appearance of spheres. Consider the system of all spheres of radius r at the points x. This system is thinned out in order to obtain a system of non-overlapping spheres, by the following thinning rule: a point (x, t) is retained if there is no other point (x', t') with

$$t' < t \quad \text{and} \quad \|x - x'\| < 2r \qquad (7.4.3)$$

where $\| \cdot \|$ denotes the Euclidean distance. All points which are not retained are eliminated.

It is not difficult to see that this construction is only another form of the construction of the system of intact spheres of the dead leaves model. It is also easy to see that the RSA model can be obtained by a natural continuation and modification of the thinning procedure above. Indeed, do not eliminate all points after the thinning step above and carry out later infinitely many analogous thinning steps. Eliminate only all points (x^*, t^*) with

$$t^* > t \quad \text{and} \quad \|x^* - x\| < 2r.$$

Apply then again rule (7.4.3) for the reduced system etc. This procedure yields an increasing sequence of point sets which tends towards the centre pattern of the RSA model.

If for the construction of the sets I and R the same sample of the Poisson process is used then clearly the dead leaves sample is a subset of the RSA sample, i.e. $I \subseteq R$.

This comparison relationship enables the use of the system of intact dead leaves as a 'lower bound' for the RSA model, which is, though, poor. Not only the volume fraction of the RSA model is bounded by the dead leaves model, but also other characteristics such as the capacity functional.

Stoyan and Stoyan (1980) attempted to compare random closed sets with respect to relations weaker than $\subseteq_{st}$; these sets are compact and can have equal shape and mean volume.

7.4.2 Comparison of Point Processes

While Section 5.4 considers point processes on the real line $\mathbb{R}$, now the case of point processes in a locally compact Polish space S is discussed. If Π is such a point process, then for a Borel set A, $\Pi(A)$ denotes the random number of points in A.

Already in the case of $S = \mathbb{R}^2$ it is clear that the relations considered in Section 5.4 cannot be generalized to general S with the exception of $\subseteq_{st}$. Therefore, in the following only this relation is considered.

Clearly, for the comparison of simple point processes (which have no multiple points) Theorem 7.4.3 can be used. For general (not necessarily simple) point processes the measure approach is used. On the space of all locally finite measures on the Borel σ-algebra of S the order $\preceq$ is introduced by

$$\nu' \preceq \nu'' \quad \text{if } \nu'(B) \leq \nu''(B) \quad \text{for all bounded Borel sets } B.$$

Denote by $\subseteq_{st}$ the corresponding stochastic order of random measures: for two random measures Φ' and Φ'' the relation

$$\Phi' \subseteq_{st} \Phi'' \tag{7.4.4}$$

means

$$Ef(\Phi') \leq Ef(\Phi'')$$

for all functions f which are increasing with respect to $\preceq$.

Generalizing an earlier result of Rolski and Szekli (1991), Norberg (1992) proved a criterion for the comparability of random capacities. This includes the following result for point processes in S.

Theorem 7.4.6. *For two point processes Π' and Π'' on S we have $\Pi' \subseteq_{st} \Pi''$ if and only if*

$$(\Pi'(K_1), \dots, \Pi'(K_n)) \leq_{st} (\Pi''(K_1), \dots, \Pi''(K_n)) \tag{7.4.5}$$

for all n and all relatively compact subsets $K_1, \dots, K_n$ of S.

Following Rolski and Szekli (1991), this result can be used to compare Cox processes (and, as a particular case, Poisson processes) on general spaces.

Theorem 7.4.7. *Let Π' and Π'' be Cox processes on S with leading random measures Λ' and Λ''. Then $\Lambda' \subseteq_{st} \Lambda''$ (in the sense of (7.4.4)) implies $\Pi' \subseteq_{st} \Pi''$ (also in the sense of (7.4.4)).*

Proof. The result is shown by proving an inequality as (7.4.5) for Π' and Π''. Let $f : \mathbb{N}_0^n \to \mathbb{R}_+$ be increasing and positive. Then κ given by

$$\kappa(x_1, \dots, x_n) = \sum_{i_1, \dots, i_n \in \mathbb{N}_0} f(i_1, \dots, i_n) \frac{x_1^{i_1}}{i_1!} e^{-x_1} \cdot \dots \cdot \frac{x_n^{i_n}}{i_n!} e^{-x_n} \tag{7.4.6}$$

is an increasing function of the x_i. This follows from the comparison properties of Poisson random variables with respect to $\leq_{st}$ given in Table 1.2 on page 63, since the right-hand side of (7.4.6) is the mean of a monotonous function of independent Poisson random variables.

Consequently, for disjoint $K_1, \ldots, K_n$

$$
\begin{aligned}
Ef(\Pi'(K_1), \ldots, \Pi'(K_n)) &= E\kappa(\Lambda'(K_1), \ldots, \Lambda'(K_n)) \\
\le E\kappa(\Lambda''(K_1), \ldots, \Lambda''(K_n)) &= Ef(\Pi''(K_1), \ldots, \Pi''(K_n)).
\end{aligned}
$$

The case of intersecting K_i can be easily reduced to the case of disjoint sets.   □

Still much work has to be done in the comparison theory for point processes. It is clear that there are processes of comparable variability. Examples of pairs of such processes are a homogeneous Poisson process and a cluster process of equal intensity or two hard-core Gibbs processes of equal intensity and different hard-core distances. It would be fine if these variability differences could be characterized by order relations of a nature similar to that of $\le_{icx}$, $\le_{rand}$ or $\le_{disp}$ and if one could conclude that comparability with respect to such a relation implies, for example, reasonable relationships for second order characteristics such as the pair correlation function.

7.5   Monotonicity and Comparison of Models of Statistical Physics

7.5.1   Monotonicity and Comparison Properties of the Ising Model

The Ising model is a famous example of a Markov random field. It is discussed here as an example for the application of results for Markov chains in Chapter 5 to a case where the state space is not a subset of $\mathbb{R}$ and rather large. Many interesting facts of the theory for this model will not be discussed here; the reader is referred to the relevant literature, see Liggett (1985), Georgii (2000), and Georgii, Häggström, and Maes (2001), where coupling methods are used.

The model is given on a discrete lattice Λ. In order to be definite assume that Λ is a subset of the quadratic lattice $\mathbb{Z}^2$ of all sites $i = (i_1, i_2)$, where i_1 and i_2 are integers. (The case of $\mathbb{Z}^d$ with $d > 2$ can be treated analogously; the case $d = 1$ is simpler.) In particular, Λ may be the set $\mathbb{Z}_r^2$ of all (i_1, i_2) with $-r \le i_1, i_2 \le r$.

A Markov random field takes values at the sites of Λ which come from a finite set of states E. In the case of the Ising model it is $E = \{0, 1\}$, while for the similar, more general, Potts model $E = \{0, 1, \ldots, q\}$. (In the literature closer to physics, for the Ising model the states $+1$ and -1 are used, corresponding to positive and negative spins in a ferromagnet. In order to treat the Potts model as a natural generalization, here the states 0 and 1 are used.)

The states of the random field are configurations $x = (x_i)$ for $i \in \Lambda$. Its state space is thus $S = E^\Lambda$. In the case of a finite Λ, the corresponding σ-algebra S is simply the power set and for general Λ it is the Borel σ-algebra corresponding to the natural product topology on E^Λ.

If Λ is finite, the distribution of the Markov random field depends on an energy function $H(x)$ and is determined by the expression

$$P(x) = \frac{1}{Z} \exp(-H(x)), \qquad (7.5.1)$$

which yields the probability of occurrence of configuration x. Here Z is a normalizing constant making P a probability distribution in $(S, \mathcal{S})$. Usually there is no closed form expression of Z and therefore it is difficult to work with the distribution P. For infinite Λ the definition is more complicated; see for the case of the Ising model Georgii (2000) and Georgii et al. (2001).

In the case of the *Ising model*

$$H(x) = \beta \sum_{\substack{\{i,j\} \cap \Lambda \neq \emptyset, \\ |i-j|=1}} \mathbf{1}_{\{x_i \neq x_j\}} \qquad (7.5.2)$$

where $|i - j| = |i_1 - j_1| + |i_2 - j_2|$ for $i = (i_1, i_2)$ and $j = (j_1, j_2)$. The model parameter β is a positive coupling coefficient, which physicists call 'inverse temperature'. The model discussed here is the 'ferromagnetic' Ising model. Also the case $\beta < 0$, the 'anti-ferromagnetic' Ising model, is of interest; see Georgii et al. (2001) and Thönnes (2000). (A further model parameter is r, if $\mathbb{Z}_r^2$ is used.) Following (7.5.2), the energy $H(x)$ of a configuration x is obtained as a sum of contributions from neighbouring sites with different states. Obviously, the distribution P prefers configurations with equal neighbours.

Note that the summation in (7.5.2) extends over pairs i and j for which at least one member belongs to Λ. In the case of $\Lambda = \mathbb{Z}_r^2$ it makes sense to include sites outside of Λ as shown below on page 255. For the moment the reader should assume a 'free boundary' where no sites outside of Λ exist; the summation is simply extended over all pairs i and j where both members are in Λ.

For the *Potts model* the energy function H in (7.5.2) is

$$H(x) = \beta \sum_{\substack{\{i,j\} \cap \Lambda \neq \emptyset, \\ |i-j|=1}} f(x_i, x_j) \mathbf{1}_{\{x_i \neq x_j\}}, \qquad (7.5.3)$$

where typical choices of the non-negative function f are

$$f(x_i, x_j) \equiv 1 \quad \text{or} \quad f(x_i, x_j) = |x_i - x_j|.$$

In the case $q = 1$ (Ising model) both forms yield the same result.

For the study of the distribution P, homogeneous Markov chains with state space $(S, \mathcal{S})$ are an important tool. They are used to simulate the model (to generate samples from P) as well as to prove distributional properties including phase transition.

A standard construction of such a Markov chain is the *Gibbs sampler* (see Gilks et al. (1996)) of the form below, which is also called 'single-site heat bath algorithm' in the physical literature. The resulting Markov chain has a monotone transition kernel for the Ising but not for the Potts model, with respect to the stochastic order relation $\leq_{st}$ based on the following natural partial order in S:

$$x \preceq y \quad \text{if} \quad x_i \leq y_i \quad \text{for all} \quad i \in \Lambda. \tag{7.5.4}$$

For the Potts model (including the Ising model, for $q = 1$) the transition function of the heat bath Markov chain (X_n) in the $\mathbb{Z}_r^2$ case is constructed as follows.

The state X_n of the chain is a random element on S, a random configuration. The new state X_{n+1} is constructed by X_n and three independent random variables U_n, M_n and T_n, which are uniform on $[0,1], \Lambda$ and $\{0, 1, \ldots, q\}$, respectively. We have

$$X_{n+1} = \varphi(X_n, U_n, M_n, T_n) \tag{7.5.5}$$

where φ is defined as

$$\varphi(x, u, m, t) = \begin{cases} x + (t - x_m)\mathbf{1}_m & \text{if } u \leq e_m(x, t) \\ x & \text{otherwise} \end{cases} \tag{7.5.6}$$

with

$$e_m(x, t) = \frac{P(\{x + (t - x_m)\mathbf{1}_m\})}{\sum_{t'=1}^{q} P(x + (t' - m)\mathbf{1}_m)}, \tag{7.5.7}$$

which cancels down to a simple expression that does not depend on Z. In (7.5.6) the term $x + (t - x_m)\mathbf{1}_m$ is the configuration obtained from x by setting $x_m = t$ and leaving the states of all other sites unchanged.

The idea behind (7.5.5) and (7.5.6) is simple. A random site M_n is chosen uniformly in $\mathbb{Z}_r^2$, and this site obtains the value t with probability $\min\{1, e_m(x, t)\}$. The term $e_m(x, t)$ denotes the conditional probability of the Potts model having a jump to t at site m given the current configuration on all other sites.

The description of the Markov chain is complete when the influence of the boundary of $\Lambda = \mathbb{Z}_r^2$ is fixed. One variant has been already introduced, the case of *free boundary*, where the sums in (7.5.2) and (7.5.3) are extended only over those pairs $\{i, j\}$ where both members belong to Λ. Furthermore, there are (at least) three ways to modify $H(x)$:

0-*boundary*: all sites $i = (i_1, i_2)$ with $|i_1| = r+1$ or $|i_2| = r+1$ obtain formally the value 0;

q-*boundary*: all sites $i = (i_1, i_2)$ with $|i_1| = r+1$ or $|i_2| = r+1$ obtain formally the value q;

periodic boundary conditions: $x_{(r+1,j)} = x_{(-r,j)}, \ x_{(i,r+1)} = x_{(i,-r)}$ etc.

Thus four Markov chains are obtained, which are in the following characterized by the indices 'f', '0', 'q', and 'p'. They are all irreducible and aperiodic and therefore ergodic. It is easy to show that the corresponding stationary distributions coincide with P given by (7.5.1) for the respective boundary regime, using the general approach in Gilks et al. (1996), p. 7. This is guaranteed by (7.5.7) and the reversibility of the Markov chains with respect to their stationary distributions. Thus the distribution P can be explored by means of the Markov chain (X_n) given by (7.5.5).

It is easy to see that for the Ising model the governing mapping φ of the Markov chain (X_n) in (7.5.5) has the following monotonicity property with respect to the partial order $\preceq$ in S:

$$x \preceq y \quad \text{implies} \quad \varphi(x, u, m, t) \preceq \varphi(y, u, m, t)$$

for all u, m and t, since $e_m(x, t)$ increases with increasing x. The Potts model with $q > 1$ does not possess this monotonicity property, neither for $f(x_i, x_j) \equiv 1$ nor for $f(x_i, x_j) = |x_i - x_j|$; but there is monotonicity for the so-called 'single-bond heat bath algorithm'; see Propp and Wilson (1996).

For all these models, there is no monotone dependence on the parameter β. Qualitatively, decreasing β (i.e. increasing temperature) induces more variability in the configurations. This will be discussed in detail in Section 7.5.2, using the randomness order $\leq_{rand}$; see also Georgii et al. (2001).

Finally, for the Ising model there is a monotone influence of the boundary assumptions. Let $\varphi_f, \varphi_0, \varphi_1$ and φ_p be the mappings (with fixed β) corresponding to free boundary, 0-boundary, 1-boundary, and periodic boundary conditions, respectively. Then

$$\varphi_0(x, u, m, t) \preceq \varphi_f(x, u, m, t) \preceq \varphi_1(x, u, m, t)$$

and

$$\varphi_0(x, u, m, t) \preceq \varphi_p(x, u, m, t) \preceq \varphi_1(x, u, m, t)$$

for all x, u, m, and t.

The results of Chapters 4 and 5 yield the following properties of the transition operators $\mathbf{T}_f$, $\mathbf{T}_0$, $\mathbf{T}_1$ and $\mathbf{T}_p$ of the corresponding Markov chains *for the Ising model*:

- All operators are monotone with respect to $\leq_{st}$.

- The following comparison statements hold:

$$\mathbf{T}_0 \leq_{st} \mathbf{T}_f \leq_{st} \mathbf{T}_1 \tag{7.5.8}$$

and

$$\mathbf{T}_0 \leq_{st} \mathbf{T}_p \leq_{st} \mathbf{T}_1. \tag{7.5.9}$$

Figure 7.2 Independent simulations of the Ising model in $\mathbb{Z}_7^2$ with $\beta = 0.44$. Left: 0-boundary, middle: free boundary, right: 1-boundary. Black squares denote 1 and white 0. The tendency of stochastic order as given by (7.5.8) is obvious.

It is easy to generalize $\mathbf{T}_0 \leq_{st} \mathbf{T}_1$ to $\mathbf{T}_{b_0} \leq_{st} \mathbf{T}_{b_1}$, where b_0 and b_1 are comparable boundaries.

The monotonicity property of the operators can be used for *perfect simulation* of the Ising model, following Section 5.2.5. The two extremal states in S are the configurations $x_i \equiv 0$ and $x_i \equiv 1$.

Perfect simulation for the Ising model was first carried out by Propp and Wilson (1996); see also the survey paper by Thönnes (2000). Note that for the Ising model similar monotonicity properties are true also for another standard MCMC simulation method, namely for the Metropolis–Hastings algorithm; see Dimakos (2001).

The comparison relationships (7.5.8) and (7.5.9) yield analogous relationships for the stationary distributions of the Markov chains; comparable boundaries produce comparable stationary distributions.

Furthermore, by Theorem 4.3.13 the Markov chains are associated. This implies positive correlations both for the states at different sites i and j and for the states at i for different times m and n.

For structural investigation of the distribution P of the Ising model on the whole lattice $S = \mathbb{Z}^2$, monotonicity and comparability is of great interest. It is useful to consider the influence of the parameter r in the case $S = \mathbb{Z}_r^2$ (with the option of letting later $r \to \infty$). Two Ising models with parameters r_1 and r_2 with 0- or 1-boundary are comparable in the following sense. In the case of 0-boundary, configurations are defined in the whole $\mathbb{Z}^2$ by assigning the value 0 to all sites $i = (i_1, i_2)$ with $|i_1| > r$ or $|i_2| > r$. Then it is possible to define Markov chains with $S = \mathbb{Z}^2$ with different parameters r.

It is clear that the corresponding transition operators satisfy

$$r_1 \leq r_2 \quad \text{implies} \quad \mathbf{T}_{r_1} \leq_{st} \mathbf{T}_{r_2}.$$

Since the initial state $(0,0, \ldots)$ produces monotone Markov chains, also the corresponding stationary distributions P_{r_1} and P_{r_2} are comparable with respect to $\leq_{st}$. With increasing r an increasing sequence (P_r) of stationary distributions is obtained. One can show that the weak limit P^0 exists, is a

Gibbs distribution and is spatially stationary (translation invariant in $\mathbb{Z}^2$). For the 1-boundary a distribution P^1 can be analogously constructed. Because of (7.5.8)

$$P^0 \leq_{st} P^1.$$

Now we are close to the famous problem of phase transition, i.e. the existence of more than one distribution P on $\mathbb{Z}^2$ for the same β, see Georgii et al. (2001). One can show that for the Ising model there is a unique distribution P if and only if $P^0 = P^1$. There is a critical value β_c of β such that for $\beta \leq \beta_c$ it is $P^0 = P^1$, while for $\beta > \beta_c$ the two distributions P^0 and P^1 are different, representing two 'phases' of the Ising model. The critical value, the 'Onsager value', is given by $\beta_c = \frac{1}{2}\ln(1 + \sqrt{2})$. (Note that in the case of dimension $d > 2$ the phase transition behaviour is different; see Georgii et al. (2001) and Liggett (1985). For $d = 1$ there is no phase transition.) For the case of the Potts model, where the random-cluster representation plays an important role; see Georgii et al. (2001). Similar investigations for disordered systems also use comparison with respect to $\leq_{st}$; see De Santis (2001) and Newman (1997).

The Markov chains in this section are particular cases of so-called *probabilistic cellular automata*; see Toom et al. (1990) and Maes (1993). Such processes are Markov chains on $S = W^E$, where E represents the set of automata and W the set of values each automaton can take. (W is finite and E countable.) Their transition operators are given by the transition probabilities $P_x(\eta, k) = P(\text{automaton } x \text{ takes value } k \text{ in the next transition}|\text{present configuration is } \eta)$ for $\eta \in S$.

López and Sanz (2000) give a comparison theorem for cellular automata with respect to $\leq_{st}$ (based on a partial order on W). Their criterion is in terms of the $P_x(\eta, k)$ and in the spirit of condition (5.2.14).

There is also a parallel continuous-time theory. For example, a 'dynamic Ising model' is studied. The corresponding models are called *interacting particle systems*. Such processes are intensity-governed continuous-time Markov processes; see Liggett (1985, 2000) and Lindvall (1992). In their study coupling and monotonicity and comparison properties play an important role. This field has partially its own terminology; for example, instead of 'monotone' the term 'attractive' is used. Forbes and Francois (1997) and López and Sanz (1998) give conditions on the transition rates that ensure comparison of more general processes with respect to $\leq_{st}$.

In the particular case $W = \{0, 1\}$ and E finite, a sufficient condition (called 'Holley's inequality') ensuring $P \leq_{st} Q$ for two distributions P and Q on S which assign positive probability to each element of S is as follows (theorem II.2.9 in Liggett (1985)):

$$P(\eta \wedge \zeta)Q(\eta \vee \zeta) \geq P(\eta)Q(\zeta) \qquad (7.5.10)$$

for all η and $\zeta \in S$, where

$$(\eta \vee \zeta)(s) = \max\{\eta(s), \zeta(s)\}, \quad (\eta \wedge \zeta)(s) = \min\{\eta(s), \zeta(s)\} \quad \text{for } s \in E.$$

Notice that this is a special case of Theorem 3.11.4 a).

P has 'positive correlations' (i.e. it is associated) if it assigns positive probability to each element of S and

$$P(\eta \wedge \zeta)P(\eta \vee \zeta) \geq P(\eta)P(\zeta). \tag{7.5.11}$$

This important result can be obtained as a special case of Theorem 3.10.14. It is called FKG inequality, since Fortuin, Kasteleyn, and Ginibre (1971) were the first to prove it.

Georgii and Küneth (1997) proved point process versions of the two inequalities above. They considered finite point processes in Polish spaces and used in their criteria the Papangelou intensities. As a particular example they studied the comparison of Gibbs distributions with distributions of finite Poisson processes with respect to $\subseteq_{st}$.

7.5.2 Comparison of Gibbs Distributions

The idea of Gibbs distributions is a very important invention of statistical physics, which is today considered as fruitful also for other areas such as spatial statistics. (There are statisticians who say that we are living in the era of Gibbs distributions after that of Gaussian distributions.) It is explained here in terms of physics.

Consider a system with different states or configurations, which appear randomly. Their number can be finite or infinite. Each state x has its energy $H(x)$, where the energy function H may depend on further parameters such as temperature. The key assumption is that the probability for the occurrence of x is proportional to $\exp(-H(x))$: states with high energy occur rarely. If the number of states is countable, then the probability of x is given by

$$P(x) = \frac{1}{Z} \exp(-H(x));$$

otherwise a probability density may describe the situation,

$$f(x) = \frac{1}{Z} \exp(-H(x)).$$

The normalizing constant Z is called partition function. (Z is called 'function' since it depends on parameters of H.) Two well-known distributions can be interpreted in an elegant way as Gibbs distributions: the Gaussian and the exponential distribution. Here x is a point on the real axis $\mathbb{R}$. In the Gaussian case the energy corresponds to a central force: $H(x) = \frac{(x-\mu)^2}{2\sigma^2}$, while in the exponential case $H(x) = \mu x$, $x \geq 0$. In both cases the constant Z is well-known, $1/\sqrt{2\pi}\sigma$ and $1/\mu$, respectively.

An important model in the context discussed here is a pair-wise interaction Gibbs point process. It describes a system of n random points in a compact convex subset $B \subset \mathbb{R}^d$ of positive Lebesgue measure. (This 'canonical' case will below be generalized to the 'grand canonical' case with a random number of points.)

Let $x = (x_1, \ldots, x_n)$ be any configuration, where x_i is the ith point. The probability density of the process has the form

$$f(x) = \frac{1}{Z} \exp \left(- \sum_{1 \leq i < j \leq n} \theta(\| x_i - x_j \|) \right). \tag{7.5.12}$$

Here $\theta : [0, \infty) \to (-\infty, \infty]$ is the 'pair-potential'. A simple but important particular case is

$$\theta(r) = \begin{cases} \infty & \text{for } r \leq \sigma, \\ 0 & \text{otherwise.} \end{cases}$$

The corresponding point process is called hard-core Gibbs process. Around each point there is no other point in a hard-core distance σ, i.e. if in y there is a point of the process then in the ball of radius σ centered at y there is no other point. Clearly, for given σ there is an upper bound $N(\sigma)$ of the possible number of points in B; assume that $n \leq N(\sigma)$.

As for other classes of distributions it is of interest to know comparison properties of Gibbs distributions. In the following three such results are given. The first of them gives a comparison property for the distributions in the preceding Section 7.5.1 with respect to the parameter β. The stochastic order used is the randomness order $\leq_{rand}$ (see Section 1.5.3). The method used is a variant of the functional method, where the order relation is particularly tailored to the distributions considered, as suggested by Uhlmann (1972); see also Ruch (1975) and Mead (1977).

Assume that a given system can take n configurations $x^1, \ldots, x^n$. The energy of configuration x is $H(x)$, and the probability of x

$$P(x) = \frac{1}{Z} \exp(-H(x)). \tag{7.5.13}$$

In contrast to Section 7.5.2, it is not assumed here that a particular form of $H(x)$ is given; the energy functions considered may be a quite general non-negative functions.

For comparison purposes two energy functions H_a and H_b are considered. Set

$$a_i = H_a(x^i) \quad \text{and} \quad b^i = H_b(x^i) \quad \text{for} \quad i = 1, 2, \ldots, n.$$

It is assumed that the configurations x^i are ordered so that the sequence (a_i) is increasing and that, with the same numbering, also (b_i) is increasing.

Let $P^a = (P_i^a)$ and $P^b = (P_i^b)$ be the distributions given by (7.5.13) with respect to H_a and H_b, where the numbering corresponds to that of the x^i. Then clearly,

$$P_1^a \geq P_2^a \geq \ldots \geq P_n^a \qquad (7.5.14)$$

and

$$P_1^b \geq P_2^b \geq \ldots \geq P_n^b. \qquad (7.5.15)$$

Theorem 7.5.1. *(Uhlmann (1972))* *If*

$$a_{i+1} - a_i \leq b_{i+1} - b_i \quad for \quad i = 1, 2, \ldots, n-1, \qquad (7.5.16)$$

then

$$\sum_{i=1}^{k} P_i^a \leq \sum_{i=1}^{k} P_i^b \quad for \quad k = 1, 2, \ldots, n, \qquad (7.5.17)$$

i.e.

$$P^b \leq_{rand} P^a. \qquad (7.5.18)$$

Proof. (See Alberti and Uhlmann (1982)). Let

$$A_k = \sum_{i=1}^{k} e^{-a_i} \quad and \quad B_k = \sum_{i=1}^{k} e^{-b_i}$$

and

$$\bar{A}_k = \sum_{i=k+1}^{n} e^{-a_i} \quad and \quad \bar{B}_k = \sum_{i=k+1}^{n} e^{-b_i} \quad for \quad k = 1, 2, \ldots, n.$$

Then

$$\sum_{i=1}^{k} P_i^a = A_k/(A_k + \bar{A}_k) \quad and \quad \sum_{i=1}^{k} P_i^b = B_k/(B_k + \bar{B}_k).$$

Consequently, it is obvious that (7.5.17) is equivalent to

$$\bar{A}_k/A_k \geq \bar{B}_k/B_k \quad for \quad k = 1, 2, \ldots n. \qquad (7.5.19)$$

Set $d_i = a_i - b_i$. The inequalities (7.5.16) imply

$$d_1 \geq d_2 \geq \ldots \geq d_n \quad for \quad k = 1, 2, \ldots, n.$$

Thus

$$A_k = \sum_{i=1}^{k} e^{-a_i} = \sum_{i=1}^{k} e^{-d_i} e^{-b_i} \le e^{-d_k} B_k$$

and analogously $\bar{A}_k \ge e^{-d_{k+1}} \bar{B}_k \ge e^{-d_k} \bar{B}_k$. This yields the inequalities (7.5.19) and the theorem. $\square$

Clearly, the assumption of equal order of configuration probabilities as in (7.5.14) and (7.5.15) is very restricting. Nevertheless, Theorem 7.5.1 has useful applications, for example the following result for Potts models.

Corollary 7.5.2. *Let there be given two Potts models on the same lattice Λ and with the same function f in (7.5.3). Then $0 < \beta^a \le \beta^b$ for the corresponding β-parameters implies*

$$P^b \le_{rand} P^a$$

for the corresponding distribution of the Potts models; i.e. the model with the smaller β is more mixed or chaotic.

In physical interpretation, the hotter system is more chaotic. The term $-\ln Z$ is called 'entropy'. The proof above shows that $A_n \le B_n$, i.e. $Z_a \le Z_b$. Thus the hotter system has also the larger entropy. Note that Theorem 7.5.1 is only one example of a more general theory, which is desribed in Alberti and Uhlmann (1982) and Alberti (1992). There, for example, also $\le_{rand}$-monotone stochastic processes are studied, processes the behavior of which becomes more and more chaotic.

In the case of a hard-core Gibbs process it is simple to give a comparison property with respect to the hard-core distance σ, for a fixed point number n and set B.

Theorem 7.5.3. *Let P^σ be the distribution of the hard-core Gibbs process with hard-core distance σ. Then $\sigma_1 \le \sigma_2$ implies $P^{\sigma_2} \le_{rand} P^{\sigma_1}$.*

Proof. Let A_σ be the set of possible configurations of the process for parameter σ. Clearly, $\sigma_1 \le \sigma_2$ implies $A_{\sigma_2} \subseteq A_{\sigma_1}$. The distribution P^σ is the uniform distribution on A_σ since for every x in A_σ, $f(x)$ takes the value $1/Z$. Theorem 1.5.41 for the randomness order then yields $P^{\sigma_2} \le_{rand} P^{\sigma_1}$. $\square$

Turn to the grand canonical case with a random number of points. Now a configuration x is a finite set of points $x_1, \ldots, x_{|x|}$. The corresponding density function is given by

$$f(x) = \frac{1}{Z}\, \alpha^{|x|} \exp\left(-\sum_{1 \le i < j \le |x|} \theta(\|\, x_i - x_j\,\|)\right) \qquad (7.5.20)$$

with a general pair-potential θ. The positive parameter α is related to a term physicists call 'chemical potential'. (A large value of α corresponds to a low potential.) For this model a comparability property with respect to α can be given.

For given α let π^α be the distribution of the point number of the process, i.e. π_k^α is the probability that the total number of points equals k. It is obtained by integration over all possible configurations consisting of k points in B. Write

$$\pi_k^\alpha = \frac{1}{Z} c_k \alpha^k \quad \text{for} \quad k = 0, 1, \ldots \tag{7.5.21}$$

and note that the c_k are independent of α; they only depend on θ and B and are given by

$$c_k = \frac{1}{k!} \int_B \cdots \int_B \exp\left(- \sum_{1 \le i < j \le k} \theta(\| x_i - x_j \|) \right) dx_1 \ldots dx_k,$$

while Z of course depends on α.

Theorem 7.5.4. $\alpha_1 \le \alpha_2$ *implies* $\pi^{\alpha_1} \le_{lr} \pi^{\alpha_2}$ *and* $\pi^{\alpha_1} \le_{st} \pi^{\alpha_2}$, *where* $\le_{lr}$ *denotes the likelihood ratio order.*

Proof. According to (7.5.21) it holds for $\alpha_1 \le \alpha_2$ that

$$\frac{\pi_k^{\alpha_2}}{\pi_k^{\alpha_1}} = \gamma \left(\frac{\alpha_2}{\alpha_1} \right)^k.$$

for some positive constant γ. This is obviously an increasing function of k. Hence $\pi^{\alpha_1} \le_{lr} \pi^{\alpha_2}$ holds. The assertion for $\le_{st}$ then follows from Theorem 1.4.4. $\qquad\square$

It is not true in general that $\alpha_1 \le \alpha_2$ ensures also the comparison of the corresponding point processes in the sense of $\subseteq_{st}$.

8

Comparing Risks

Almost any human activity is related to some sort of risk. These risks typically involve an economic factor. Therefore the comparison of risks is an important topic of economic research, especially in the financial and actuarial literature. However, there is no clear cut definition of the word *risk*. In the actuarial literature a risk is usually defined to be a non-negative random variable describing a potential loss of a customer or a company. In contrast, classical economic theory typically deals with gains instead of losses. Here the word *risk* is used to describe the situation where the gains are random variables with a known distribution. In order to avoid confusion we will use the notation *prospects* when dealing with random gains and their utilities. This is the usual notation in classical economic theory and in finance. The word *risk* will be reserved for random losses which arise in the context of actuarial applications.

8.1 Economics of Uncertainty

8.1.1 Basics of Stochastic Dominance

Frequently in every-day life one has to choose an action from a given set of alternatives with uncertain consequences. Consider, for example, an investor who has to allocate his resources to different investment opportunities, or an individual who has to decide whether or not to buy a lottery ticket. In economics the uncertain consequences are called *prospects* or *lotteries*.

We will assume here that the prospects can be described by random variables with distributions that are known to the decision maker. This is sometimes called the situation of *decision under risk* in contrast to *decision under uncertainty* where the decision maker is not able to assign distributions to the prospects.

Moreover, it will be assumed that the decision maker's actions only depend on the distributions and not on the concrete realizations of the random variables.

In their famous book *Theory of Games and Economic Behavior*, von Neumann and Morgenstern (1947) described some axioms for a rational decision maker, which imply the so-called *expected utility hypothesis*. This hypothesis says that for a rational decision maker there exists a utility function u such that she prefers prospect Y to prospect X if and only if $Eu(X) \le Eu(Y)$.

In practice, however, it is almost impossible to elicit the utility function of a decision maker explicitly. Therefore it is a question of interest, whether there are criteria for the distributions of X and Y that enable prediction of the decision maker's choice when there is only partial knowledge of his utility function, say, that his utility function belongs to some prescribed class $\mathcal{F}$ of functions. A similar situation occurs if there is a group of decision makers with different utility functions, and one asks whether all members of the group will come to the same decision.

Both situations lead to stochastic order relations of the type

$$X \preceq Y \quad \text{if } Eu(X) \le Eu(Y) \text{ for all } u \in \mathcal{F}.$$

Stochastic orders of this type have been studied extensively in Chapter 2 under the name *integral stochastic orders*. In economics they are usually called *stochastic dominance rules*.

8.1.2   First- and Second-Order Stochastic Dominance

Suppose now that the prospects can be measured in real-valued units (e.g. monetary units, number of items sold, etc.) so that the prospects are real-valued random variables, and assume that the decision maker is rational in the sense of preferring more to less. Then the utility function is necessarily increasing.

If every decision maker with an increasing utility function prefers prospect Y to X then Y is said to dominate X with respect to *first order stochastic dominance* (FSD), written $X \le_{FSD} Y$. According to Theorem 1.2.8 $X \le_{FSD} Y$ is just another notation for $X \le_{st} Y$. Therefore it is not necessary here to describe properties of this stochastic order. The reader can be referred to Section 1.2.

However, we mention the fact that $X \le_{FSD} Y$ holds if and only if $P(X > t) \le P(Y > t)$ for all real t. In the notation used in Chapter 2 this means that there is a small generator consisting of increasing indicator functions (see Example 2.5.2). This has an interesting economic interpretation: for checking FSD it is sufficient to consider decision makers who follow a so-called *aspiration level* strategy, i.e. they have an aspiration level t and their preferences are determined by the probabilities of exceeding this aspiration level. Such decision makers are sometimes called *satisficer*; see Simon (1957).

The FSD rule does not take into account the decision maker's attitude to risk. Most decision makers are assumed to be *risk averse* in the sense of preferring the certain yield of EX to the risky prospect X whatever the distribution of X, i.e. the utility function satisfies $Eu(X) \le u(EX)$ for all X.

It is a simple consequence of Jensen's inequality that this holds if and only if u is concave.

If every risk averse decision maker with an increasing concave utility function prefers Y to X then Y is said to dominate X with respect to *second order stochastic dominance* (SSD), written $X \leq_{SSD} Y$. This order relation was introduced in Section 1.5 under the notation $\leq_{icv}$. We mainly considered the stochastic orders $\leq_{cx}$ and $\leq_{icx}$ in Section 1.5, since these orders are more important in other fields. Nevertheless, it is not necessary to study the concept of SSD in detail here. Recall from page 17 that $X \leq_{icv} Y$ is equivalent to $-Y \leq_{icx} -X$. Thus all results for $\leq_{icx}$ can easily be translated into results for $\leq_{icv}$. As a counterpart to the Theorems 1.5.7 and 1.5.21 the following result was shown by Rothschild and Stiglitz (1970).

Theorem 8.1.1. *The following statements are equivalent:*

(i) $X \leq_{SSD} Y$*;*

(ii) $E \min\{X, t\} \leq E \min\{Y, t\}$ *for all real t;*

(iii) *there are random variables $\hat{X}$ and $\hat{Y}$ with the same distributions as X and Y such that $\hat{Y} \geq E[\hat{X}|\hat{Y}]$ almost surely.*

Whitmore (1970) introduced the notion of *third order stochastic dominance* (TSD). In his notation $X \leq_{TSD} Y$ means that $Eu(X) \leq Eu(Y)$ holds for all functions u which are three times differentiable with $u' \geq 0$, $u'' \leq 0$ and $u''' \geq 0$. Notice that this is one of the orders considered in Section 1.6, namely the order $\leq_{3-icv}$.

Extensions to higher order stochastic dominance, even for fractional order, were given by Fishburn (1976, 1980). Taking the limit to infinity yields the Laplace transform order introduced in Definition 1.6.4. The corresponding utility functions are infinitely differentiable with derivatives alternating in sign. They are sometimes said to have *mixed risk aversion*; see Pratt and Zeckhauser (1987) and Caballé and Pomansky (1996).

In a famous paper, Pratt (1964) considered the problem of comparative risk aversion: under what conditions is a decision maker with a utility function v more risk averse than one with a utility function u? One possible definition of this concept is in terms of *risk premiums*. For that it is convenient to assume that all involved utility functions are strictly increasing and at least twice differentiable. Let w be the current wealth of a decision maker with utility function u facing a risky prospect X. The risk premium $\pi_u(w, X)$ is defined as the value of exchanging the risky prospect X by a sure payment of EX, i.e. it is the solution of the equation

$$Eu(w + X) = u(w + EX - \pi_u(w, X)). \tag{8.1.1}$$

For strictly increasing and continuous u this quantity is well defined. The utility function v is said to be more risk averse than u, if

$$\pi_u(w, X) \le \pi_v(w, X) \quad \text{for all } w \text{ and all } X.$$

Pratt (1964) showed the following result.

Theorem 8.1.2. *For all strictly increasing and twice differentiable utility functions u and v the following statements are equivalent:*

(i) *v is more risk averse than u;*

(ii) *$v(x) = k(u(x))$ for some increasing concave function k;*

(iii) *$r_u(x) \le r_v(x)$ for all x, where r_u is the* absolute risk aversion *defined as*

$$r_u(x) = -\frac{u''(x)}{u'(x)} \tag{8.1.2}$$

and similarly for v.

Meyer (1977) considered *second order stochastic dominance with respect to a function v*, i.e. the stochastic dominance rule obtained by considering all decision makers with a utility function that is more risk averse than a fixed utility function v. Let us denote this stochastic order by $\le_{SSD,v}$. The equivalence of (i) and (ii) in Theorem 8.1.1 can be extended to this case.

Theorem 8.1.3. *The following statements are equivalent:*

(i) *$X \le_{SSD,v} Y$;*

(ii) *$E \min\{v(X), t\} \le E \min\{v(Y), t\}$ for all real t.*

Not all results, however, can easily be generalized from $\le_{SSD}$ to $\le_{SSD,v}$. For strictly concave v it is not true in general that

$$X \le_{SSD,v} Y \quad \text{implies } EX \le EY. \tag{8.1.3}$$

The implication (8.1.3) holds only if v is increasing and convex. Similarly, closure under convolution (i.e. property (C) from Definition 2.4.1) does not hold for all v. It will be shown below that closure under convolution only holds if v is a CARA utility function, which means that it has *constant absolute risk aversion*. The only CARA utility functions are $v(x) = \exp(\alpha x)$ with $\alpha > 0$, $v(x) = -\exp(\alpha x)$ with $\alpha < 0$ (in both cases $r_v(x) \equiv \alpha$)), and $v(x) = x$ and hence $r_v(x) \equiv 0$, i.e. in the latter case the decision maker is risk neutral.

Theorem 8.1.4. *The following statements are equivalent:*

(i) *$X \le_{SSD,v} Y$ implies $X + Z \le_{SSD,v} Y + Z$ for all Z independent of X and Y;*

(ii) *v is a CARA utility function.*

Proof. It follows from Corollary 2.3.9 that the maximal generator of $\leq_{SSD,v}$ is given by the set of all functions $\{k \circ v : k$ increasing concave$\}$. Hence, if v is a CARA utility function then the maximal generator is invariant under translations since then for all real t it is $k(v(x+t)) = k(\alpha_t v(x) + \beta_t)$ for some non-negative α_t and some real β_t. On the other hand, if v is not a CARA utility function, i.e. there are real numbers a and b with $r_v(a) < r_v(b)$, then the function $w(x) = v(x + a - b)$ is not in the maximal generator of $\leq_{SSD,v}$, since $r_w(b) = r_v(a) < r_v(b)$.

The equivalence of (i) and (ii) thus follows from Theorem 2.4.2 e).   □

8.1.3   Stochastic Dominance with DARA Utility Functions

It is frequently supposed that a decision maker's aversion to risk is decreasing with his wealth. This is reflected by a utility function that exhibits *decreasing absolute risk aversion*, shortly denoted by DARA.

Mostly it is assumed, in addition, that u is concave (and hence r_u non-negative), so that the decision maker really is risk averse. Therefore the stochastic dominance rule based on DARA utility functions is typically defined as follows.

Definition 8.1.5. $X \leq_{DARA} Y$ if $Eu(X) \leq Eu(Y)$ for all increasing concave functions with decreasing absolute risk aversion r_u.

This order relation was considered in Vickson (1974, 1975). We want to give a characterization of its maximal generator. To do so, we need a pointwise characterization of DARA utility functions. First observe that r_u is the derivative of $-\ln u'$, so that u is DARA if $\ln u'$ is decreasing and convex. In other words, u' is decreasing and log-convex. The set of all log-convex functions is a closed convex cone; see Roberts and Varberg (1973). Therefore the following characterization of functions with a log-convex derivative holds.

Lemma 8.1.6. *Assume that $u : \mathbb{R} \to \mathbb{R}$ is differentiable. Then u' is log-convex if and only if $x \mapsto u(x + h) - u(x)$ is log-convex for all $h > 0$.*

This yields the following characterization of the maximal generator.

Theorem 8.1.7. *The maximal generator of $\leq_{DARA}$ is the set of all increasing concave functions u with the property that $x \mapsto u(x + h) - u(x)$ is log-convex for all $h > 0$.*

Proof. It is a consequence of theorem 13F in Roberts and Varberg (1973) that this set of functions is a convex cone closed under pointwise convergence. Therefore the assertion is an immediate consequence of Lemma 8.1.6 and Corollary 2.3.9, since an arbitrary function u with these properties can be approximated by a sequence of infinite differentiable DARA utility functions via $u_n(x) = \int u(x-t)\phi(nt) \, dt$, where ϕ is the density of the standard normal distribution. $\square$

Remark 8.1.8. We conjecture that all functions in this maximal generator are differentiable, but we are not able to prove this.

There is no hope to find a reasonable small generator for $\leq_{DARA}$, which would make it easy to check this order relation. Similarly as in the case of multivariate convex order, the extreme points are dense in the set of all DARA utility functions. This was shown by Gollier and Kimball (1997). This may be a reason why this stochastic dominance rule is not very often considered in the literature. Since we could not even find closure properties of this order relation in the literature, we will collect them here for completeness.

Theorem 8.1.9. *The stochastic order relation $\leq_{DARA}$ has the properties (R), (E), (M), (T), (C), and (MI), but is not closed with respect to weak convergence.*

Proof. The mentioned properties are easy consequences of Theorem 2.4.2. Property (W) cannot hold, since it does not hold for the stronger order $\leq_{SSD}$. $\square$

An important reason for the consideration of third order stochastic dominance is that it is only a little stronger than $\leq_{DARA}$ but easier to check.

Theorem 8.1.10. *If $X \leq_{TSD} Y$ then $X \leq_{DARA} Y$.*

Proof. It is sufficient to show that the third derivative of a (three times differentiable) DARA utility function is non-negative. Taking the derivatives on both sides of the equation $u''(x) = -u'(x)r_u(x)$ yields

$$u'''(x) = -u''(x)r_u(x) - u'(x)r_u'(x),$$

and the latter expression is non-negative if u is increasing concave and r_u is non-negative and decreasing. $\square$

On the other hand, $\leq_{DARA}$ is stronger than $\leq_{Lt}$. This was shown in Caballé and Pomansky (1996) for non-negative random variables, but the result also holds for arbitrary real random variables. Indeed, the proof of this more general version is almost trivial.

Theorem 8.1.11. *If $X \leq_{DARA} Y$ then $X \leq_{Lt} Y$.*

Proof. It is sufficient to show that the functions $x \mapsto -\exp(-tx)$, $t > 0$, are DARA, and indeed they have constant risk aversion. $\square$

The proof of Caballé and Pomansky (1996) uses the non-trivial fact that every function with a completely monotone derivative is DARA. Note that this result is a corollary to Theorem 8.1.11.

It is well known that not all decision makers are risk averse in all situations. The typical example where a wide-spread risk seeking behavior can be observed is buying lottery tickets. Therefore we think that it is natural to consider also stochastic orders which are defined by monotone risk aversion, but where the utility functions are not necessarily concave. A stronger order than $\leq_{DARA}$ is obtained, if the requirement of concavity is abandoned in Definition 8.1.5. This relation will be denoted as $\leq'_{DARA}$. Then the following results hold.

Theorem 8.1.12. *a) The maximal generator of $\leq'_{DARA}$ is the set of all increasing functions u with the property that $x \mapsto u(x + h) - u(x)$ is log-convex for all $h > 0$.*
b) The stochastic order relation $\leq'_{DARA}$ has the properties (R), (E), (M), (T), (C), and (MI), but (W) does not hold.

Proof. This can be shown as in the Theorems 8.1.7 and 8.1.9, except for (W). But (W) cannot hold here, since the maximal generator of this order relation does not contain any non-trivial bounded function. This follows from the fact that monotonicity of r_u implies that u is either convex or concave or first concave and then convex. But such a function cannot be bounded unless it is constant. ☐

As far as we know, the stochastic orders that can be defined by utility functions with increasing risk aversion have not yet been considered in the literature. This might seem surprising, but it will be shown in Theorem 8.1.14 that these orders do not lead to new concepts.

Definition 8.1.13. Let X, Y be real random variables. Then define
a) $X \leq_{IARA} Y$ if $Eu(X) \leq Eu(Y)$ for all increasing concave functions with increasing absolute risk aversion r_u;
b) $X \leq'_{IARA} Y$ if $Eu(X) \leq Eu(Y)$ for all increasing functions with increasing absolute risk aversion r_u.

At first sight one may think that these order relations have a maximal generator that can be characterized as in Theorem 8.1.7, just by replacing log-convexity by log-concavity. But this is wrong. The problem is that the set of log-*concave* functions is not a convex cone. In fact, if we consider the closed convex cone generated by the increasing functions with increasing absolute risk aversion, then we get the set of *all* increasing functions, and a similar result holds for increasing concave utility functions.

Theorem 8.1.14.
a) $X \leq_{IARA} Y$ if and only if $X \leq_{icv} Y$.
b) $X \leq'_{IARA} Y$ if and only if $X \leq_{st} Y$.

Proof. a) According to Lemma 2.3.4 it is sufficient to show that the maximal generator of $\leq_{IARA}$ contains the functions $x \mapsto \min\{x - a, 0\}$ for all real a. It is clear that it is sufficient to verify this for $a = 0$. To do so, define a sequence (u_n) of utility functions by

$$u_n(x) = \begin{cases} x & \text{for } x \leq 0 \\ \frac{1}{n}(1 - e^{-nx}) & \text{for } x > 0 \end{cases}.$$

Then

$$r_{u_n}(x) = \begin{cases} 0 & \text{for } x < 0 \\ n & \text{for } x > 0 \end{cases}.$$

Hence u_n is an increasing concave function with non-negative increasing risk aversion. Since (u_n) converges uniformly to $u(x) = \min\{x, 0\}$, we are done.

b) In this case it is sufficient to approximate the function $u(x) = \mathbf{1}_{[0,\infty)}(x)$. Here choose

$$u_n(x) = \begin{cases} e^{nx} & \text{for } x \leq 0 \\ \frac{1}{n}(n + 1 - e^{-nx}) & \text{for } x > 0 \end{cases}.$$

Then

$$r_{u_n}(x) = \begin{cases} -n & \text{for } x < 0 \\ n & \text{for } x > 0 \end{cases}.$$

and thus u_n is an increasing function with non-decreasing risk aversion. Moreover (u_n) converges pointwise to u, and since all these functions are uniformly bounded, it follows from the theorem of dominated convergence that u belongs to the maximal generator. $\quad\Box$

Many people buy both insurance and lotteries at the same time. This means that they show risk averse and risk seeking behavior simultaneously. There are numerous attempts to find utility functions that reflect this behavior. Friedman and Savage (1948) suggested the use of so called concave-convex-concave utility functions. Landsberger and Meilijson (1990b) derived axiomatically a class of utility function which is suited for this situation. Their approach is based on restrictions of the mean preserving spreads defined in Definition 1.5.25. The idea is to assume that the decision maker is only averse to mean preserving spreads with intervals which contain a fixed real number ν. This was generalized in Müller (1998b), where the restriction to an arbitrary set M of intervals (a, b) is considered.

We say that a distribution function G differs from a distribution function F by a mean preserving (a, b)-spread, if (a, b) is the interval in the definition of an MPS in Definition 1.5.25. (a, b)-spreads are connected with functions which are called (a, b)-concave.

Definition 8.1.15. A real function f is said to be (a, b)-*concave* if $f(x) \geq \ell(x)$ for $a \leq x \leq b$ and $f(x) \leq \ell(x)$ otherwise, where ℓ is the affine function the graph of which is the line through $(a, f(a))$ and $(b, f(b))$.

Indeed, the following result holds, see Müller (1998b).

Theorem 8.1.16. *For all distribution functions F and G and all $a < b$ it holds: G differs from F by a mean preserving (a, b)-spread if and only if $\int f \, dF \geq \int f \, dG$ for all (a, b)-concave functions f.*

Now consider the case where the distribution G can be obtained from the distribution F by a sequence of (a, b)-spreads, where the intervals (a_n, b_n) are restricted to an arbitrary subset M of the set of all intervals. We will denote this as a *mean preserving M-increase in risk*. The following characterization generalizes Theorem 1.5.27.

Theorem 8.1.17. *If a distribution function G differs from a distribution function F by a mean preserving M-increase in risk, then*

$$\int u \, dF \geq \int u \, dG \quad \text{for all } u \in \mathcal{U}_M, \tag{8.1.4}$$

where $\mathcal{U}_M$ is the set of all functions that are (a, b)-concave for all $(a, b) \in M$; and $\mathcal{U}_M$ is the largest class of functions satisfying (8.1.4).

Landsberger and Meilijson (1990a) considered the special case $M = L_\nu = \{(a, b) : a \leq \nu \leq b\}$ for some fixed ν, what they called mean preserving spread about ν. Their main result is stated in the next theorem, which is an easy consequence of Theorem 8.1.17.

Theorem 8.1.18. *If $M = L_\nu = \{(a, b) \in S : a \leq \nu \leq b\}$, then $\mathcal{U}_M$ is the set of all functions that are antistarshaped at ν, i.e. $\mathcal{U}_M$ is the set of all functions u for which*

$$x \mapsto \frac{u(x) - u(\nu)}{x - \nu}, \quad x \neq \nu \tag{8.1.5}$$

is decreasing.

These functions are called antistarshaped since it is easy to see that f is antistarshaped if and only if the epigraph of $-f$ is a starshaped set with respect to the point $(\nu, -f(\nu))$. See also p. 9 for the definition of a starshaped function.

There are, however, other interesting examples for choices of M. Of course, if M is the set of *all* intervals, then $\mathcal{U}_M$ is the set of all concave functions. This is the famous result of Rothschild and Stiglitz (1970) (see Theorem 1.5.27). Müller (1998b) considered the set $M_c = \{(a, b) : b - a \geq c\}$ for $c > 0$. There an example is given for an unfair lottery, an insurance contract and a utility function in $\mathcal{U}_{M_c}$ such that it is optimal for the decision maker to buy both insurance and lottery at the same time.

8.2 Financial Applications

8.2.1 Consistency of Mean-deviation Rules

The most popular approach to the comparison of risky investments is based on the comparison of mean and variance of the prospects. It was introduced in the famous work of Markowitz on portfolio optimization, see Markowitz (1959). In this approach a risk averse decision maker is assumed to assign to the prospect X the 'utility'

$$U(X) = EX - \alpha \operatorname{Var}(X),$$

where $\alpha > 0$ characterizes the degree of risk aversion. It is well known that this approach has some drawbacks (see Example 8.2.1). Therefore there are many attempts to generalize this concept to other mean-risk approaches, where

$$U(X) = EX - \alpha R(X),$$

and $R(X)$ is some other measure of risk like lower semi-variance (defined below), some other (partial) moment, a quantile, or a function of one or more of these measures.

It is a natural question to ask whether or not such a decision rule is consistent with stochastic dominance rules, i.e. does $X \leq_{SD} Y$ imply $U(X) \leq U(Y)$, where $\leq_{SD}$ is one of the stochastic dominance rules described in the previous section, and U is one of the functionals described above?

One of the main drawbacks of the mean-variance approach is that it is not even consistent with FSD. This also holds if variance is replaced by lower semi-variance.

Example 8.2.1. (Mean-(semi)variance decision rules are not consistent with FSD.) Suppose that X has a two point distribution with $P(X = 0) = P(X = x) = 1/2$ for some $x > 0$ and that $Y = 2X$. Then obviously $X \leq_{FSD} Y$. Notice that X and Y can be considered as lotteries with the same probability of winning, but in case of a win Y pays double the price compared to X. However, if U is given by $U(X) = EX - \alpha \operatorname{Var}(X)$ then $U(X) > U(Y)$ for $x > 2/3\alpha$. Hence a decision maker with utility U will have the irrational behavior of preferring X to Y for x large enough.

This relation remains true if variance is replaced by lower semi-variance $\operatorname{Var}^-(X) = E(\max\{EX - X, 0\})^2$. This follows from the fact that X and Y are symmetric about their mean and therefore $\operatorname{Var}^-(X) = 1/2 \operatorname{Var}(X)$ and similarly for Y. Therefore in this example the mean-semi-variance decision rule leads to the same decision as the mean-variance rule with α replaced by $\alpha/2$.

For some important families of distributions, however, mean-variance decision rules are consistent with second order stochastic dominance. In the case of normally distributed random variables X and Y

$$X \leq_{SSD} Y \quad \text{if and only if } EX \leq EY \text{ and } \text{Var}(X) \geq \text{Var}(Y). \quad (8.2.1)$$

Relation (8.2.1) also holds for some other classes of distributions; see the discussion in Bigelow (1993).

There are other mean-semi-deviation decision rules that are always consistent with first or second order stochastic dominance. An example is the *absolute semi-deviation* defined as

$$\delta^{(1)}(X) = E(X - EX)_- = E(\max\{EX - X, 0\}) = \frac{1}{2}E|X - EX|.$$

Notice that no symmetry is required for the last equality. Ogryczak and Ruszczyński (1999) have shown the following result.

Theorem 8.2.2. *Let $U(X) = EX - \alpha\delta^{(1)}(X)$ with $\alpha \leq 1$. Then $X \leq_{SSD} Y$ implies $U(X) \leq U(Y)$.*

Proof. It is easy to see that

$$0 \leq E(X - t)_- - E(X - s)_- \leq t - s \quad \text{for all } s \leq t. \quad (8.2.2)$$

Moreover, $X \leq_{SSD} Y$ implies $EX \leq EY$ and $E(X - t)_- \geq E(Y - t)_-$, since the function $x \mapsto -(x - t)_-$ is increasing concave. Therefore, choosing in (8.2.2) $s = EX$ and $t = EY$ yields

$$\begin{aligned}
EY - EX &\geq E(X - EY)_- - E(X - EX)_- \\
&\geq \alpha(E(X - EY)_- - E(X - EX)_-) \\
&\geq \alpha(E(Y - EY)_- - E(X - EX)_-).
\end{aligned}$$

Hence $U(X) \leq U(Y)$.   □

This result remains true if the absolute semi-deviation is replaced by a general lower partial moment

$$\delta^{(k)}(X) = \left(E(\max\{EX - X, 0\}^k)\right)^{1/k}$$

for some $k \in \mathbb{N}$; see Ogryczak and Ruszczyński (2001) for details.

8.2.2 Portfolio Optimization

The most important problem of portfolio optimization can be described as follows. Consider an investor who has the possibility of investment in n different stocks. Investing one unit of money into stock i yields a (random) return X_i. Thus there is a vector of returns $\mathbf{X} = (X_1, \ldots, X_n)$. The investor has to allocate her budget m to the different stocks. If she has the utility function u then she faces the optimization problem

$$\max_{a_1,\ldots,a_n} Eu\left(\sum_{i=1}^n a_i X_i\right) \tag{8.2.3}$$

$$\text{subject to } \sum_{i=1}^n a_i = m, \quad a_i \geq 0, \quad i = 1,\ldots,n.$$

Sometimes already partial knowledge of the utility function is sufficient to find the optimal allocation. This is the case if there is an optimal allocation $(a_1^*,\ldots,a_n^*)$ such that $\sum_{i=1}^n a_i^* X_i$ stochastically dominates $\sum_{i=1}^n a_i X_i$ for all other feasible allocations $(a_1,\ldots,a_n)$.

For example, it can be shown that for independent and identically distributed random variables $X_1,\ldots,X_n$ and a risk averse investor with a concave utility function the strategy of maximal diversification is always optimal. Indeed this even holds in the more general case of an exchangeable vector $\mathbf{X}$.

Theorem 8.2.3. *Let* $\mathbf{X} = (X_1,\ldots,X_n)$ *be exchangeable and* u *concave. Then the optimization problem* (8.2.3) *has the optimal solution* $\mathbf{a}^* = (m/n,\ldots,m/n)$.

Proof. This is a simple application of the Strassen theorem. Indeed, the result follows easily from Theorem 1.5.23 by choosing there $f_i(x) = a_i x$. $\qquad\square$

Next consider the case that $X_1,\ldots,X_n$ are independent, but with different distributions. One might conjecture that a rational decision maker (having an increasing utility function) will optimally choose a larger proportion of those shares that yield stochastically larger returns. This is not true, however, as the following example shows.

Example 8.2.4. (A rational decision maker may buy larger shares of investments with smaller returns.) Suppose that $n = 2$ and that $P(X_1 = 0) = P(X_1 = 20) = 1/2$ and that $P(X_2 = 10) = P(X_2 = 90) = 1/2$. Thus obviously $X_1 \leq_{FSD} X_2$. Consider a decision maker who is a satisficer with aspiration level 18, i.e. her utility function is $u(x) = 1$ if $x \geq 18$ and $u(x) = 0$ otherwise. Let $g(\alpha) = Eu(\alpha X_1 + (1-\alpha)X_2)$. A simple calculation yields that $g(0.8) = 3/4$ and $g(\alpha) = 1/2$ for all $0 \leq \alpha \leq 1$ with $\alpha \neq 0.8$. Hence the optimal decision of the decision maker is to spend 80 % of her budget for the investment with the stochastically smaller return, and only 20 % for the investment with the stochastically larger return.

However, it is true that the optimal allocation gives a higher proportion to the investments with larger returns, if either the utility function satisfies additional properties or if the returns can be compared with respect to an order that is stronger than FSD, for example if FSD is replaced by the stronger notion of likelihood ratio order introduced in Section 1.4. The following result is essentially due to Landsberger and Meilijson (1990a).

Theorem 8.2.5. *Assume that $X_1, \ldots, X_n$ are independent with*

$$X_1 \geq_{lr} X_2 \geq_{lr} \ldots \geq_{lr} X_n,$$

and that u is increasing. Then the optimization problem (8.2.3) has an optimal solution $a_1^*, \ldots, a_n^*$ with $a_1^* \geq a_2^* \geq \ldots \geq a_n^*$.

Proof. Assume that (8.2.3) has an optimal solution $\mathbf{a}^* = (a_1^*, \ldots, a_n^*)$ with $a_j^* < a_k^*$ for some $j < k$. Define $\mathbf{a}' = (a_1', \ldots, a_n')$ by exchanging a_j^* and a_k^*, i.e. $a_j' = a_k^*$, $a_k' = a_j^*$ and $a_i' = a_i^*$ for all other i. It will be shown that $\mathbf{a}'$ is also an optimal solution of (8.2.3).

To do so, define the bivariate function $g(x, y) = a_j^* y + a_k^* x$. Then

$$\Delta g(x, y) = g(x, y) - g(y, x) = (a_k^* - a_j^*)(x - y) \geq 0$$

for all $x \geq y$. Thus it follows from Theorem 1.9.4 a) that $g(X, Y) \geq_{st} g(Y, X)$, i.e. $a_j^* X_k + a_k^* X_j \geq_{st} a_j^* X_j + a_k^* X_k$. Closure under convolution of $\leq_{st}$ (shown in Theorem 1.2.17) then yields

$$Eu\left(\sum_{i=1}^n a_i^* X_i\right) \leq Eu\left(\sum_{i=1}^n a_i' X_i\right)$$

for any increasing function u. $\qquad\square$

If the decision maker is assumed to be risk averse, then the condition of likelihood ratio order can be relaxed to reversed hazard rate order. Taking into account Theorem 1.9.4 b), the proof of the following result is similar to that of Theorem 8.2.5. Details are given in Kijima and Ohnishi (1996).

Theorem 8.2.6. *Assume that $X_1, \ldots, X_n$ are independent with*

$$X_1 \geq_{rh} X_2 \geq_{rh} \ldots \geq_{rh} X_n,$$

and that u is increasing and concave. Then the optimization problem (8.2.3) has an optimal solution $a_1^*, \ldots, a_n^*$ with $a_1^* \geq a_2^* \geq \ldots \geq a_n^*$.

Müller and Scarsini (2001) extend the portfolio optimization problem in (8.2.3) to include the possibility of buying call options on the stocks yielding a return $Y_i = (X_i - K_i)_+/p_i$ per unit of money, where $p_i = E(X_i - K_i)_+/EX_i$ is the relative price for an option for stock i (so that $EY_i = EX_i$) and K_i is the strike price. Then a risk averse decision maker with increasing concave utility function u and budget m faces the following decision problem:

$$\max_{a_1,\ldots,a_n,b_1,\ldots,b_n} Eu\left(\sum_{i=1}^{n}(a_iX_i + b_iY_i)\right) \qquad (8.2.4)$$

$$\text{subject to } \sum_{i=1}^{n}(a_i + b_i) = m, \quad a_i, b_i \geq 0, \quad i = 1,\ldots,n.$$

It seems to be intuitively clear that investing money in an option involves a higher risk than investing the same amount in the corresponding stock. One may conjecture that a risk averse decision maker will never buy an option, and the optimal solution of the above problem therefore should have the feature $b_1 = \cdots = b_n = 0$. If the components of $\mathbf{X}$ are independent, then this is in fact easy to show. However, if they exhibit negative dependence then this may not be the case. This is not surprising. A put option on X_i can be seen as a call option on $-X_i$, so every investor who buys put options to *hedge* her risks is utilizing this phenomenon. On the other hand, if the components of $\mathbf{X}$ are positively dependent, we are again lead to conjecture that a risk averse decision maker should not buy options. It will be shown now that this is indeed a consequence of Theorem 3.12.14, when $\mathbf{X}$ is conditionally increasing.

Theorem 8.2.7. *If the random vector $\mathbf{X}$ is CI, then there is an optimal solution of the problem (8.2.4) with $b_1 = \cdots = b_n = 0$.*

Proof. It will be shown that

$$Eu\left(\sum_{i=1}^{n}(a_iX_i + b_iY_i)\right) \leq Eu\left(\sum_{i=1}^{n}(a_i + b_i)X_i\right) \qquad (8.2.5)$$

for all $a_1,\ldots,a_n,b_1,\ldots,b_n \geq 0$. To do so, let $\mathbf{X}'$ be the vector with components $X'_i = a_iX_i + b_iY_i$, and let $\mathbf{Y}'$ be the vector with components $Y'_i = (a_i + b_i)X_i$, $i = 1,\ldots,n$. Both X'_i and Y'_i are monotone transformations of X_i, namely $X'_i = f_i(X_i) = a_iX_i + b_i(X_i - K_i)_+/p_i$ and $Y'_i = g_i(X_i) = (a_i + b_i)X_i$. Hence $\mathbf{X}'$ and $\mathbf{Y}'$ have a common copula. Moreover, $EX'_i = EY'_i = a_i + b_i$ and, since $p_i < 1$, the functions f_i and g_i cross exactly once at the point $x_i = K_i/(1 - p_i)$. Thus it follows from the cut criterion of Theorem 1.5.17 that $X'_i \geq_{cx} Y'_i$. Therefore Theorem 3.12.14 yields $\mathbf{X}' \geq_{dcx} \mathbf{Y}'$ and thus $Eu(\sum X'_i) \geq Eu(\sum Y'_i)$ for all convex u, which is equivalent to inequality (8.2.4) for all concave u. $\qquad \square$

8.3   Ordering of Actuarial Risks

8.3.1   Bounds for Aggregate Claims of Dependent Risks

Actuarial sciences is one of the most important fields of research where stochastic orders are applied. The problem of comparing risks lies at the heart of insurance business. A risk born by an insurance company is most naturally described by a random variable S. If the insurance company does not want to bear all of this risk then it will pass on parts of it to a so-called reinsurance company. The part that remains with the first company is usually called *retention*. If the contract is such that there is a fixed retention t then it is called a *stop-loss* contract. This means that the first company bears the whole risk, as long as it is less than t. If, however, $S > t$ then the reinsurance company will take over the amount of $S - t$ so that only the amount of t remains with the first company. Hence the expected cost for the reinsurance company, called the net premium, is $\pi_S(t) = E(S - t)_+$.

The function π_S is called the *stop-loss transform* of the risk S and there is a simple one-to-one correspondence between the distribution function F_S and the stop-loss transform π_S, namely

$$\pi_S(t) = E(S - t)_+ = \int_t^\infty \bar{F}_S(x)\mathrm{d}x,$$

This is also the reason why π_S in other applications is called the *integrated survival function*. Under that name it has been discussed in detail around Theorem 1.5.7.

It is natural to consider the following stochastic order for the comparison of actuarial risks.

Definition 8.3.1. If S and S' are two risks, then S precedes S' in stop-loss order, written $S \leq_{sl} S'$, if $\pi_S(t) \leq \pi_{S'}(t)$ for all real t.

This means that S' is called riskier than S if it leads to higher net premiums for the reinsurance company, independent of the choice of the retention level. According to Theorem 1.5.7 this order is by no means new. It is just the increasing convex order $\leq_{icx}$. On the other hand, due to the derivation given above it is natural that it is used in actuarial sciences under the name stop-loss order, and therefore we will stick to this notation throughout this section.

The typical situation of an insurance company is that they have a number n of clients who all have insured their individual risks $X_1, \ldots, X_n$, and hence the company bears the risk $S = \sum_{i=1}^n X_i$. This is the view point of the so called *individual model of risk theory*.

Very often it is assumed that the risks $X_1, \ldots, X_n$ are independent, since this is very convenient for the mathematical analysis. In the following example, however, it will be shown that this assumption can lead to dramatic errors in the conclusions if it is not true.

Example 8.3.2. (Dramatic effect of dependence.) Suppose that an insurance company dealing with the business of property insurance has $n = 10^6$ clients, each of them owning a house. For simplicity, assume further that each house is worth 1 unit of money (which may be millions of Euro) and will be damaged (and then completely destroyed) in the considered period with a probability of $1/10\,000$. Hence $P(X_i = 1) = 1/10\,000 = 1 - P(X_i = 0)$ for $i = 1, \dots, 10^6$. Thus the expected risk in that period for the insurance company is $ES = 100$, irrespective of the dependence structure between the risks. Let the retention level be $t = 150$ and suppose that the individual risks are stochastically independent. Since n is rather large, the distribution of the aggregate claim S is very well approximated by the normal distribution with $\mu = 100$ and $\sigma = 10$. Thus $t = \mu + 5\sigma$, and it is well known that $P(S > t)$ is then extremely small. Indeed, the net premium is approximately $E(S - t)_+ = 2.8 \times 10^{-8}$.

However, the assumption of independence is very unrealistic in the given context. It may happen for instance that the damages are caused by a storm, a flood or an earthquake. In these cases, many or even all houses can be destroyed simultaneously. Assume that such a 'bad event' happens in the considered period with a probability of $1/100$, i.e. there is an exogenous variable Θ which describes the occurrence of this event with $P(\Theta = 1) = 1 - P(\Theta = 0) = 1/100$. Given Θ, the risks $X_1, \dots, X_n$ shall be independent with $P(X_i = 1 | \Theta = 1) = 1/1\,000$ and $P(X_i = 1 | \Theta = 0) = 1/11\,000$. Then unconditionally each individual still faces the same risk as before, namely $P(X_i = 1) = 1/10\,000 = 1 - P(X_i = 0)$ for $i = 1, \dots, 10^6$. However, there is some sort of positive dependence between the individuals. Though this dependence seems to be very weak (indeed the correlation coefficient is $\mathrm{Corr}(X_i, X_j) = 1/10\,000$ for all $i \neq j$), the situation for the insurance company changes dramatically. It this case the sum S of all the claims is approximately a mixture of two normal distributions, namely

$$S \sim 0.01 \times \mathcal{N}(1\,000, 1\,000) + 0.99 \times \mathcal{N}\left(\frac{1\,000}{11}, \frac{1\,000}{11}\right).$$

An easy calculation shows that $E(S_n - t)_+ \approx 8.5$. Comparing this with the value 2.8×10^{-8} obtained under the assumption of independence shows the big difference which is caused here by assuming only a very small correlation.

The example clearly indicates the practical relevance of investigating the dependence structure of the random vector $\mathbf{X} = (X_1, \dots, X_n)$.

Therefore it is an interesting task to consider stochastic orders $\preceq$ between random vectors $\mathbf{X} = (X_1, \dots, X_n)$ and $\mathbf{X}' = (X_1', \dots, X_n')$ such that $\mathbf{X} \preceq \mathbf{X}'$ implies

$$\sum_{i=1}^{n} X_i \leq_{sl} \sum_{i=1}^{n} X_i'.$$

In earlier chapters several orders with this property have already been encountered. It is clear that the multivariate versions of $\leq_{st}$ and $\leq_{icx}$ have this property. However, more interesting are dependence orders with this property. Müller (1997c) showed that $\leq_{sm}$ is an example for such an order.

Theorem 8.3.3. *Let* $\mathbf{X} = (X_1, \ldots, X_n)$ *and* $\mathbf{X}' = (X_1', \ldots, X_n')$ *be random vectors with* $\mathbf{X} \leq_{sm} \mathbf{X}'$ *and let*

$$S = \sum_{i=1}^{n} X_i \quad and \quad S' = \sum_{i=1}^{n} X_i'.$$

Then $S \leq_{sl} S'$.

Proof. Since $S \leq_{sl} S'$ holds if and only if $Ef(S) \leq Ef(S')$ for all increasing convex functions f, it is sufficient to show that for such a function f the function $\mathbf{x} \mapsto g(\mathbf{x}) = f(x_1 + \ldots + x_n)$ is supermodular. According to the definition of supermodularity given in (3.9.1), g is supermodular if and only if

$$f(t + \varepsilon + \delta) + f(t) \geq f(t + \varepsilon) + f(t + \delta)$$

for all real t and all $\varepsilon, \delta > 0$, which is equivalent to the convexity of f. $\quad\square$

Choosing $\mathbf{X}$ in Theorem 8.3.3 as a vector with independent components shows that a portfolio of positively supermodular dependent risks (in the sense of Definition 3.10.2) is more risky with respect to stop-loss order than a corresponding portfolio of independent risks with the same marginal distributions.

The following theorem shows a similar result for associated random variables. For a proof we refer to Denuit, Dhaene, and Ribas (2001).

Theorem 8.3.4. *Let* $\mathbf{X} = (X_1, \ldots, X_n)$ *and* $\mathbf{X}' = (X_1', \ldots, X_n')$ *be random vectors with the same marginals, and assume that the components of* $\mathbf{X}$ *are independent, while the components of* $\mathbf{X}'$ *are associated. Then*

$$\sum_{i=1}^{n} X_i \leq_{sl} \sum_{i=1}^{n} X_i'.$$

The proof of Theorem 8.3.3 shows that there are also stochastic orders weaker than $\leq_{sm}$ with this property. Indeed it is obvious that any integral stochastic order whose generator contains all functions $(x_1, \ldots, x_n) \mapsto f(x_1 + \ldots + x_n)$ with f increasing convex will do the job. Thus Theorem 8.3.3 remains valid if $\leq_{sm}$ is replaced by $\leq_{idcx}$ or $\leq_{iplcx}$. On the other hand, we are not aware of any interesting stochastic model where one of these weaker orders can be shown but supermodular order cannot be shown. Therefore we will mainly consider $\leq_{sm}$ here, since it is better known and since it has some properties with important meanings in the actuarial context, like the properties of a multivariate dependence order mentioned on page 110 and shown in Theorem 3.9.8.

Property (P9) for example states that supermodular order is invariant with respect to increasing transformations of the marginals. This is important in the actuarial context, since in many cases insurance companies do not offer full compensation for the losses suffered by the policy holder. Instead, if individual i bears a risk X_i, then it will be specified in the insurance contract that the company will pay a compensation $\phi_i(X_i)$. Here ϕ_i is an increasing function with $0 \leq \phi_i(x) \leq x$. A typical example is $\phi_i(x) = \min\{(x - d)_+, M\}$, where M is the maximal payment and d is the deductible. Hence the loss for the insurance company is

$$S = \sum_{i=1}^{n} \phi_i(X_i).$$

But according to property (P9) of supermodular order it is still true that $\mathbf{X} \leq_{sm} \mathbf{X}'$ implies $S \leq_{sl} S'$. This result, however, does not remain true if $\leq_{sm}$ is replaced by a weaker order like $\leq_{idcx}$ or $\leq_{iplcx}$.

Another important property is (P5). It states that within a given Fréchet class $\Gamma(F_1, \dots, F_n)$, i.e. in the set of all distributions with fixed marginals $F_1, \dots, F_n$, there is a maximal element with respect to supermodular order, namely the upper Fréchet bound given by $F(\mathbf{x}) = \min_i F_i(x_i)$. Hence if we know the distributions of the individual risks, then we can immediately give an upper bound for the net reinsurance premium of a stop-loss contract for the portfolio, derived from the upper Fréchet bound. This upper bound may be more realistic than it seems at first sight. For earthquake risks this may be a quite realistic model.

On the other hand, it was shown in Theorem 3.9.15 that the lower Fréchet bound F^- defined by

$$F^-(\mathbf{x}) = \max\left\{0, \sum_{i=1}^{n} F_i(x_i) - (n - 1)\right\} \quad \text{for } \mathbf{x} \in \mathbb{R}^n,$$

is the minimal element in $\Gamma(F_1, \dots, F_n)$ with respect to supermodular order, whenever it defines a proper distribution function. Thus the following result is obtained by combining Theorem 3.9.15 with Theorem 8.3.3.

Theorem 8.3.5. *Let $\mathbf{X} = (X_1, \dots, X_n)$ be a random vector with the lower Fréchet bound as its distribution, and let $\mathbf{X}' = (X'_1, \dots, X'_n)$ be any vector of risks with the same marginal distributions. Then*

$$\sum_{i=1}^{n} X_i \quad \leq_{sl} \quad \sum_{i=1}^{n} X'_i.$$

In actuarial sciences there is a very important special case where the lower Fréchet bound is a proper distribution function, namely that of so-called *mutually exclusive* risks. The following terminology was introduced by Dhaene and Denuit (1999).

Definition 8.3.6. The vector $\mathbf{X} = (X_1, \dots, X_n)$ of non-negative risks is said to be *mutually exclusive* if

$$P(X_i > 0, X_j > 0) = 0 \quad \text{for all } 1 \leq i < j \leq n.$$

This means that when one of the risks is positive, then all others must be zero. This can only hold if

$$\sum_{i=1}^{n} P(X_i > 0) \le 1, \quad \text{or equivalently, if} \quad \sum_{i=1}^{n} P(X_i = 0) \ge n - 1.$$

Thus the marginal distributions of a vector of mutually exclusive risks fulfill condition (ii) of Theorem 3.1.2. In fact, it is easy to see that a vector of $n \ge 3$ risks is mutually exclusive if and only if the following holds: the marginals fulfill condition (ii) of Theorem 3.1.2 with zero as the minimal point of the support, and their joint distribution is given by the lower Fréchet bound. Therefore the following corollary is a direct consequence of Theorem 8.3.5.

Corollary 8.3.7. Let $\mathbf{X} = (X_1, \dots, X_n)$ be a vector of mutually exclusive risks, and let $\mathbf{X}' = (X_1', \dots, X_n')$ be a vector of risks with the same marginal distributions $F_1, \dots, F_n$. Then

$$\sum_{i=1}^{n} X_i \quad \le_{sl} \quad \sum_{i=1}^{n} X_i'.$$

Dhaene and Denuit (1999) gave an alternative direct proof of this result. It is based on the fact that the distribution of the aggregate claim S can be derived easily from the marginal distributions $F_1, \dots, F_n$, if the risks are mutually exclusive. Indeed it is obvious that then for all $t > 0$

$$\bar{F}_S(t) = P(S > t) = P(\sum_{i=1}^{n} X_i > t) = \sum_{i=1}^{n} P(X_i > t) = \sum_{i=1}^{n} \bar{F}_i(t). \quad (8.3.1)$$

This implies

$$\pi_S(t) = E(S - t)_+ = \sum_{i=1}^{n} E(X_i - t)_+.$$

Since it is clear that

$$E\left(\sum_{i=1}^{n} X_i - t\right)_+ \ge \sum_{i=1}^{n} E(X_i - t)_+$$

for any random vector $\mathbf{X} = (X_1, \dots, X_n)$, this yields again Corollary 8.3.7.

Similar explicit expressions hold for the distribution of the aggregate claim in all cases where the lower Fréchet bound exists.

Theorem 8.3.8. Let $\mathbf{X} = (X_1, \dots, X_n)$, $n \ge 3$, have the lower Fréchet bound as its distribution, and let $S = \sum_{i=1}^{n} X_i$. Then the distribution function of S is given either as

a)

$$F_S(t) = \sum_{i=1}^{n} F_i(t + \xi_i - \xi),$$

if condition (i) of Theorem 3.1.2 is fulfilled with $\xi_i, 1 \leq i \leq n$, the maximum of the support of X_i, and $\xi = \sum_{i=1}^{n} \xi_i$; or
b)

$$F_S(t) = 1 - \sum_{i=1}^{n} \left(1 - F_i(t + \xi_i - \xi)\right),$$

if condition (ii) of Theorem 3.1.2 is fulfilled with $\xi_i, 1 \leq i \leq n$, the minimum of the support of X_i, and $\xi = \sum_{i=1}^{n} \xi_i$.

Proof. To show part b), assume first that $\xi_i = 0$ for all $i = 1, \ldots, n$. Then we have the case of mutually exclusive risks and the assertion follows from (8.3.1). Now assume that $\xi_i \neq 0$ for at least one $i \in \{1, \ldots, n\}$. Then the risks $\tilde{X}_i = X_i - \xi_i, 1 \leq i \leq n$, are mutually exclusive and from $\tilde{S} = \sum_{i=1}^{n} \tilde{X}_i = S - \xi$ it follows that

$$F_S(t) = F_{\tilde{S}}(t - \xi) = 1 - \sum_{i=1}^{n} \left(1 - F_{\tilde{X}_i}(t - \xi)\right) = 1 - \sum_{i=1}^{n} \left(1 - F_i(t + \xi_i - \xi)\right).$$

Part a) can be shown analogously.   □

In general, the lower Fréchet bound is not a proper distribution function. Therefore it may happen that there are different minimal elements with respect to supermodular order within a given Fréchet class $\Gamma(F_1, \ldots, F_n)$. Nevertheless, it is sometimes possible to determine a dependence structure with minimal risk of the aggregate claim. This in particular is the case when there is a dependence structure such that the aggregate claim $S = \sum X_i$ is constant. This happens more often than one might think at first sight. In the following we will discuss some cases where this holds. We start with an example.

Example 8.3.9. (Risks with constant sum.) Consider the situation described in Example 8.2.1, where there were $n = 10^6$ clients, and each claim had the distribution $P(X_i = 1) = 10^{-4} = 1 - P(X_i = 0)$. In this case the lower Fréchet bound is not a proper distribution function, since

$$\sum_{i=1}^{n} P(X_i > 0) = 100,$$

and therefore none of the necessary conditions in Theorem 3.1.2 is fulfilled. On the other hand, there are possible dependence structures such that the aggregate claim S is equal to its expectation $ES = 100$ with certainty.

A simple example with an intuitive explanation is as follows. Assume that the clients can be divided into 100 regions, each consisting of 10 000 clients. Suppose further that we have in each region a fire-raiser who will choose at random one house in his region that he will set on fire, but he will be caught with certainty when doing so. This obviously implies that exactly 100 houses will be destroyed, so that the total claim will be $ES = 100$ with certainty. But each individual house will be hit with a probability of $1/10\,000$. In fact, it is the case that the risks within one region are mutually exclusive, and the risks in different regions are independent. Since there are many different possibilities of dividing the risks into such groups, there are a plenty of different joint distributions that lead to this safest dependence structure.

Also in more general cases it is possible to find a dependence structure such that the distribution of the aggregate claim is constant.

Theorem 8.3.10. *Assume that there are $m, n \in \mathbb{N}$, $a_j \in \mathbb{R}$ and $p_j \in \mathbb{N}_0$, $j = 1, \ldots, m$, such that for all $i = 1, \ldots, n$ the random variable X_i has a discrete distribution given by*

$$P(X_i = a_j) = \frac{p_j}{n} \quad \text{for } j = 1, \ldots, m.$$

Then there is a random vector $\mathbf{X} = (X_1, \ldots, X_n)$ with these marginal distributions such that

$$S = \sum_{i=1}^{n} X_i = ES = \sum_{j=1}^{m} p_j a_j \quad \text{with certainty.}$$

Proof. We give a constructive proof of this result. Let U be a real random variable which is uniformly distributed on $[0, 1]$, and let F and F^{-1} be the distribution function and the quantile function of the random variable X_1 (and thus of any X_i), respectively. Let

$$X_i = F^{-1}\left(\text{frac}\left(U + \frac{i-1}{n} \right) \right) \quad \text{for } i = 1, \ldots, n,$$

where as usual $\text{frac}(u) = u$ if $0 \leq u < 1$, and $\text{frac}(u) = u - 1$ if $1 \leq u < 2$. Since F^{-1} is constant on any interval $[(i-1)/n, i/n)$, this implies

$$S = \sum_{i=1}^{n} X_i = \sum_{i=1}^{n} F^{-1}\left(\frac{i-1}{n} \right) = ES$$

with certainty. $\square$

In the case of continuous distributions there are many cases where the sum of $n = 2$ identically distributed random variables can be constant, and this then immediately generalizes to arbitrary even n. A particularly simple case is contained in the next result.

Theorem 8.3.11. *Assume that the distribution of X_1 is symmetric, i.e. there is some real x_0 such that $X_1 - x_0 =_{st} x_0 - X_1$. Then there is for any natural n a vector $\mathbf{X} = (X_1, \ldots, X_{2n})$ with identically distributed components such that*

$$S = \sum_{i=1}^{2n} X_i = ES \quad \text{with certainty.} \tag{8.3.2}$$

Proof. Here again an easy constructive proof is available. Set

$$X_{2k-1} = X_1 \quad \text{and} \quad X_{2k} = 2x_0 - X_1 \quad \text{for } k = 1, \ldots, n.$$

Then the random variables $X_1, \ldots, X_{2n}$ all have the same distribution as X_1, since X_1 is assumed to be symmetric with relation to x_0. As $X_{2k-1} + X_{2k} = 2x_0$, it is obviously

$$S = \sum_{i=1}^{2n} X_i = 2nx_0 = ES \quad \text{with certainty.}$$

$\square$

For symmetric unimodal distributions it is also possible to find three random variables with the same distribution and a constant sum; see Rüschendorf and Uckelmann (2002). Thus for symmetric unimodal distributions (8.3.2) is possible for arbitrary n.

Note that the assumptions of Theorem 8.3.11 also hold for distributions with unbounded support like the normal distribution. Indeed, it even holds for distributions not having an expectation like the Cauchy distribution.

On the other hand, there are also marginal distributions where it is not possible to find a dependence structure such that the aggregate claim is constant with certainty. An example is the case where the individual risks are non-negative but unbounded above, as, for instance, for the exponential distribution. Then it is obvious that $P(S > ES) \geq P(X_1 > ES) > 0$.

Nevertheless, the construction in the proof of Theorem 8.3.10 leads to a small variability of S, namely to bounds for the support of S. Recall that

$$\text{ess inf } S = \inf\{t : P(S < t) > 0\} \quad \text{and} \quad \text{ess sup } S = \sup\{t : P(S > t) > 0\}.$$

Theorem 8.3.12. *Let F be an arbitrary distribution function. Then there is a random vector $\mathbf{X} = (X_1, \ldots, X_n) \in \Gamma(F, \ldots, F)$ (i.e. all marginals have the distribution F) such that the following bounds hold for the support of $S = \sum X_i$:*

$$\text{ess inf } S \geq \sum_{i=0}^{n-1} F^{-1}\left(\frac{i}{n}\right) \geq ES - (\text{ess sup } X_1 - \text{ess inf } X_1)$$

and

$$\text{ess sup } S \le \sum_{i=1}^{n} F^{-1}\left(\frac{i}{n}\right) \le ES + \text{ess sup } X_1 - \text{ess inf } X_1.$$

Moreover, these bounds are the best possible ones if the distribution F has bounded support.

Proof. We use the same construction as in the proof of Theorem 8.3.10, i.e. let U be uniformly distributed on $(0, 1)$, and define

$$X_i = F^{-1}\left(\text{frac}\left(U + \frac{i-1}{n}\right)\right) \quad \text{for } i = 1, \dots, n. \tag{8.3.3}$$

Thus S is a function of U, say $S = g(U)$, where g is periodic with period $1/n$ (i.e. $g(x + k/n) = g(x)$ for all $k = 1, 2, \dots$) and g is obviously increasing on $[0, 1/n)$ since F^{-1} is increasing. Hence

$$\text{ess inf } S \ge g(0) = \sum_{i=0}^{n-1} F^{-1}\left(\frac{i}{n}\right) \tag{8.3.4}$$

and

$$\text{ess sup } S \le \lim_{x \uparrow 1/n} g(x) \le \sum_{i=1}^{n} F^{-1}\left(\frac{i}{n}\right). \tag{8.3.5}$$

Since F^{-1} is increasing, these sums can be considered respectively as lower and upper Riemann sums for the corresponding integrals. Taking into account that $EX_1 = \int_0^1 F^{-1}(x)\mathrm{d}x$ and that $\text{ess inf } X_1 \le F^{-1} \le \text{ess sup } X_1$, the right-hand side bounds can be derived as follows.

$$
\begin{aligned}
\sum_{i=0}^{n-1} F^{-1}\left(\frac{i}{n}\right) &= F^{-1}(0) + n \sum_{i=1}^{n-1} \frac{1}{n} F^{-1}\left(\frac{i}{n}\right) \\
&\ge \text{ess inf } X_1 + n \int_0^{\frac{n-1}{n}} F^{-1}(x)\mathrm{d}x \\
&= \text{ess inf } X_1 + ES - n \int_{\frac{n-1}{n}}^{1} F^{-1}(x)\mathrm{d}x \\
&\ge ES + \text{ess inf } X_1 - \text{ess sup } X_1,
\end{aligned}
$$

and similarly

$$\sum_{i=1}^{n} F^{-1}\left(\frac{i}{n}\right) \le F^{-1}(1) + n \int_{\frac{1}{n}}^{1} F^{-1}(x)\mathrm{d}x \le ES + \text{ess sup } X_1 - \text{ess inf } X_1.$$

It remains to demonstrate that these bounds are tight. Let $a = \text{ess inf } X_1 < \text{ess sup } X_1 = b$. In the case of inequality (8.3.4) suppose that $P(X_1 = a) = 1/(kn) = 1 - P(X_1 = b)$, where k is arbitrary. Then $ES = nb - (b - a)/k$ and since there is a positive probability that X_1 assumes the lower value a,

$$\text{ess inf } S \le a + (n-1)b = ES - \frac{k-1}{k}(\text{ess sup } X_1 - \text{ess inf } X_1).$$

Since k was arbitrary, this shows the tightness of the bounds for ess inf S. Similarly, if $P(X_1 = b) = 1/(kn) = 1 - P(X_1 = a)$, then $ES = na + (b-a)/k$ and hence

$$\text{ess sup } S \ge b + (n-1)a = ES + \frac{k-1}{k}(\text{ess sup } X_1 - \text{ess inf } X_1).$$

$\square$

In the practically relevant case of non-negative and unbounded risks, most of the bounds given in Theorem 8.3.12 are trivial. However, the first inequality in (8.3.4) remains interesting also in that case. The construction used in the proof leads to an aggregate claim S with a quite small variability.

Example 8.3.13. (Bounds for ess inf S and $\text{Var}(S)$ for exponentially distributed claims.) For the case of exponentially distributed claims with mean $\mu = 1$ the infimum of the support and the variance of S obtained from the construction in the proof of Theorem 8.3.12 are given in the table below for various n. Notice that $ES = n$.

n	$Var(S)$	ess inf S
1	1	0
2	1.394	0.693
3	1.660	1.504
4	1.865	2.367
5	2.032	3.259
10	2.606	7.921
15	1.978	12.721
20	3.258	17.579
50	4.242	47.123
100	5.079	96.777

The following general upper bound for $ES - \text{ess inf } S$ can be derived in this case from Theorem 8.3.12.

$$ES - \text{ess inf } S \;=\; ES - \sum_{i=0}^{n-1} F^{-1}\left(\frac{i}{n}\right) \;\le\; n \int_{\frac{n-1}{n}}^{1} F^{-1}(x)\,dx$$

$$=\; n \int_{0}^{\frac{1}{n}} -\ln(x)\,dx \;=\; 1 + \ln(n).$$

Hence the variability of S grows asymptotically at the order $O(\log(n))$, whereas it grows with $O(n)$ if we assume the individual claims to be independent.

It is possible to improve these results by combining the ideas of Theorem 8.3.12 with those of Theorem 8.3.11. Assume that n is even, and that

$$X_{2k-1} = F^{-1}\left(\mathrm{frac}\left(U + \frac{i-1}{n}\right)\right) \quad \text{for } k = 1,\dots,n/2,$$

whereas

$$X_{2k} = F^{-1}\left(\mathrm{frac}\left(1 - U + \frac{i-1}{n}\right)\right) \quad \text{for } k = 1,\dots,n/2.$$

Then the variability of S is even smaller. In this case a lower bound for ess inf S is given by

$$\operatorname{ess\,inf} S \ \geq \ 2\sum_{k=1}^{n/2} F^{-1}\left(\frac{2k-1}{n}\right)$$

For $X \sim \mathrm{Exp}(1)$ and $n = 100$ this method yields $S - \operatorname{ess\,inf} S = 0.6914$, whereas $ES - \operatorname{ess\,inf} S = 3.223$ when using the construction of (8.3.3).

Denuit, Genest, and Marceau (1999) derive bounds for the distribution of S with respect to $\leq_{st}$ for fixed marginals of $\mathbf{X}$. It follows from Theorem 8.3.12 and Example 8.3.13 that these bounds cannot be of good quality for large n, since it has been demonstrated in Theorem 8.3.12 and Example 8.3.13 that $F_S(t) = 0$ is possible for all $0 \leq t \leq ES - \varepsilon$ for some quite small ε.

8.3.2   Some Models for Dependent Risks

This section presents some models for risk portfolios which exhibit positive dependence, and it will be shown how they can be compared by suitable stochastic orders. Most of the following material is taken from Bäuerle and Müller (1998).

In the first model it is assumed that the portfolio consists of different groups such that there is a strong dependence between the members of one group, but much less dependence between members of different groups. As a typical example where this is a realistic setting, consider catastrophic risks like earthquakes or hurricanes where the groups are specified by geographic regions. There is certainly a strong dependence between the expected losses of clients from the same region, but the losses will be nearly independent for clients living far from each other. For such situations the following model can be suggested. It was introduced by Tong (1989) and was further considered by Bäuerle (1997a).

Model 1

Consider a portfolio $\mathbf{X} = (X_1, \ldots, X_n)$ consisting of n risks $X_1, \ldots, X_n$. We assume that the risks can be divided into $r \leq n$ groups according to an n-dimensional vector $\mathbf{k} = (k_1, \ldots, k_r, 0, \ldots, 0)$ with $k_\nu \in \mathbb{N}$ and $\sum_{\nu=1}^r k_\nu = n$, where risk X_i is in group ν if and only if $k_1 + \ldots + k_{\nu-1} < i \leq k_1 + \ldots + k_\nu$. Each of the risks in the portfolio is influenced by three risk factors which will be modeled as independent random variables V, G_ν and Z_i:

1. a global risk factor V which concerns all of the risks in the portfolio in the same fashion;
2. a group specific risk factor G_ν which influences only the risks in group ν, $1 \leq \nu \leq r$, and has no effect on other risks in the portfolio;
3. an individual risk factor Z_i which reflects the individual share of risk X_i, $i = 1, \ldots, n$.

Moreover, it is assumed that there exists a function $g : \mathbb{R}^3 \to \mathbb{R}$ such that the ith risk is given by $X_i = g(V, G_\nu, Z_i)$ whenever i is in group ν. Since higher outcomes of a risk factor should be associated with higher risk in the portfolio, it is assumed that g is increasing. This situation is typical for many insurance businesses. In private health insurance, for example, the risk caused by an individual person depends on an overall risk factor which collects environmental aspects (e.g. pollution, epidemics etc.), on a group specific factor like profession, and on an individual risk factor which summarizes health conditions. In car insurance, the group risk factor could be connected with the local area of the policy holder. Assuming this kind of dependence within a portfolio, it is now interesting to investigate the effect which the constellation of group sizes has on the aggregate claim of the portfolio. It is well known that positive correlations in a risk portfolio increase the payable amount of the insurance company; see e.g. Dhaene and Goovaerts (1997) or Müller (1997c). Obviously it is quite hard to compare two risky portfolios when, for example, the number and sizes of the groups change. However, in some cases this is possible as will be shown in Theorem 8.3.15 below. In order to state it, let $\mathbf{k}$ and $\mathbf{k}'$ be two n-dimensional vectors with

$$\mathbf{k} = (k_1, \ldots, k_r, 0, \ldots, 0), \quad \mathbf{k}' = (k_1', \ldots, k_l', 0, \ldots, 0)$$

$1 \leq r, l \leq n$, $k_i, k_i' \in \mathbb{N}$ for all i and $\sum_{i=1}^n k_i = \sum_{i=1}^n k_i' = n$. Let two

n-dimensional risky portfolios $\mathbf{X}$ and $\mathbf{X}'$ be given by

$$
\begin{aligned}
X_1 &= g(Z_1, G_1, V) & X_1' &= g(Z_1', G_1, V) \\
&\vdots & &\vdots \\
X_{k_1} &= g(Z_{k_1}, G_1, V) & X_{k_1'}' &= g(Z_{k_1'}', G_1, V) \\
X_{k_1+1} &= g(Z_{k_1+1}, G_2, V) & X_{k_1'+1}' &= g(Z_{k_1'+1}', G_2, V) \\
&\vdots & &\vdots \\
X_{k_1+k_2} &= g(Z_{k_1+k_2}, G_2, V) & X_{k_1'+k_2'}' &= g(Z_{k_1'+k_2'}', G_2, V) \\
&\vdots & &\vdots \\
X_n &= g(Z_n, G_r, V) & X_n' &= g(Z_n', G_l, V)
\end{aligned}
$$

where the individual risk factors $Z_1, \ldots, Z_n, Z_1', \ldots, Z_n'$ are i.i.d. random variables, the group specific risk factors $G_1, \ldots, G_{\max\{r,l\}}$ are i.i.d. random variables and the global risk factor V is a random variable independent of $\{Z_i\}, \{Z_i'\}$ and $\{G_\nu\}$; $g : \mathbb{R}^3 \to \mathbb{R}$ is an increasing function. Denote $S = \sum_{i=1}^n X_i$ and $S' = \sum_{i=1}^n X_i'$, respectively.

As an appropriate order relation for the comparison of group structures $\mathbf{k}$ and $\mathbf{k}'$, the majorization order $\leq_M$ introduced in Definition 1.5.31 will be used.

It is not possible in this situation to show that $\mathbf{X}$ and $\mathbf{X}'$ can be compared with respect to supermodular order, but they can be compared with respect to the slightly weaker symmetric supermodular order also introduced in Definition 3.9.4.

The following analogue to Theorem 8.3.3 holds in this case. The easy proof is similar to that of Theorem 8.3.3, and therefore omitted.

Theorem 8.3.14. *Let* $\mathbf{X} = (X_1, \ldots, X_n)$ *and* $\mathbf{X}' = (X_1', \ldots, X_n')$ *be random vectors with* $\mathbf{X} \leq_{symsm} \mathbf{X}'$. *Then*

$$
\sum_{i=1}^n X_i \leq_{sl} \sum_{i=1}^n X_i'.
$$

The main result for this model is as follows; for the proof the reader is referred to Bäuerle and Müller (1998).

Theorem 8.3.15. *Under the assumptions of model 1* $\mathbf{k} \leq_M \mathbf{k}'$ *implies*

$$
\mathbf{X} \leq_{symsm} \mathbf{X}'
$$

and hence

$$
S \leq_{sl} S'.
$$

In this setting it is easy to determine the riskiest and the safest portfolio with respect to the stop-loss order of aggregate claims. The minimum and the maximum with respect to majorization under all vectors $\mathbf{k}$ with $\sum k_i = n$ are given by $\mathbf{k}^s = (1, 1, \ldots, 1)$ and $\mathbf{k}^r = (n, 0, \ldots, 0)$, respectively. This yields the following corollary.

Corollary 8.3.16. *Let* $\mathbf{k}^r = (n, 0, \ldots, 0)$ *and* $\mathbf{k}^s = (1, \ldots, 1)$ *be two* n-*dimensional vectors and denote by* S^r *and* S^s *the aggregate claims of the corresponding risk portfolios as in model 1. Then for arbitrary* $\mathbf{k}$ *with* $\sum_{i=1}^n k_i = n$ *and respective aggregate claim* S,

$$S^s \leq_{sl} S \leq_{sl} S^r.$$

Thus the most risky portfolio is given when there is only one group and the safest portfolio is obtained when each individual forms his own group.

Model 1 is strongly related to the *component models* considered in Wang (1998). As another important class of models, *common mixture models* were considered there, which will be investigated now.

Model 2

Here two so-called common mixture models will be compared. This means that there is some external mechanism, described by a random variable W (and random variables V and W, respectively), which has influence on all the risks. This environmental parameter can describe some state of nature (weather conditions, earthquakes, etc.) as well as economic or legal conditions (inflation, court rules etc.) which have an common impact on all risks. Given this environmental parameter, the individual risks are independent. In the probability literature models of the type described below have been considered by Shaked and Tong (1985) and Bäuerle (1997a).

Suppose that there are two n-dimensional random vectors $\mathbf{X}$ and $\mathbf{X}'$ with the structure

$$(X_1, \ldots, X_n) = (g_1(Z_1, W), \ldots, g_n(Z_n, W)) \qquad (8.3.6)$$
$$(X_1', \ldots, X_n') = (\tilde{g}_1(Z_1', V, W), \ldots, \tilde{g}_n(Z_n', V, W)) \qquad (8.3.7)$$

where $Z_1, \ldots, Z_n$ and $Z_1', \ldots, Z_n'$ are i.i.d. random variables and (V, W) is a random vector independent of $Z_1, \ldots, Z_n$ and $Z_1', \ldots, Z_n'$. Moreover, the functions $g_i : \mathbb{R}^2 \to \mathbb{R}$ and $\tilde{g}_i : \mathbb{R}^3 \to \mathbb{R}$ are such that for every fixed w and all $i = 1, \ldots, n$

$$g_i(Z_i, w) =_{st} \tilde{g}_i(Z_i', V, w), \qquad (8.3.8)$$

i.e. they have the same distribution.

In this case it can be shown that the portfolio $\mathbf{X}' = (X_1', \ldots, X_n')$ is more risky than the portfolio $\mathbf{X} = (X_1, \ldots, X_n)$, if the functions $\tilde{g}_i$ are increasing in the second argument. In fact, let $S = \sum_{i=1}^n X_i$ and $S' = \sum_{i=1}^n X_i'$. Then the following result has been shown in Bäuerle and Müller (1998).

Theorem 8.3.17. *If the functions $\tilde{g}_i$ are increasing in the second argument then $\mathbf{X} \leq_{sm} \mathbf{X}'$ and hence $S \leq_{sl} S'$.*

The model for $\mathbf{X}'$ contains an additional environmental variable V which has influence on all random variables $X_1', \ldots, X_n'$ in the same direction. Hence there is more dependence in $\mathbf{X}'$ than in $\mathbf{X}$, since the external mechanism which has a common influence on all risks is more important in the model for $\mathbf{X}'$. This will become more explicit in the special case treated next.

Assume that W is degenerate. Hence $X_i' = \tilde{g}_i(Z_i', V)$ and $X_i = g_i(Z_i)$. This means that $X_1', \ldots, X_n'$ are conditionally independent given $V = v$, and the monotonicity of $\tilde{g}_i$ in the second argument means that the conditional distribution of X_i' given $V = v$ is stochastically increasing in v for all $i = 1, \ldots, n$. Moreover, $X_1, \ldots, X_n$ are independent random variables which by (8.3.8) have the same marginal distributions as $X_1', \ldots, X_n'$. This special case yields the following corollary to Theorem 8.3.17.

Corollary 8.3.18. *Let V be any random variable and let $\mathbf{X}' = (X_1', \ldots, X_n')$ be a random vector such that $X_1', \ldots, X_n'$ are conditionally independent given $V = v$ and such that the conditional distributions $P(X_i' \in \cdot \,|V = v)$ are $\leq_{st}$-increasing in v for all $i = 1, \ldots, n$. Moreover, let $\mathbf{X} = (X_1, \ldots, X_n)$ be a vector of independent random variables with the same marginal distributions as $\mathbf{X}'$. Then*

$$\mathbf{X} \leq_{sm} \mathbf{X}' \quad and \quad \sum_{i=1}^{n} X_i \leq_{sl} \sum_{i=1}^{n} X_i'.$$

Another application of Theorem 8.3.17 will be given in the next subsection. Further examples can be found in Wang (1998).

Consider the important special case of portfolios consisting of risks X_i having a two point distribution with masses in 0 and α_i with $P(X_i = 0) = p_i$. This occurs, for example, in the individual life model or in models for credit default risk. Dhaene and Goovaerts (1997) determined the most risky portfolio with given marginals for this case and especially considered portfolios with dependencies only between couples.

In the actuarial context the riskiest portfolio has the following interpretation. It has the property that if a policy holder with a low mortality dies in the considered period, then all policy holders with higher mortality also die in the same period with probability 1. Since this assumption is not very realistic, it would be desirable to have a parametric model with a dependence parameter ρ which continuously varies between independence and maximal dependence as described above.

We investigate here two such models, one for the case of indistinguishable individuals and one for the case that the probability for the occurrence of a claim differs between the individuals.

8.3.3　Indistinguishable Individuals

The individuals in a portfolio are indistinguishable if the joint distribution of the random vector of their risks is not affected by permutations of the risks. Usually such a sequence of random variables is said to be exchangeable (or interchangeable); see Feller (1971) or Chow and Teicher (1978). Of course this implies that all risks have the same marginal distribution. Thus, under the stated assumptions of a two-point distribution with masses in zero and some other point there is a $p \in (0,1)$ and some $\alpha > 0$ such that $P(X_i = 0) = p = 1 - P(X_i = \alpha)$ for all $i = 1, \ldots, n$. Without loss of generality it will be assumed that $\alpha = 1$ so that the random variables $X_1, X_2, \ldots$ form a sequence of exchangeable Bernoulli variables.

Thus we suppose that S_n is the total claim amount of a portfolio of n risks, which stem from a sequence of exchangeable Bernoulli variables. A well-known Theorem of de Finetti (see Feller (1971), p. 228) states that in this case the distribution of S_n is a mixture of binomial distributions, i.e.

$$P(S_n = k) = \int_0^1 \binom{n}{k} \vartheta^k (1 - \vartheta)^{n-k} F(\mathrm{d}\vartheta)$$

for some mixing distribution F. Thus, the distribution of S_n is completely determined by F. In fact, it is already completely determined by the first n moments of F. (For a survey on exchangeable Bernoulli variables, including many examples and methods for estimating their parameters, Madsen (1993) is a good reference.)

The following result shows how the mixing distribution F affects the riskiness of the portfolio S_n. It is a direct consequence of corollary 3.7 in Lefèvre and Utev (1996).

Theorem 8.3.19. *Let S_n (S'_n) be the total claim amount of a portfolio of n risks which stem from a sequence of exchangeable Bernoulli variables with mixing distribution F (F'). Then $F \leq_{sl} F'$ implies $S_n \leq_{sl} S'_n$.*

From Theorem 8.3.19 it follows easily that the least risky portfolio of exchangeable Bernoulli variables with given marginals is that which consists of independent risks and the riskiest portfolio is that with mixing distribution concentrated on $\{0, 1\}$, which means that the risks are comonotone. Then the portfolio consists of identical risks $\mathbf{X} = (X_1, X_1, \ldots, X_1)$ and the distribution of the total claim amount $S_n = nX_1$ is a two point distribution with $P(S_n = 0) = p = 1 - P(S_n = n)$. Comparing the stop-loss premiums of this portfolio with an arbitrary other portfolio of Bernoulli risks, Theorem 8.3.19 can be strengthened as follows.

Theorem 8.3.20. *Let $\mathbf{X} = (X_1, \ldots, X_n)$ be a portfolio of identically distributed Bernoulli risks with an arbitrary dependence structure and let $\mathbf{Y} = (Y_1, \ldots, Y_1)$ be a portfolio of identical risks with the same distribution.*

Let $\pi_{\mathbf{X}}(t) = E(\sum X_i - t)_+$ be the net stop-loss reinsurance premium of portfolio $\mathbf{X}$ and define $\pi_{\mathbf{Y}}(t)$ analogously. Then the ratio $\pi_{\mathbf{Y}}(t)/\pi_{\mathbf{X}}(t)$ is increasing on its range $[0, n)$.

Proof. For $t < n$, we have $\pi_{\mathbf{Y}}(t) = p(n - t)$ and hence

$$\frac{\pi_{\mathbf{X}}(t)}{\pi_{\mathbf{Y}}(t)} = \frac{E\left(\sum_{i=1}^{n} X_i - t\right)_+}{p(n - t)} = \frac{1}{p} E\left(\sum_{i=1}^{n} \frac{X_i - 1}{n - t} - 1\right)_+.$$

The last term is obviously decreasing in t. Thus taking the reciprocal value yields the result. □

Example 8.3.21. (Theorem 8.3.20 is not true for general distributions.) It was conjectured in Bäuerle and Müller (1998) that Theorem 8.3.20 may be true for general distributions. This, however, is wrong as can be seen from the following bivariate example. Assume that $P((X_1, X_2) = (3, 3)) = 1/2$ and

$$P((X_1, X_2) = (0, 0)) = P((X_1, X_2) = (0, 2)) = P((X_1, X_2) = (2, 0))$$
$$= P((X_1, X_2) = (2, 2)) = \frac{1}{8}.$$

Then X_1 and X_2 have the same distribution, namely

$$P(X_i = 3) = \frac{1}{2} \text{ and } P(X_i = 0) = P(X_i = 2) = \frac{1}{4} \quad \text{for } i = 1, 2.$$

The vector (X_1, X_2) can be derived from a common mixture model with $P(\Theta = 0) = P(\Theta = 1) = 1/2$ and X_1, X_2 conditionally independent given Θ with

$$P(X_i = 3|\Theta = 1) = 1 \text{ and } P(X_i = 2|\Theta = 0) = P(X_i = 0|\Theta = 0) = \frac{1}{2}$$

for $i = 1, 2$. In this case $\pi_{\mathbf{Y}}(t)/\pi_{\mathbf{X}}(t) = 1$ for $t = 0$ and $t = 4$, whereas $\pi_{\mathbf{Y}}(t)/\pi_{\mathbf{X}}(t) = 10/9$ for $t = 2$. Hence the ratio $\pi_{\mathbf{Y}}(t)/\pi_{\mathbf{X}}(t)$ is not increasing. It will become clear in equation (8.3.11) that this remains true for the same common mixture model with large n.

Suppose $Y, X_1, \ldots, X_n$ are i.i.d. random variables (without loss of generality we assume that they are concentrated on $[0,1]$) and we are interested in the stop-loss premiums of the safest portfolio $\pi_{\mathbf{X}}^n(t) = E(\sum_{i=1}^{n} X_i - nt)_+$ and the riskiest one $\pi_{\mathbf{Y}}^n(t) = E(nY - nt)_+$, where $t \in (0, 1)$ gives the retention percentage. In this setting we obtain the following result.

Theorem 8.3.22. Suppose $Y, X_1, \ldots, X_n$ are i.i.d. random variables and let $\mathbf{X} = (X_1, \ldots, X_n)$ and $\mathbf{Y} = (Y, \ldots, Y)$. Then the ratio $\pi_{\mathbf{Y}}^n(t)/\pi_{\mathbf{X}}^n(t)$ is increasing in the number n of risks and the limit is equal to $E(Y - t)_+/(EY - t)$ if $t < EY$ and $+\infty$ if $t \geq EY$.

Proof. We have

$$\frac{\pi_{\mathbf{Y}}^n(t)}{\pi_{\mathbf{X}}^n(t)} = \frac{E(nY - nt)_+}{E(\sum_{i=1}^n X_i - nt)_+} = \frac{E(Y - t)_+}{E(\frac{1}{n}\sum_{i=1}^n X_i - t)_+}.$$

Hence it suffices to show that $E(\frac{1}{n}\sum_{i=1}^n X_i - t)_+$ is decreasing in n. Since $X_1, \ldots, X_n$ are i.i.d. it follows from Theorem 1.5.24 that

$$\frac{1}{n+1}\sum_{i=1}^{n+1} X_i \leq_{sl} \frac{1}{n}\sum_{i=1}^n X_i \tag{8.3.9}$$

and thus monotonicity follows.

Since the random variables $X_1, X_2, \ldots$ are independent and identically distributed with a finite mean, the strong law of large numbers can be applied and yields

$$\lim_{n \to \infty} \frac{1}{n}\sum_{i=1}^n X_i = EX_1 = EY. \tag{8.3.10}$$

Hence the stated limit follows.   □

Remark 8.3.23. Theorem 1.5.24 shows that equation (8.3.9) holds if the $X_1, X_2, \ldots$ are exchangeable. Hence the monotonicity part of Theorem 8.3.22 remains true for the more general case of exchangeable random variables, but in that case the limit is different. It is obtained by a version of the strong law of large numbers for sequences of exchangeable random variables, which yields

$$\lim_{n \to \infty} \frac{1}{n}\sum_{i=1}^n X_i = E[X_1|\Theta],$$

where Θ is the random variable which describes the mixing mechanism in de Finetti's Theorem; see Feller (1971) or Chow and Teicher (1978) for details.

Hence in this case

$$\lim_{n \to \infty} \frac{\pi_{\mathbf{Y}}^n(t)}{\pi_{\mathbf{X}}^n(t)} = \frac{E(Y - t)_+}{E(E[Y|\Theta] - t)_+}. \tag{8.3.11}$$

It follows from Theorem 8.3.22 that the relative stop-loss premium can be arbitrarily high when the retention exceeds the expected aggregate claim. This again demonstrates how dramatic the effect of neglecting dependencies of risks can be.

8.3.4 Distinguishable Individuals

In the next model individuals in the portfolio may have different probabilities for claims and different claim amounts. We want to construct a portfolio of risks X_i with $P(X_i = 0) = p_i$ and $P(X_i = \alpha_i) = q_i = 1 - p_i$ where $0 < p_i < 1$ and $\alpha_i > 0$ are arbitrary. Moreover, a dependence parameter $\rho \in [0,1]$ will be introduced such that $\rho = 0$ corresponds to independence and $\rho = 1$ corresponds to comonotonicity. A very simple model with this property would be to take some mixture of the independent and the comonotone case. This, however, is probably not very realistic. Instead, some sort of an additive damage model is proposed, which is well known in reliability theory. Assume that there are two sources that cause some normally distributed damage. One source influences all individuals in the same manner, while the other source depends on the individual behavior. A claim of amount α_i occurs if the sum of these two damages exceeds some level z_i.

The formal construction is based on model 2 (see page 292) with normal distributions and functions which assume only two values. As usual denote by $\mathcal{N}(\mu, \sigma^2)$ the univariate normal distribution with mean μ and variance $\sigma^2 > 0$. For convenience the definition is extended to the case $\sigma^2 = 0$, where $\mathcal{N}(\mu, 0)$ denotes the one point distribution in μ. The p-quantile of the standard normal distribution will be denoted by z_p, i.e. if $X \sim \mathcal{N}(0,1)$, then $P(X \le z_p) = p$. Now assume that $0 \le \rho^2 < \tau^2 \le 1$ and consider model 2 with $W \sim \mathcal{N}(0, \rho^2)$, $V \sim \mathcal{N}(0, \tau^2 - \rho^2)$, $Z_i \sim \mathcal{N}(0, 1 - \rho^2)$ and $Z_i' \sim \mathcal{N}(0, 1 - \tau^2)$. All random variables are independent. Define

$$g_i(z, w) = \alpha_i \cdot \mathbf{1}\{z + w \ge z_{p_i}\} = \begin{cases} \alpha_i & \text{for } z + w \ge z_{p_i} \\ 0 & \text{otherwise} \end{cases}$$

and

$$\tilde{g}_i(z, v, w) = \alpha_i \cdot \mathbf{1}\{z + v + w \ge z_{p_i}\}.$$

Recall that $X_i = g_i(Z_i, W)$ and $X_i' = \tilde{g}(Z_i', V, W)$ for $i = 1, \dots, n$. Since $Z_i' + V =_{st} Z_i \sim \mathcal{N}(0, 1 - \rho^2)$, condition (8.3.8) is fulfilled. Moreover, $Z_i + W =_{st} Z_i' + V + W \sim \mathcal{N}(0,1)$ so that $P(X_i = \alpha_i) = P(Z_i + W \ge z_{p_i}) = q_i$ and $P(X_i = 0) = P(Z_i + W \le z_{p_i}) = p_i$. Similarly $P(X_i' = 0) = p_i = 1 - P(X_i' = \alpha_i)$. By Theorem 8.3.17, it is $\mathbf{X} \le_{sm} \mathbf{X}'$ and hence $\mathbf{X}$ is less risky than $\mathbf{X}'$.

Now let us write $\mathbf{X}(\rho) = (X_1(\rho), \dots, X_n(\rho))$ for the above defined portfolio $\mathbf{X}$ to make the dependence on ρ explicit. The definition of $\mathbf{X}'$ implies that $\mathbf{X}' =_{st} \mathbf{X}(\tau^2)$ which can be seen by interchanging the roles of Z_i and Z_i' as well as those of W and $V + W$. The following result is obtained.

Theorem 8.3.24. Let $0 \le \rho < \rho' \le 1$. Then $\mathbf{X}(\rho) \le_{sm} \mathbf{X}(\rho')$ and hence

$$\sum_{i=1}^n X_i(\rho) \le_{sl} \sum_{i=1}^n X_i(\rho').$$

A detailed look at the assumptions which led to the construction of the vector $\mathbf{X}(\rho)$ shows that $\mathbf{X}(\rho)$ was obtained by taking the copula of an exchangeable normal vector with correlation coefficients $\text{Corr}(X_i, X_j) = \rho$, and transforming it monotonically into Bernoulli random variables, which of course still have the same copula (though this is not unique for discrete random vectors). Thus there is an alternative way to derive the result of Theorem 8.3.24. It can also be derived from Theorem 3.13.5 in combination with Theorem 8.3.3.

Such models become more and more popular in the modeling of credit risk; see, for example, Bäuerle (2002) and the references therein.

Frostig (2001) compares a heterogeneous portfolio of independent risks with a homogeneous portfolio of independent risks with respect to symmetric supermodular order. She shows that an upper bound for the heterogeneous portfolio is given by the homogeneous portfolio in which the distribution of each risk is a mixture with equal weights of the risks in the heterogeneous portfolio. A lower bound is obtained by assuming that the risks in the homogeneous portfolio are the average of the individual risks in the heterogeneous portfolio.

There are many other applications of stochastic orderings in actuarial sciences so that we are not able to treat here all of them. The interested reader is referred to van Heerwaarden (1991), Sundt (1991) and Rolski, Schmidli, Schmidt, and Teugels (1999) for general results on ordering risks and to Asmussen (2000), Müller and Pflug (2001) and Willmot and Lin (2001) for recent applications in the context of bounds and approximations of ruin probabilities.

List of Symbols

Basic Notation

A^c	complement of set A
$\mathbf{A}^T$	transposition of matrix $\mathbf{A}$
$\mathbf{1}_A, \mathbf{1}\{A\}$	indicator function of set A
$\mathbf{1}$	vector $\mathbf{1} = (1, \dots, 1)$
$\mathcal{A}, \mathcal{B}, \mathcal{F}$	σ-algebras
Corr	correlation coefficient
Cov	covariance
δ_x	one point distribution in x
E	expectation
$E[X|Y]$	conditional expectation
$\text{Exp}(\alpha)$	exponential distribution with parameter α
$\mathbf{e}_i$	base vector $\mathbf{e}_i = (0, \dots, 0, 1, 0, \dots, 0)$
$F, F_X, F_\mathbf{X}$	distributions, distribution functions
F^{-1}, F_X^{-1}	generalized inverse distribution functions
$\bar{F}, \bar{F}_X, \bar{F}_\mathbf{X}$	survival functions
F^{*k}	kth convolution of F, $F^{*0} = \delta_0$
f, f_X	density functions
$\hat{f}, \hat{f}_X$	Laplace transforms
$\mathbb{N}$	$= \{1, 2, \dots\}$
$\mathbb{N}_0$	$= \mathbb{N} \cup \{0\}$
N_t	counting process
$P, P_X, P_\mathbf{X}$	probability measures, distributions
$\mathbb{R}$	real line
$\mathbb{R}_+$	$= [0, \infty)$
Var	variance
x_+	$= \max\{x, 0\}$
$\mathbf{x}, \mathbf{y}$	vectors
$\mathbf{x} \vee \mathbf{y}$	$= (\max\{x_1, y_1\}, \dots, \max\{x_n, y_n\})$
$\mathbf{x} \wedge \mathbf{y}$	$= (\min\{x_1, y_1\}, \dots, \min\{x_n, y_n\})$

X	random variable
$X_{(i:n)}$	ith order statistic of $X_1, \ldots, X_n$
$\mathbf{X}$	random vector
(X_n)	discrete time stochastic process
(X_t)	continuous time stochastic process
$\mathbb{Z}$	set of all integers
$(\Omega, \mathcal{A}, P)$	probability space
$\leq, \preceq$	order relations
$=_{st}$	equality in distribution

Special Symbols

$\mathcal{B}_b$	set of b-bounded functions, 67
c_Σ	system quantity, 152
$\mathcal{C}^\infty$	set of infinite differentiable functions, 79
$\mathcal{C}_b$	set of bounded continuous functions, 74
C^+	upper Fréchet bound copula, 86
$\Delta g(x, y)$	$= g(x, y) - g(y, x)$, 51
Δ_i^ε	difference operator, 95
F_e	equilibrium distribution, 49
F^+	upper Fréchet bound, 86
F^-	lower Fréchet bound, 86
$\mathcal{F}$	generator of integral stochastic order, 65
$\mathcal{F}_{\mathfrak{I}}$	see p. 105
$\mathcal{G}_{hr}, \mathcal{G}_{rh}$	see p. 53
$\mathcal{G}_{lr}$	see p. 52
$\mathbb{M}_b, \mathbb{M}_b^N$	sets of signed measures, 67
$\mathbb{P}, \mathbb{P}_b$	sets of probability measures, 67
$\pi_X(t)$	integrated survival function, stop-loss transform, 20, 279
$\Phi_{F,G}$	relative inverse distribution function, 3
$r_X(t)$	failure rate function, 9
$\mathcal{R}_\mathcal{F}$	maximal generator, 70
$\tilde{\mathcal{R}}_\mathcal{F}$	extended maximal generator, 70
ϱ	traffic intensity, 217
$Var^-(X)$	lower semi-variance, 274
$\mathbf{x}_\downarrow$	decreasing rearrangement of $\mathbf{x}$, 31
$\mathbf{x}_\uparrow$	increasing rearrangement of $\mathbf{x}$, 31
$\mathbf{X}^\perp$	random vector with independent components, 121
Ξ	random closed set, 247
$\| \cdot \|_b$	weighted supremum norm, 67
$\uparrow_{st}$	stochastically increasing, 125

List of Stochastic Orders

List of Stochastic Orders. (continuation)

References

Ahmed, A.-H. N., Leon, R., and Proschan, F. (1981). Generalized association, with applications in multivariate statistics. *Ann. Stat.* **9**, 168-176.

Alberti, P. M. (1992). Geometry of completely positive maps over certain injective vN-algebras. In M. Mathieu (Ed.), *Elementary operators and applications (Blaubeuren, 1991)* (pp. 117–141). River Edge, NJ: World Scientific Publishing.

Alberti, P. M. (2000). Monotone operators with respect to majorization order. *Personal Communication.*

Alberti, P. M. and Crell, B. (1984). Nonlinear evolution equations and H theorems. *J. Statist. Phys.* **35**, 131–149.

Alberti, P. M. and Crell, B. (1986). A note on the rate of convergence for certain nonlinear evolution equations. *Math. Methods Appl. Sci.* **8**, 479–491.

Alberti, P. M. and Uhlmann, A. (1982). *Stochasticity and partial order: Doubly stochastic maps and unitary mixing.* Dordrecht: D. Reidel Publishing Co.

Alzaid, A., Kim, J. S., and Proschan, F. (1991). Laplace ordering and its applications. *J. Appl. Probab.* **28**, 116–130.

Anderson, W. J. (1972). Local behaviour of solutions of stochastic integral equations. *Trans. Amer. Math. Soc.* **164**, 309–321.

Anderson, W. J. (1991). *Continuous-time Markov chains.* New York: Springer-Verlag.

Araujo, A. and Giné, E. (1980). *The central limit theorem for real and Banach valued random variables.* Chichester: John Wiley & Sons, Ltd.

Arnold, B. C. and Villaseñor, J. A. (1986). Lorenz ordering of means and medians. *Statist. Probab. Lett.* **4**, 47–49.

Asmussen, S. (2000). *Ruin probabilities.* River Edge, NJ: World Scientific Publishing.

Atkinson, J. B. (2000). Some related paradoxes of queuing theory: new cases and a unifying explanation. *J. Oper. Res. Soc.* **51**, 921–935.

Baccelli, F. and Brémaud, P. (1994). *Elements of queueing theory.* Berlin: Springer-Verlag.

Baccelli, F. and Liu, Z. (1992). Comparison properties of stochastic decision free Petri nets. *IEEE Trans. Automat. Control* **37**, 1905–1920.

Barlow, R. E. and Proschan, F. (1975). *Statistical theory of reliability and life testing*. New York: Holt.

Barlow, R. E. and Proschan, F. (1976). Theory of maintained systems: distribution of time to first system failure. *Math. Oper. Res.* **1**, 32–42.

Bassan, B. and Scarsini, M. (1991). Convex orderings for stochastic processes. *Comment. Math. Univ. Carolin.* **32**, 115–118.

Bäuerle, N. (1997a). Inequalities for stochastic models via supermodular orderings. *Comm. Statist. Stochastic Models* **13**, 181–201.

Bäuerle, N. (1997b). Monotonicity results for $MR/GI/1$ queues. *J. Appl. Probab.* **34**, 514–524.

Bäuerle, N. (2002). Risk management in credit risk portfolios with correlated assets. *Insurance Math. Econom.*, to appear.

Bäuerle, N. and Müller, A. (1998). Modeling and comparing dependencies in multivariate risk portfolios. *ASTIN Bulletin* **28**, 59–76.

Bäuerle, N. and Rolski, T. (1998). A monotonicity result for the workload in Markov-modulated queues. *J. Appl. Probab.* **35**, 741–747.

Belzunce, F., Lillo, R. E., Ruiz, J.-M., and Shaked, M. (2001). Stochastic comparisons of nonhomogeneous processes. *Probab. Engrg. Inform. Sci.* **15**, 199–224.

Belzunce, F., Ortega, E., and Ruiz, J. M. (1999). The Laplace order and ordering of residual lives. *Statist. Probab. Lett.* **42**, 145–156.

Bergmann, R. (1978). Some classes of semi-ordering relations for random vectors and their use for comparing covariances. *Math. Nachr.* **82**, 103–114.

Bergmann, R. (1979). Qualitative properties and bounds for the serial covariances of waiting times in single-server queues. *Oper. Res.* **27**, 1168–1179.

Bergmann, R. (1991). Stochastic orders and their application to a unified approach to various concepts of dependence and association. In K. Mosler and M. Scarsini (Eds.), *Stochastic orders and decision under risk* (pp. 48–73). IMS Lecture Notes Vol. 19.

Bergmann, R., Daley, D. J., Rolski, T., and Stoyan, D. (1979). Bounds for cumulants of waiting-times in $GI/GI/1$ queues. *Math. Operationsforsch. Statist. Ser. Optim.* **10**, 257–263.

Bergmann, R. and Stoyan, D. (1978). Monotonicity properties of second order characteristics of stochastically monotone Markov chains. *Math. Nachr.* **82**, 99–102.

Bertsekas, D. P. and Shreve, S. E. (1978). *Stochastic optimal control: The discrete time case*. New York: Academic Press.

Bickel, P. J. and Freedman, D. A. (1981). Some asymptotic theory for the bootstrap. *Ann. Statist.* **9**, 1196–1217.

Bigelow, J. P. (1993). Consistency of mean-variance analysis and expected utility analysis: a complete characterization. *Econom. Lett.* **43**, 187–192.

Billingsley, P. (1999). *Convergence of probability measures* (Second ed.). New

York: John Wiley & Sons Inc.

Blackwell, D. (1953). Equivalent comparisons of experiments. *Ann. Math. Statistics* **24**, 265–272.

Bloch-Mercier, S. (2001). Monotone Markov processes with respect to the reversed hazard rate ordering: an application to reliability. *J. Appl. Probab.* **38**, 195–208.

Block, H. W., Langberg, N. A., and Savits, T. H. (1990). Comparisons for maintenance policies involving complete and minimal repair. In H. W. Block (Ed.), *Topics in statistical dependence (Somerset, PA, 1987)* (pp. 57–68). Hayward, CA: IMS Lecture Notes Vol. 16.

Block, H. W., Savits, T. H., and Shaked, M. (1982). Some concepts of negative dependence. *Ann. Probab.* **10**, 765–772.

Block, H. W., Savits, T. H., and Shaked, M. (1985). A concept of negative dependence using stochastic ordering. *Statist. Probab. Lett.* **3**, 81–86.

Boland, P. J., El-Neweihi, E., and Proschan, F. (1994). Applications of the hazard rate ordering in reliability and order statistics. *J. Appl. Probab.* **31**, 180–192.

Border, K. C. (1991). Functional analytic tools for expected utility theory. In C. D. Aliprantis (Ed.), *Positive operators, Riesz spaces, and economics (Pasadena, CA, 1990)* (pp. 69–88). Berlin: Springer.

Borovkov, A. A. and Foss, S. G. (1992). Stochastically recursive sequences and their generalizations. *Siberian Adv. Math.* **2**, 16–81.

Brandt, A. (1986). The stochastic equation $Y_{n+1} = A_n Y_n + B_n$ with stationary coefficients. *Adv. in Appl. Probab.* **18**, 211–220.

Brandt, A., Franken, P., and Lisek, B. (1990). *Stationary stochastic models*. Chichester: John Wiley & Sons Ltd.

Brandt, A. and Last, G. (1994). On the pathwise comparison of jump processes driven by stochastic intensities. *Math. Nachr.* **167**, 21–42.

Bullen, P. S. (1971). A criterion for n-convexity. *Pacific J. Math.* **36**, 81–98.

Caballé, J. and Pomansky, A. (1996). Mixed risk aversion. *J. Econom. Theory* **71**, 485–513.

Cambanis, S. and Simons, G. (1982). Probability and expectation inequalities. *Z. Wahrsch. Verw. Gebiete* **59**, 1–25.

Cambanis, S., Simons, G., and Stout, W. (1976). Inequalities for $Ek(X,Y)$ when the marginals are fixed. *Z. Wahrsch. Verw. Gebiete* **36**, 285–294.

Carlsson, H. and Nerman, O. (1986). An alternative proof of Lorden's renewal inequality. *Adv. in Appl. Probab.* **18**, 1015–1016.

Chang, C.-S., Chao, X., and Pinedo, M. (1991). Monotonicity results for queues with doubly stochastic Poisson arrivals: Ross's conjecture. *Adv. in Appl. Probab.* **23**, 210–228.

Chang, C.-S. and Yao, D. D. (1993). Rearrangement, majorization and stochastic scheduling. *Math. Oper. Res.* **18**, 658–684.

Chao, X. and Scott, C. (2000). Several results on the design of queueing systems. *Oper. Res.* **48**, 965–970.

Chong, K. M. (1975). Spectral orders, conditional expectations and martingales. *Z. Wahrsch. Verw. Gebiete* **31**, 329–331.

Choquet, G. (1969). *Lectures on analysis. Vol. II: Representation theory.* W. A. Benjamin, Inc., New York, Amsterdam.

Chow, Y. S., Robbins, H., and Siegmund, D. (1971). *Great expectations: the theory of optimal stopping.* Boston: Houghton Mifflin Co.

Chow, Y. S. and Teicher, H. (1978). *Probability theory.* New York: Springer-Verlag.

Cohen, A. and Sackrowitz, H. B. (1995). On stochastic ordering of random vectors. *J. Appl. Probab.* **32**, 960–965.

Cohen, A., Sackrowitz, H. B., and Samuel-Cahn, E. (1995). Cone order association. *J. Multivariate Anal.* **55**, 320–330.

Cohen, J. E., Kemperman, J. H. B., and Zbăganu, G. (1998). *Comparisons of stochastic matrices.* Boston: Birkhäuser Boston Inc.

Cohen, J. W. (1982). *The single server queue* (Second ed.). Amsterdam: North-Holland Publishing Co.

Cottle, R. W., Habetler, G. J., and Lemke, C. E. (1970). On classes of copositive matrices. *Linear Algebra and Appl.* **3**, 295–310.

Cover, T. M. and Thomas, J. A. (1991). *Elements of information theory.* New York: John Wiley & Sons Inc.

Coyle, A. J. and Taylor, P. G. (1995). Tight bounds on the sensitivity of generalised semi-Markov processes with a single generally distributed lifetime. *J. Appl. Probab.* **32**, 63–73.

Daduna, H. and Szekli, R. (1996). A queueing theoretical proof of increasing property of Pólya frequency functions. *Statist. Probab. Lett.* **26**, 233–242.

Daley, D. J. (1968). Stochastically monotone Markov chains. *Z. Wahrsch. Verw. Gebiete* **10**, 305–317.

Daley, D. J. (1987). Certain optimality properties of the first-come first-served discipline for $G/G/s$ queues. *Stochastic Process. Appl.* **25**, 301–308.

Daley, D. J. and Rolski, T. (1984). Some comparability results for waiting times in single- and many-server queues. *J. Appl. Probab.* **21**, 887–900.

Daley, D. J. and Vere-Jones, D. (1988). *An introduction to the theory of point processes.* New York: Springer-Verlag.

Dall'Aglio, G. (1972). Fréchet classes and compatibility of distribution functions. In *Symposia mathematica, vol ix* (pp. 131–150). London: Academic Press.

Dalton, H. (1920). The measurement of inequality of incomes. *Econom. J.* **30**, 348–361.

Day, M. V. (1983). Comparison results for diffusions conditioned on positivity. *J. Appl. Probab.* **20**, 766–777.

Deng, Y. L. (1985). On the comparison of point processes. *J. Appl. Probab.* **22**, 300–313.

Denuit, M., Dhaene, J., and Ribas, C. (2001). Does positive dependence between individual risks increase stop-loss premiums? *Insurance Math.*

Econom. **28**, 305–308.

Denuit, M., Genest, C., and Marceau, É. (1999). Stochastic bounds on sums of dependent risks. *Insurance Math. Econom.* **25**, 85–104.

Denuit, M. and Lefèvre, C. (2001). Stochastic *s*-(increasing) convexity. In *Generalized convexity and generalized monotonicity (Karlovassi, 1999)* (pp. 167–182). Berlin: Springer.

Denuit, M., Lefevre, C., and Shaked, M. (1998). The *s*-convex orders among real random variables, with applications. *Math. Inequal. Appl.* **1**, 585–613.

Denuit, M., Lefevre, C., and Shaked, M. (2000). On *s*-convex approximations. *Adv. Appl. Probab.* **32**, 994-1010.

Denuit, M. and Müller, A. (2001). Smooth generators of integral stochastic orders. *Preprint.*

Denuit, M. and Vermandele, C. (1998). Optimal reinsurance and stop-loss order. *Insurance Math. Econom.* **22**, 229–233.

De Santis, E. (2001). Strict inequality for phase transition between ferromagnetic and frustrated systems. *Electron. J. Probab.* **6**, no. 6, 27 pp.

De Vylder, E. (1996). *Advanced risk theory.* Bruxelles: Editions de l'Université Libre Bruxelles.

De Vylder, F. and Goovaerts, M. (1982). Upper and lower bounds on stop-loss premiums in case of known expectation and variance of the risk variable. *Mitt. Verein. Schweiz. Versicherungsmath.* **1**, 149–164.

Dhaene, J. and Denuit, M. (1999). The safest dependence structure among risks. *Insurance Math. Econom.* **25**, 11–21.

Dhaene, J. and Goovaerts, M. J. (1997). On the dependency of risks in the individual life model. *Insurance Math. Econom.* **19**, 243–253.

Dimakos, X. (2001). A guide to exact simulation. *Internat. Statist. Review* **69**, 27–48.

Doksum, K. (1969). Starshaped transformations and the power of rank tests. *Ann. Math. Statist.* **40**, 1167–1176.

Dudley, R. M. (1989). *Real analysis and probability.* Pacific Grove, CA: Wadsworth & Brooks/Cole Advanced Books & Software.

Duflo, M. (1997). *Random iterative models.* Berlin: Springer-Verlag.

Dyckerhoff, R. and Mosler, K. (1997). Orthant orderings of discrete random vectors. *J. Statist. Plann. Inference* **62**, 193–205.

Ebrahimi, N., Maasoumi, E., and Soofi, E. S. (1999). Ordering univariate distributions by entropy and variance. *J. Econometrics* **90**, 317–336.

Edmundson, H. P. (1956). *Bounds on the expectation of a convex function of a random variable* (Technical Report Paper No. 982). RAND Corporation.

Efron, B. (1965). Increasing properties of Pólya frequency functions. *Ann. Math. Statist.* **36**, 272–279.

Elton, J. and Hill, T. P. (1992). Fusions of a probability distribution. *Ann. Probab.* **20**, 421–454.

Elton, J. and Hill, T. P. (1998). On the basic representation theorem for convex domination of measures. *J. Math. Anal. Appl.* **228**, 449–466.

Embrechts, P., Klüppelberg, C., and Mikosch, T. (1997). *Modelling extremal events.* Berlin: Springer-Verlag.

Esary, J. D. and Proschan, F. (1963). Coherent structures of non-identical components. *Technometrics* **5**, 191–209.

Esary, J. D., Proschan, F., and Walkup, D. W. (1967). Association of random variables, with applications. *Ann. Math. Statist.* **38**, 1466–1474.

Evans, J. W. (1993). Random and cooperative adsorption. *Rev. Modern Phys.* **65**, 1281–1329.

Feller, W. (1971). *An introduction to probability theory and its applications. Vol. II.* New York: John Wiley & Sons Inc.

Fill, J. A. (1998). An interruptible algorithm for perfect sampling via Markov chains. *Ann. Appl. Probab.* **8**, 131–162.

Fill, J. A. and Machida, M. (2001). Stochastic monotonicity and realizable monotonicity. *Ann. Probab.* **29**, 938–978.

Fishburn, P. C. (1970). *Utility theory for decision making.* New York: John Wiley & Sons Inc.

Fishburn, P. C. (1976). Continua of stochastic dominance relations for bounded probability distributions. *J. Math. Econom.* **3**, 295–311.

Fishburn, P. C. (1980). Continua of stochastic dominance relations for unbounded probability distributions. *J. Math. Econom.* **7**, 271–285.

Fishburn, P. C. and Lavalle, I. H. (1995). Stochastic dominance on unidimensional grids. *Math. Oper. Res.* **20**, 513–525.

Forbes, F. and Francois, O. (1997). Stochastic comparison for Markov processes on a product of partially ordered sets. *Statist. Probab. Lett.* **33**, 309–320.

Fortuin, C. M., Kasteleyn, P. W., and Ginibre, J. (1971). Correlation inequalities on some partially ordered sets. *Comm. Math. Phys.* **22**, 89–103.

Foss, S. G. (1980). Approximation of multichannel queueing systems. *Siberian Math. J.* **21**, 851–857.

Foss, S. G. and Chernova, N. I. (2001). On the optimality of the FCFS discipline in multiserver systems and queueing networks. *Siberian Math. J.* **42**, 372–385.

Franco, M., Ruiz, J. M., and Ruiz, M. C. (2001). On closure of the IFR(2) and NBU(2) classes. *J. Appl. Probab.* **38**, 235–241.

Franken, P. and Kirstein, B. M. (1977). Zur Vergleichbarkeit zufälliger Prozesse. *Math. Nachr.* **78**, 197–205.

Franken, P., König, D., Arndt, U., and Schmidt, V. (1982). *Queues and point processes.* Chichester: John Wiley & Sons, Ltd.

Franken, P. and Stoyan, D. (1975). Einige Bemerkungen über monotone und vergleichbare Markowsche Prozesse. *Math. Nachr.* **66**, 201–209.

Friedman, M. and Savage, L. (1948). The utility analysis of choices involving risk. *Journal of Political Economy* **56**, 279–304.

Frostig, E. (2001). A comparison between homogeneous and heterogeneous portfolios. *Insurance Math. Econom.* **29**, 59–71.

Gaede, K.-W. (1965). Konfidenzgrenzen bei Warteschlangen- und Lagerhaltungsproblemen. *Z. Angew. Math. Mech.* **45**, T91–T92.

Gaede, K.-W. (1973). Einige Abschätzungen in der Bedienungstheorie. In *Proc. Oper. Res. 2, DGOR Ann. Meet. Hamburg 1972.* (pp. 241–255). Physica-Verlag.

Geman, D. and Horowitz, J. (1973). Remarks on Palm measures. *Ann. Inst. H. Poincaré Sect. B (N.S.)* **9**, 215–232.

Georgii, H.-O. (2000). Phase transition and percolation in Gibbsian particle models. In K. Mecke and D. Stoyan (Eds.), *Statistical physics and spatial statistics.* Berlin: Springer Verlag, (pp. 267–294).

Georgii, H.-O., Häggström, O., and Maes, C. (2001). The random geometry of equilibrium phases. In C. Domb and J. Lebowitz (Eds.), *Phase transitions and critical phenomena.* London: Academic Press, (pp. 1–142).

Georgii, H.-O. and Küneth, T. (1997). Stochastic comparison of point random fields. *J. Appl. Probab.* **34**, 868–881.

Gilks, W. R., Richardson, S., and Spiegelhalter, D. J. (Eds.). (1996). *Markov chain Monte Carlo in practice.* London: Chapman & Hall.

Glasserman, P. and Yao, D. D. (1994). *Monotone structure in discrete-event systems.* New York: John Wiley & Sons Inc.

Gollier, C. and Kimball, M. S. (1997). *New methods in the classical economics of uncertainty: comparing risks.* (Discussion paper, University of Michigan)

Grandell, J. (1976). *Doubly stochastic Poisson processes.* Berlin: Springer-Verlag. (Lecture Notes in Mathematics, Vol. 529)

Gupta, R. D. and Richards, D. S. P. (1992). Multivariate Liouville distributions. III. *J. Multivariate Anal.* **43**, 29–57.

Hanisch, K.-H. and Stoyan, D. (1978). Abschätzungen für die mittlere Gesamtvorgangsdauer von Pert-Netzwerken. *Wiss. Z. Hochschule Architektur Bauwesen Weimar* **25**, 32-33.

Hansel, G. and Troallic, J.-P. (1978). Mesures marginales et théorème de Ford-Fulkerson. *Z. Wahrsch. Verw. Gebiete* **43**, 245–251.

Hardy, G. H., Littlewood, J. E., and Pólya, G. (1934). *Inequalities.* Cambridge: Cambridge University Press.

Harris, B. (1962). Determining bounds on expected values of certain functions. *Ann. Math. Statist.* **33**, 1454–1457.

Hartley, H. O. and David, H. A. (1954). Universal bounds for mean range and extreme observation. *Ann. Math. Statist.* **25**, 85–99.

Heyman, D. P. and Sobel, M. J. (1984). *Stochastic models in operations research. Vol. II.* New York: McGraw-Hill.

Hickey, R. J. (1982). A note on the measurement of randomness. *J. Appl. Probab.* **19**, 229–232.

Hickey, R. J. (1983). Majorisation, randomness and some discrete distributions. *J. Appl. Probab.* **20**, 897–902.

Hickey, R. J. (1984). Continuous majorisation and randomness. *J. Appl. Probab.* **21**, 924–929.

Hickey, R. J. (1986). Concepts of dispersion in distributions: a comparative note. *J. Appl. Probab.* **23**, 914–921.

Hinderer, K. (1970). *Foundations of non-stationary dynamic programming with discrete time parameter.* Berlin: Springer-Verlag.

Hinderer, K. (1984). On the structure of solutions of stochastic dynamic programs. In M. Iosifescu (Ed.), *Proceedings of the seventh conference on probability theory (Braşov, 1982).* Utrecht: VNU Sci. Press, (pp. 173–182).

Hoeffding, W. (1955). The extrema of the expected value of a function of independent random variables. *Ann. Math. Statist.* **26**, 268–275.

Hu, T. (1996). Stochastic comparisons of order statistics under multivariate imperfect repair. *J. Appl. Probab.* **33**, 156–163.

Hu, T. (2000). Negatively superadditive dependence of random variables with applications. *China J. Appl. Prob. Stat.* **16**, 133–144.

Hu, T. and Joe, H. (1995). Monotonicity of positive dependence with time for stationary reversible Markov chains. *Probab. Engrg. Inform. Sci.* **9**, 227–237.

Hu, T. and Pan, X. (2000). Comparisons of dependence for stationary Markov processes. *Probab. Engrg. Inform. Sci.* **14**, 299–315.

Huang, Z. Y. (1984). A comparison theorem for solutions of stochastic differential equations and its applications. *Proc. Amer. Math. Soc.* **91**, 611–617.

Hürlimann, W. (1997a). Best bounds for expected financial payoffs. I. Algorithmic evaluation. *J. Comput. Appl. Math.* **82**, 199–212.

Hürlimann, W. (1997b). Best bounds for expected financial payoffs. II. Applications. *J. Comput. Appl. Math.* **82**, 213–227.

Ikeda, N. and Watanabe, S. (1981). *Stochastic differential equations and diffusion processes.* Amsterdam: North-Holland Publishing Co.

Iosifescu, M. and Theodorescu, R. (1969). *Random processes and learning.* New York: Springer-Verlag.

Jacobs, D. R., Jr. and Schach, S. (1972). Stochastic order relationships between $GI/G/k$ systems. *Ann. Math. Statist.* **43**, 1623–1633.

Jansen, K., Haezendonck, J., and Goovaerts, M. J. (1986). Upper bounds on stop-loss premiums in case of known moments up to the fourth order. *Insurance Math. Econom.* **5**, 315–334.

Jeulin, D. (Ed.). (1997). *Proceedings of the International Symposium on Advances in Theory and Applications of Random Sets.* River Edge, NJ: World Scientific Publishing.

Joag-Dev, K. and Proschan, F. (1983). Negative association of random variables, with applications. *Ann. Statist.* **11**, 286–295.

Joe, H. (1990). Multivariate concordance. *J. Multivariate Anal.* **35**, 12–30.

Joe, H. (1997). *Multivariate models and dependence concepts.* London:

Chapman & Hall.

Johansen, S. (1972). A representation theorem for a convex cone of quasi convex functions. *Math. Scand.* **30**, 297–312.

Johansen, S. (1974). The extremal convex functions. *Math. Scand.* **34**, 61–68.

Johnson, N. L., Kotz, S., and Balakrishnan, N. (1994). *Continuous univariate distributions. Vol. 1* (Second ed.). New York: John Wiley & Sons Inc.

Jonathan, D. and Plenio, M. B. (1999). Entanglement-assisted local manipulation of pure quantum states. *Phys. Rev. Lett.* **83**, 3566–3569.

Kaas, R. and Heerwaarden, A. E. van. (1992). Stop-loss order, unequal means, and more dangerous distributions. *Insurance Math. Econom.* **11**, 71–77.

Kall, P. and Wallace, S. W. (1994). *Stochastic programming.* Chichester: John Wiley & Sons, Ltd.

Kalmykov, G. I. (1962). On the partial ordering of Markov processes on the real line. *Teor. Verojatnost. i Primenen* **7**, 466–469.

Kamae, T. and Krengel, U. (1978). Stochastic partial ordering. *Ann. Probab.* **6**, 1044–1049.

Kamae, T., Krengel, U., and O'Brien, G. L. (1977). Stochastic inequalities on partially ordered spaces. *Ann. Probab.* **5**, 899–912.

Kamburowski, J. (1989). PERT networks under incomplete probabilistic information. In R. Słowiński (Ed.), *Advances in project scheduling.* Amsterdam: Elsevier, (pp. 433–466).

Kamburowski, J. (1992). Bounding the distribution of project duration in PERT networks. *Oper. Res. Lett.* **12**, 17–22.

Karlin, S. (1960). Dynamic inventory policy with varying stochastic demands. *Management Sci.* **6**, 231–258.

Karlin, S. and Novikoff, A. (1963). Generalized convex inequalities. *Pacific J. Math.* **13**, 1251–1279.

Karlin, S. and Rinott, Y. (1980a). Classes of orderings of measures and related correlation inequalities. I. Multivariate totally positive distributions. *J. Multivariate Anal.* **10**, 467–498.

Karlin, S. and Rinott, Y. (1980b). Classes of orderings of measures and related correlation inequalities. II. Multivariate reverse rule distributions. *J. Multivariate Anal.* **10**, 499–516.

Karpelevich, F. I. and Rybko, A. N. (2000). Thermodynamic limit for the mean field model of simple symmetrical closed queueing network. *Markov Process. Related Fields* **6**, 89–105.

Keilson, J. and Kester, A. (1977). Monotone matrices and monotone Markov processes. *Stochastic Processes Appl.* **5**, 231–241.

Kella, O. and Sverchkov, M. (1994). On concavity of the mean function and stochastic ordering for reflected processes with stationary increments. *J. Appl. Probab.* **31**, 1140–1142.

Kellerer, H. G. (1972). Markov-Komposition und eine Anwendung auf Martingale. *Math. Ann.* **198**, 99–122.

Kellerer, H. G. (1984). Duality theorems for marginal problems. *Z. Wahrsch.*

Verw. Gebiete **67**, 399–432.

Kellerer, H. G. (1997). Representation of Markov kernels by random mappings under order conditions. In V. Beneš and J. Štěpán (Eds.), *Distributions with given marginals and moment problems (Prague, 1996)*. Dordrecht: Kluwer, (pp. 143–160).

Kemperman, J. H. B. (1977). On the FKG-inequality for measures on a partially ordered space. *Indag. Math.* **39**, 313–331.

Kendall, W. S. and Møller, J. (2000). Perfect simulation using dominating processes on ordered spaces, with application to locally stable point processes. *Adv. in Appl. Probab.* **32**, 844–865.

Kertz, R. P. and Rösler, U. (1992). Stochastic and convex orders and lattices of probability measures, with a martingale interpretation. *Israel J. Math.* **77**, 129–164.

Kiefer, J. and Wolfowitz, J. (1955). On the theory of queues with many servers. *Trans. Amer. Math. Soc.* **78**, 1–18.

Kiefer, J. and Wolfowitz, J. (1956). On the characteristics of the general queueing process, with applications to random walk. *Ann. Math. Statist.* **27**, 147–161.

Kifer, Y. (1986). *Ergodic theory of random transformations*. Boston: Birkhäuser.

Kijima, M. (1997). *Markov processes for stochastic modeling*. London: Chapman & Hall.

Kijima, M. (1998). Hazard rate and reversed hazard rate monotonicities in continuous-time Markov chains. *J. Appl. Probab.* **35**, 545–556.

Kijima, M. and Ohnishi, M. (1996). Portfolio selection problems via the bivariate characterization of stochastic dominance relations. *Math. Finance* **6**, 237–277.

Kijima, M. and Ohnishi, M. (1999). Stochastic orders and their applications in financial optimization. *Math. Methods Oper. Res.* **50**, 351–372.

Kimeldorf, G. and Sampson, A. R. (1987). Positive dependence orderings. *Ann. Inst. Statist. Math.* **39**, 113–128.

Kimeldorf, G. and Sampson, A. R. (1989). A framework for positive dependence. *Ann. Inst. Statist. Math.* **41**, 31–45.

Kirstein, B. M. (1976). Monotonicity and comparability of time-homogeneous Markov processes with discrete state space. *Math. Operationsforsch. Statist.* **7**, 151–168.

Klar, B. (2000). A class of tests for exponentiality against HNBUE alternatives. *Statist. Probab. Lett.* **47**, 199–207.

Klefsjö, B. (1982). The HNBUE and HNWUE classes of life distributions. *Naval Res. Logist. Quart.* **29**, 331–344.

Klefsjö, B. (1983a). Testing exponentiality against HNBUE. *Scand. J. Statist.* **10**, 65–75.

Klefsjö, B. (1983b). A useful ageing property based on the Laplace transform. *J. Appl. Probab.* **20**, 615–626.

Koshevoy, G. and Mosler, K. (1998). Lift zonoids, random convex hulls and the variability of random vectors. *Bernoulli* **4**, 377–399.

Kotz, S., Balakrishnan, N., and Johnson, N. L. (2000). *Continuous multivariate distributions*. *Vol. 1* (Second ed.). Wiley-Interscience, New York.

Kulik, R. and Szekli, R. (2001). Sufficient conditions for long-range count dependence of stationary point processes on the real line. *J. Appl. Probab.* **38**, 570–581.

Kwieciński, A. and Szekli, R. (1991). Compensator conditions for stochastic ordering of point processes. *J. Appl. Probab.* **28**, 751–761.

Kwieciński, A. and Szekli, R. (1996). Some monotonicity and dependence properties of self-exciting point processes. *Ann. Appl. Probab.* **6**, 1211–1231.

Landsberger, M. and Meilijson, I. (1990a). Demand for risky financial assets: a portfolio analysis. *J. Econom. Theory* **50**, 204–213.

Landsberger, M. and Meilijson, I. (1990b). Lotteries, insurance, and star-shaped utility functions. *J. Econom. Theory* **52**, 1–17.

Last, G. (1993). On dependent marking and thinning of point processes. *Stochastic Process. Appl.* **45**, 73–94.

Last, G. and Brandt, A. (1995). *Marked point processes on the real line.* New York: Springer-Verlag.

Last, G. and Szekli, R. (1998). Asymptotic and monotonicity properties of some repairable systems. *Adv. in Appl. Probab.* **30**, 1089–1110.

Lefèvre, C. and Utev, S. (1996). Comparing sums of exchangeable Bernoulli random variables. *J. Appl. Probab.* **33**, 285–310.

Lehmann, E. L. (1955). Ordered families of distributions. *Ann. Math. Statist.* **26**, 399–419.

Lehmann, E. L. (1966). Some concepts of dependence. *Ann. Math. Statist.* **37**, 1137–1153.

Li, H. and Shaked, M. (1997). Ageing first-passage times of Markov processes: a matrix approach. *J. Appl. Probab.* **34**, 1–13.

Li, H. and Xu, S. (2000). On the dependence structure and bounds of correlated parallel queues and their applications to synchronized stochastic systems. *J. Appl. Probab.* **37**, 1020–1043.

Li, H. J. and Zhu, H. (1994). Stochastic equivalence of ordered random variables with applications in reliability theory. *Statist. Probab. Lett.* **20**, 383–393.

Liggett, T. M. (1985). *Interacting particle systems.* New York: Springer-Verlag.

Liggett, T. M. (1999). *Stochastic interacting systems: contact, voter and exclusion processes.* Berlin: Springer-Verlag.

Liggett, T. M. (2000). Monotonicity of conditional distributions and growth models on trees. *Ann. Probab.* **28**, 1645–1665.

Lillo, R. E., Nanda, A. K., and Shaked, M. (2001). Preservation of some likelihood ratio stochastic orders by order statistics. *Statist. Probab. Lett.* **51**, 111–119.

Lindley, D. V. (1952). The theory of queues with a single server. *Proc. Cambridge Philos Soc.* **48**, 277–289.

Lindqvist, B. H. (1988). Association of probability measures on partially ordered spaces. *J. Multivariate Anal.* **26**, 111–132.

Lindvall, T. (1992). *Lectures on the coupling method.* New York: John Wiley & Sons Inc.

Lippman, S. A. and McCall, J. J. (1976). Job search in a dynamic economy. *J. Econom. Theory* **12**, 365–390.

Lisek, B. (1978). Comparability of special distributions. *Math. Operationsforsch. Statist. Ser. Statist.* **9**, 587–598.

López, F. J., Martinez, S., and Sanz, G. (2000). Stochastic domination and Markovian couplings. *Adv. in Appl. Probab.* **32**, 1064–1076.

López, F. J. and Sanz, G. (1998). Stochastic comparisons and couplings for interacting particle systems. *Statist. Probab. Lett.* **40**, 93–102.

López, F. J. and Sanz, G. (2000). Stochastic comparisons for general probabilistic cellular automata. *Statist. Probab. Lett.* **46**, 401–410.

Lorden, G. (1970). On excess over the boundary. *Ann. Math. Statist.* **41**, 520–527.

Lorenz, M. O. (1905). Methods of measuring concentration of wealth. *J. Amer. Statist. Assoc.* **9**, 209–219.

Loynes, R. M. (1962). The stability of a queue with non-independent interarrival and service times. *Proc. Cambridge Philos. Soc.* **58**, 497–520.

Lund, R. B., Meyn, S. P., and Tweedie, R. L. (1996). Computable exponential convergence rates for stochastically ordered Markov processes. *Ann. Appl. Probab.* **6**, 218–237.

Lund, R. B. and Tweedie, R. L. (1996). Geometric convergence rates for stochastically ordered Markov chains. *Math. Oper. Res.* **21**, 182–194.

MacDonald, I. L. and Zucchini, W. (1997). *Hidden Markov and other models for discrete-valued time series.* London: Chapman & Hall.

Machina, M. J. and Pratt, J. W. (1997). Increasing risk: Some direct constructions. *J. Risk Uncertain.* **14**, 103-127.

Madansky, A. (1959). Bounds on the expectation of a convex function of a multivariate random variable. *Ann. Math. Statist.* **30**, 743–746.

Madsen, R. W. (1993). Generalized binomial distributions. *Comm. Statist. Theory Methods* **22**, 3065–3086.

Maes, C. (1993). Coupling interacting particle systems. *Rev. Math. Phys.* **5**, 457–475.

Makowski, A. M. (1994). On an elementary characterization of the increasing convex ordering, with an application. *J. Appl. Probab.* **31**, 834–840.

Mann, H. B. and Whitney, D. R. (1947). On a test of whether one of two random variables is stochastically larger than the other. *Ann. Math. Statistics* **18**, 50–60.

Markowitz, H. M. (1959). *Portfolio selection: Efficient diversification of investments*. New York: John Wiley & Sons Inc.

Marshall, A. W. (1991). Multivariate stochastic orderings and generating cones of functions. In K. Mosler and M. Scarsini (Eds.), *Stochastic orders and decision under risk (Hamburg, 1989)*. Hayward, CA: IMS Lecture Notes Vol. 19, (pp. 231–247).

Marshall, A. W. and Olkin, I. (1979). *Inequalities: theory of majorization and its applications*. New York: Academic Press.

Marshall, K. T. (1973). Linear bounds on the renewal function. *SIAM J. Appl. Math.* **24**, 245–250.

Massey, W. A. (1985). Asymptotic analysis of the time dependent $M/M/1$ queue. *Math. Oper. Res.* **10**, 305–327.

Massey, W. A. (1987). Stochastic orderings for Markov processes on partially ordered spaces. *Math. Oper. Res.* **12**, 350–367.

Matheron, G. (1975). *Random sets and integral geometry*. New York: John Wiley & Sons, Inc.

Mead, C. A. (1977). Mixing characters and its applications to irreversible processes in macroscopic systems. *J. Chem. Phys.* **66**, 459.

Meester, L. E. and Shanthikumar, J. G. (1993). Regularity of stochastic processes: A theory of directional convexity. *Prob. Eng. Inform. Sci.* **7**, 343–360.

Meester, L. E. and Shanthikumar, J. G. (1999). Stochastic convexity on general space. *Math. Oper. Res.* **24**, 472–494.

Menich, R. and Serfozo, R. F. (1991). Optimality of routing and servicing in dependent parallel processing systems. *Queueing Systems Theory Appl.* **9**, 403–418.

Meyer, J. (1977). Choice among distributions. *J. Econom. Theory* **14**, 326–336.

Meyer, P.-A. (1966). *Probability and potentials*. Waltham: Blaisdell Publishing Co.

Milgrom, P. and Weber, R. J. (1982). A theory of auctions and competitive bidding. *Econometrica* **50**, 1089-1122.

Mira, A. and Geyer, C. J. (2000). On non-reversible Markov chains. In N. Madras (Ed.), *Monte Carlo methods (Toronto, 1998)*. Providence, RI: American Mathematical Society, (pp. 95–110).

Mira, A., Møller, J., and Roberts, G. O. (2001). Perfect slice samplers. *J. R. Stat. Soc. Ser. B* **63**, 593–606.

Møller, J. (1999). Markov chain Monte Carlo and spatial point processes. In O. Barndorff-Nielsen, W. Kendall, and M. N. M. Lieshout (Eds.), *Stochastic geometry (Toulouse, 1996)* (pp. 141–172). Boca Raton: Chapman & Hall.

Møller, J. and Schladitz, K. (1999). Extensions of Fill's algorithm for perfect simulation. *J. R. Stat. Soc. Ser. B Stat. Methodol.* **61**, 955–969.

Mosler, K. (1982). *Entscheidungsregeln bei Risiko: Multivariate stochastische*

Dominanz. Berlin: Springer-Verlag.

Mosler, K. and Scarsini, M. (1993). *Stochastic orders and applications: A classified bibliography.* Berlin: Springer-Verlag.

Müller, A. (1996). Orderings of risks: a comparative study via stop-loss transforms. *Insurance Math. Econom.* **17**, 215–222.

Müller, A. (1997a). How does the value function of a Markov decision process depend on the transition probabilities? *Math. Oper. Res.* **22**, 872–885.

Müller, A. (1997b). Stochastic orders generated by integrals: a unified study. *Adv. in Appl. Probab.* **29**, 414–428.

Müller, A. (1997c). Stop-loss order for portfolios of dependent risks. *Insurance Math. Econom.* **21**, 219–223.

Müller, A. (1998a). Comparing risks with unbounded distributions. *J. Math. Econom.* **30**, 229–239.

Müller, A. (1998b). Another tale of two tails: On characterizations of comparative risk. *J. Risk Uncertain.* **16**, 187-197.

Müller, A. (2000). On the waiting times in queues with dependency between interarrival and service times. *Oper. Res. Lett.* **26**, 43–47.

Müller, A. (2001a). Bounds for optimal stopping values of dependent random variables with given marginals. *Stat. Probab. Lett.* **52**, 73 – 78.

Müller, A. (2001b). Stochastic ordering of multivariate normal distributions. *Ann. Inst. Statist. Math.* **53**, 567–575.

Müller, A. and Pflug, G. (2001). Asymptotic ruin probabilities for risk processes with dependent increments. *Insurance Math. Econom.* **28**, 381–392.

Müller, A. and Rüschendorf, L. (2001). On the optimal stopping values induced by general dependence structures. *J. Appl. Probab.* **38**, 672–684.

Müller, A. and Scarsini, M. (2000). Some remarks on the supermodular order. *J. Multivariate Anal.* **73**, 107–119.

Müller, A. and Scarsini, M. (2001). Stochastic comparison of random vectors with a common copula. *Math. Oper. Res.* **26**, 723–740.

Nachbin, L. (1965). *Topology and order.* London: Van Nostrand.

Nanda, A. K. and Shaked, M. (2002). The hazard rate and the reversed hazard rate orders, with applications to order statistics. *Ann. Inst. Statist. Math.*, to appear.

Nawrotzki, K. (1962). Eine Monotonieeigenschaft zufälliger Punktfolgen. *Math. Nachr.* **24**, 193–200.

Nelsen, R. B. (1999). *An introduction to copulas.* New York: Springer-Verlag.

Newman, C. M. (1997). *Topics in disordered systems.* Basel: Birkhäuser Verlag.

Norberg, T. (1992). On the existence of ordered couplings of random sets – with applications. *Israel J. Math.* **77**, 241–264.

O'Brien, G. L. (1980). A new comparison theorem for solutions of stochastic differential equations. *Stochastics* **3**, 245–249.

Ogryczak, W. and Ruszczyński, A. (1999). From stochastic dominance to mean-risk models: Semideviations as risk measures. *European J. Oper. Res.* **116**, 33–50.

Ogryczak, W. and Ruszczyński, A. (2001). On consistency of stochastic dominance and mean-semideviation models. *Math. Program.* **89**, 217–232. (Mathematical programming and finance)

Ohlin, J. (1969). On a class of measures for dispersion with applications to optimal insurance. *ASTIN Bulletin* **5**, 249–266.

Oja, H. (1981). On location, scale, skewness and kurtosis of univariate distributions. *Scand. J. Statist.* **8**, 154–168.

Olkin, I. and Tong, Y. L. (1994). Positive dependence of a class of multivariate exponential distributions. *SIAM J. Control Optim.* **32**, 965–974.

Pečarić, J. E., Proschan, F., and Tong, Y. L. (1992). *Convex functions, partial orderings, and statistical applications.* Boston: Academic Press Inc.

Pellerey, F. and Shaked, M. (1997). Characterizations of the IFR and DFR aging notions by means of the dispersive order. *Statist. Probab. Lett.* **33**, 389–393.

Perlman, M. D. (1974). Jensen's inequality for a convex vector-valued function on an infinite-dimensional space. *J. Multivariate Anal.* **4**, 52–65.

Peskun, P. H. (1973). Optimum Monte-Carlo sampling using Markov chains. *Biometrika* **60**, 607–612.

Pigou, A. C. (1912). *Wealth and welfare.* New York: Macmillan.

Pinedo, M. (1995). *Scheduling: theory, algorithms, and systems.* Englewood Cliffs: Prentice Hall.

Pledger, G. and Proschan, F. (1973). Stochastic comparisons of random processes, with applications in reliability. *J. Appl. Probab.* **10**, 572–585.

Popoviciu, T. (1944). *Les fonctions convexes.* Paris: Hermann et Cie.

Pratt, J. W. (1964). Risk aversion in the small and in the large. *Econometrica* **32**, 122–136.

Pratt, J. W. and Zeckhauser, R. J. (1987). Proper risk aversion. *Econometrica* **55**, 143–154.

Propp, J. G. and Wilson, D. B. (1996). Exact sampling with coupled Markov chains and applications to statistical mechanics. *Random Structures Algorithms* **9**, 223–252.

Proschan, F. and Sethuraman, J. (1976). Stochastic comparisons of order statistics from heterogeneous populations, with applications in reliability. *J. Multivariate Anal.* **6**, 608–616.

Puterman, M. L. (1994). *Markov decision processes: discrete stochastic dynamic programming.* New York: John Wiley & Sons Inc.

Quirk, J. P. and Saposnik, R. (1962). Admissibility and measurable utility functions. *Rev. Econ. Stud.* **29**, 140–146.

Rachev, S. T. (1991). *Probability metrics and the stability of stochastic models.* Chichester: John Wiley & Sons, Ltd.

Ramachandran, D. and Rüschendorf, L. (1995). A general duality theorem for marginal problems. *Probab. Theory Related Fields* **101**, 311–319.

Raqab, M. Z. and Amin, W. A. (1996). Some ordering results on order statistics and record values. *IAPQR Trans.* **21**, 1–8.

Reuter, H. and Riedrich, T. (1981). On maximal sets of functions compatible with a partial ordering for distribution functions. *Math. Operationsforsch. Statist. Ser. Optim.* **12**, 597–605.

Rieder, U. and Zagst, R. (1994). Monotonicity and bounds for convex stochastic control models. *Z. Oper. Res.* **39**, 187–207.

Righter, R. and Shanthikumar, J. G. (1992). Extremal properties of the FIFO discipline in queueing networks. *J. Appl. Probab.* **29**, 967–978.

Roberts, A. W. and Varberg, D. E. (1973). *Convex functions.* Academic Press, New York-London.

Roberts, G. O. and Tweedie, R. L. (2000). Rates of convergence of stochastically monotone and continuous time Markov models. *J. Appl. Probab.* **37**, 359–373.

Robertson, T., Wright, F. T., and Dykstra, R. L. (1988). *Order restricted statistical inference.* Chichester: John Wiley & Sons, Ltd.

Rogers, L. C. G. and Williams, D. (1987). *Diffusions, Markov processes, and martingales. Vol. 2.* New York: John Wiley & Sons Inc.

Rogozin, B. A. (1966). Certain extremal problems of queueing theory. *Teor. Verojatnost. i Primenen* **11**, 161–169.

Rolski, T. (1975). Mean residual life. *Bull. Inst. Internat. Statist.* **46**, 266–270.

Rolski, T. (1976). Order relations in the set of probability distribution functions and their applications in queueing theory. *Dissertationes Math. (Rozprawy Mat.)* **132**, 52.

Rolski, T. (1977). On some classes of distribution functions determined by an order relation. In R. Bartoszyński, E. Fidelis, and W. Klonecki (Eds.), *Proceedings of the symposium to honour Jerzy Neyman (Warsaw, 1974).* Warsaw: Państw. Wydawn. Nauk., (pp. 293–302).

Rolski, T. (1981). Queues with nonstationary input stream: Ross's conjecture. *Adv. in Appl. Probab.* **13**, 603–618.

Rolski, T. (1984). Comparison theorems for queues with dependent inter-arrival times. In F. Baccelli and G. Fayolle (Eds.), *Modelling and performance evaluation methodology (Paris, 1983).* Berlin: Springer, (pp. 42–67).

Rolski, T. (1986). Upper bounds for single server queues with doubly stochastic Poisson arrivals. *Math. Oper. Res.* **11**, 442–450.

Rolski, T., Schmidli, H., Schmidt, V., and Teugels, J. (1999). *Stochastic processes for insurance and finance.* Chichester: John Wiley & Sons, Ltd.

Rolski, T. and Stoyan, D. (1974). Two classes of semi-orderings and their application in the queuing theory. *Z. Angew. Math. Mech.* **54**, 127–128.

Rolski, T. and Szekli, R. (1991). Stochastic ordering and thinning of point processes. *Stochastic Process. Appl.* **37**, 299–312.

Rosenblatt, M. (1952). Remarks on a multivariate transformation. *Ann. Math. Statistics* **23**, 470–472.

Ross, S. M. (1978). Average delay in queues with non-stationary Poisson arrivals. *J. Appl. Probab.* **15**, 602–609.

Ross, S. M. (1983). *Stochastic processes.* New York: John Wiley & Sons Inc.

Rossberg, H.-J. (1968). Optimale Eigenschaften einiger Wartesysteme bei regelmäßigem Eingang bzw. konstanten Bedienungszeiten. *Z. Angew. Math. Mech.* **48**, 395–403.

Rothschild, M. and Stiglitz, J. E. (1970). Increasing risk. I. A definition. *J. Econom. Theory* **2**, 225–243.

Rubinstein, R. Y. and Melamed, B. (1998). *Modern simulation and modeling.* New York: John Wiley & Sons Inc.

Rubinstein, R. Y., Samorodnitsky, G., and Shaked, M. (1985). Antithetic variates, multivariate dependence and simulation of stochastic systems. *Management Sci.* **31**, 66–77.

Ruch, E. (1975). The diagram lattice as structural principle. *Theoret. Chim. Acta* **38**, 167–183.

Rudin, W. (1987). *Real and complex analysis* (Third ed.). New York: McGraw-Hill.

Rüschendorf, L. (1980). Inequalities for the expectation of Δ-monotone functions. *Z. Wahrsch. Verw. Gebiete* **54**, 341–349.

Rüschendorf, L. (1981a). Characterization of dependence concepts in normal distributions. *Ann. Inst. Statist. Math.* **33**, 347–359.

Rüschendorf, L. (1981b). Stochastically ordered distributions and monotonicity of the OC-function of sequential probability ratio tests. *Math. Operationsforsch. Statist. Ser. Statist.* **12**, 327–338.

Rüschendorf, L. (1991). On conditional stochastic ordering of distributions. *Adv. Appl. Probab.* **23**, 46-63.

Rüschendorf, L. and Uckelmann, L. (2002). Variance minimization and random variables with constant sum. In *Proceedings of the fourth conference on distributions with given marginals (Barcelona 2000).* Dordrecht: Kluwer, to appear.

Rüschendorf, L. and Valk, V. de. (1993). On regression representations of stochastic processes. *Stochastic Process. Appl.* **46**, 183–198.

Ryff, J. V. (1963). On the representation of doubly stochastic operators. *Pacific J. Math.* **13**, 1379–1386.

Scarsini, M. (1988). Multivariate stochastic dominance with fixed dependence structure. *Oper. Res. Lett.* **7**, 237–240.

Scarsini, M. (1998). Multivariate convex orderings, dependence, and stochastic equality. *J. Appl. Probab.* **35**, 93–103.

Scarsini, M. and Shaked, M. (1990). Some conditions for stochastic equality. *Naval Res. Logist.* **37**, 617–625.

Scarsini, M. and Shaked, M. (1997). Fréchet classes and nonmonotone dependence. In V. Beneš and J. Štěpán (Eds.), *Distributions with given marginals and moment problems (Prague, 1996).* Dordrecht: Kluwer, (pp. 59–71).

Schassberger, R. (1977). Insensitivity of steady-state distributions of generalized semi-Markov processes. I. *Ann. Probab.* **5**, 87–99.

Schassberger, R. (1978a). Insensitivity of steady-state distributions of generalized semi-Markov processes with speeds. *Adv. in Appl. Probab.* **10**, 836–851.

Schassberger, R. (1978b). Insensitivity of steady-state distributions of generalized semi-Markov processes. II. *Ann. Probab.* **6**, 85–93.

Schur, I. (1923). Über eine Klasse von Mittelbildungen mit Anwendungen in der Determinantentheorie. *Sitzungsber. Berlin. Math. Gesellschaft* **22**, 9–20.

Sempi, C. (1982). On the space of distribution functions. *Riv. Mat. Univ. Parma* **8**, 243–250.

Shaked, M. (1982). Dispersive ordering of distributions. *J. Appl. Probab.* **19**, 310–320.

Shaked, M. and Shanthikumar, J. G. (1987a). Characterization of some first passage times using log-concavity and log-convexity as aging notions. *Prob. Eng. Inform. Sci.* **1**, 279–291.

Shaked, M. and Shanthikumar, J. G. (1987b). Multivariate hazard rates and stochastic ordering. *Adv. in Appl. Probab.* **19**, 123–137.

Shaked, M. and Shanthikumar, J. G. (1990). Parametric stochastic convexity and concavity of stochastic processes. *Ann. Inst. Statist. Math.* **42**, 509–531.

Shaked, M. and Shanthikumar, J. G. (1994). *Stochastic orders and their applications.* Boston: Academic Press Inc.

Shaked, M. and Shanthikumar, J. G. (1997). Supermodular stochastic orders and positive dependence of random vectors. *J. Multivariate Anal.* **61**, 86–101.

Shaked, M. and Szekli, R. (1995). Comparison of replacement policies via point processes. *Adv. in Appl. Probab.* **27**, 1079–1103.

Shaked, M. and Tong, Y. L. (1985). Some partial orderings of exchangeable random variables by positive dependence. *J. Multivariate Anal.* **17**, 333–349.

Shanthikumar, J. G. (1987). On stochastic comparison of random vectors. *J. Appl. Probab.* **24**, 123–136.

Shanthikumar, J. G. (1988). DFR property of first-passage times and its preservation under geometric compounding. *Ann. Probab.* **16**, 397–406.

Shanthikumar, J. G. and Yao, D. D. (1991). Bivariate characterization of some stochastic order relations. *Adv. in Appl. Probab.* **23**, 642–659.

Simon, H. A. (1957). *Models of man, social and rational. Mathematical essays on rational human behavior in a social setting.* New York: John Wiley & Sons Inc.

Skala, H. J. (1993). The existence of probability measures with given marginals. *Ann. Probab.* **21**, 136–142.

Sklar, A. (1959). Fonctions de répartition à n dimensions et leurs marges.

Publications de l'Institut de Statistique de l'Université de Paris **8**, 229–231.

Skorokhod, A. V. (1965). *Studies in the theory of random processes.* Reading: Addison-Wesley.

Sonderman, D. (1980). Comparing semi-Markov processes. *Math. Oper. Res.* **5**, 110–119.

Stidham, S., Jr. and Weber, R. (1993). A survey of Markov decision models for control of networks of queues. *Queueing Systems Theory Appl.* **13**, 291–314.

Stoyan, D. (1972a). Halbordnungsrelationen für Verteilungsgesetze. *Math. Nachr.* **52**, 315–331.

Stoyan, D. (1972b). Über einige Eigenschaften monotoner stochastischer Prozesse. *Math. Nachr.* **52**, 21–34.

Stoyan, D. (1973). Bounds for the extrema of the expected value of a convex function of independent random variables. *Studia Sci. Math. Hungar.* **8**, 153–159.

Stoyan, D. (1983). *Comparison methods for queues and other stochastic models.* Chichester: John Wiley & Sons, Ltd.

Stoyan, D. (1998). Random sets: Models and statistics. *Int. Stat. Rev.* **66**, 1–27.

Stoyan, D., Kendall, W. S., and Mecke, J. (1995). *Stochastic geometry and its applications.* Chichester: John Wiley & Sons, Ltd.

Stoyan, D. and Schlather, M. (2000). Random sequential adsorption: Relatioship to dead leaves and characterization of variability. *J. Stat. Phys.* **100**, 969–979.

Stoyan, H. (1973). Monotonie- und Stetigkeitseigenschaften mehrliniger Wartesysteme der Bedienungstheorie. *Math. Operationsforsch. Statist.* **4**, 155–163.

Stoyan, H. and Stoyan, D. (1969). Monotonieeigenschaften der Kundenwartezeiten im Modell $GI/G/1$. *Z. Angew. Math. Mech.* **49**, 729–734.

Stoyan, H. and Stoyan, D. (1980). On some partial orderings of random closed sets. *Math. Operationsforsch. Statist. Ser. Optim.* **11**, 145–154.

Strassen, V. (1965). The existence of probability measures with given marginals. *Ann. Math. Statist.* **36**, 423–439.

Sundt, B. (1991). *An introduction to non-life insurance mathematics* (Second ed.). Karlsruhe: Verlag Versicherungswirtschaft e.V.

Szekli, R. (1995). *Stochastic ordering and dependence in applied probability.* New York: Springer-Verlag.

Szekli, R., Disney, R. L., and Hur, S. (1994). $MR/GI/1$ queues with positively correlated arrival stream. *J. Appl. Probab.* **31**, 497–514.

Taylor, J. M. (1983). Comparisons of certain distribution functions. *Math. Operationsforsch. Statist. Ser. Statist.* **14**, 397–408.

Tchen, A. H. (1980). Inequalities for distributions with given marginals. *Ann.*

Probab. **8**, 814–827.

Thirring, W. (1983). *A course in mathematical physics. Vol. 4.* New York: Springer-Verlag.

Thönnes, E. (1999). Perfect simulation of some point processes for the impatient user. *Adv. in Appl. Probab.* **31**, 69–87.

Thönnes, E. (2000). A primer in perfect simulation. In K. Mecke and D. Stoyan (Eds.), *Statistical physics and spatial statistics.* Berlin: Springer Verlag, (pp. 349–378).

Thorisson, H. (2000). *Coupling, stationarity, and regeneration.* New York: Springer-Verlag.

Tong, Y. L. (1980). *Probability inequalities in multivariate distributions.* New York: Academic Press.

Tong, Y. L. (1989). Inequalities for a class of positively dependent random variables with a common marginal. *Ann. Statist.* **17**, 429–435.

Tong, Y. L. (1990). *The multivariate normal distribution.* New York: Springer-Verlag.

Toom, A. L., Vasilyev, N. B., Stavskaya, O. N., Mityushin, L. G., Kurtyumov, G. L., and Pirogov, S. A. (1990). Discrete lokal Markov systems. In R. L. Dobrushin, V. I. Kryukov, and A. L. Toom (Eds.), *Stochastic cellular systems: Ergodicity, memory, morphogenesis.* Manchester: Manchester University Press, (pp. 1–182).

Uhlmann, A. (1972). Endlich-dimensionale Dichtematrizen. I. *Wiss. Z. Karl-Marx-Univ. Leipzig Math.-Natur. Reihe* **21**, 421–452.

van Doorn, E. A. (1981). *Stochastic monotonicity and queueing applications of birth-death processes.* New York: Springer-Verlag.

van Heerwaarden, A. E. (1991). *Ordering of risks.* Amsterdam: Tinbergen Institute research series.

van Zwet, W. R. (1964). *Convex transformations of random variables.* Amsterdam: Mathematisch Centrum.

Veinott, A. F., Jr. (1965). Optimal policy in a dynamic, single product, nonstationary inventory model with several demand classes. *Operations Res.* **13**, 761–778.

Vervaat, W. (1979). On a stochastic difference equation and a representation of nonnegative infinitely divisible random variables. *Adv. in Appl. Probab.* **11**, 750–783.

Vickson, R. G. (1974). Stochastic dominance tests for decreasing absolute risk aversion. I. Discrete random variables. *Management Sci.* **21**, 1438–1446.

Vickson, R. G. (1975). Stochastic dominance for decreasing absolute risk aversion. *J. Finan. Quant. Anal.* **10**, 799–811.

von Neumann, J. and Morgenstern, O. (1947). *Theory of Games and Economic Behavior (Second ed.).* Princeton: Princeton University Press.

Wang, S. (1998). Aggregation of correlated risk portfolios: Models and algorithms. *Proceedings of the Casualty Actuarial Society* **LXXXV**, 848–939.

Wheeden, R. L. and Zygmund, A. (1977). *Measure and integral.* New York: Marcel Dekker Inc.

Whitmore, G. A. (1970). Third degree stochastic dominance. *Amer. Econ. Rev.* **60**, 457 - 459.

Whitt, W. (1974). The continuity of queues. *Adv. Appl. Probab.* **6**, 175–183.

Whitt, W. (1980a). Uniform conditional stochastic order. *J. Appl. Probab.* **17**, 112–123.

Whitt, W. (1980b). The effect of variability in the $GI/G/s$ queue. *J. Appl. Probab.* **17**, 1062–1071.

Whitt, W. (1982). Multivariate monotone likelihood ratio and uniform conditional stochastic order. *J. Appl. Probab.* **19**, 695–701.

Whitt, W. (1986). Stochastic comparisons for non-Markov processes. *Math. Oper. Res.* **11**, 609–618.

Willmot, G. E. and Lin, X. S. (2001). *Lundberg approximations for compound distributions with insurance applications.* New York: Springer-Verlag.

Wilson, D. B. (2000). How to couple from the past using a read-once source of randomness. *Random Structures Algorithms* **16**, 85–113.

Wolff, R. W. (1977a). The effect of service time regularity on system performance. In K. M. Chandy and M. Reiser (Eds.), *Computer performance.* Amsterdam: North Holland, (pp. 297–304).

Wolff, R. W. (1977b). An upper bound for multi-channel queues. *J. Appl. Probab.* **14**, 884–888.

Wolff, R. W. (1987). Upper bounds on work in system for multichannel queues. *J. Appl. Probab.* **24**, 547–551.

Wolff, R. W. (1989). *Stochastic modeling and the theory of queues.* Englewood Cliffs: Prentice-Hall Inc.

Yamada, T. (1973). On a comparison theorem for solutions of stochastic differential equations and its applications. *J. Math. Kyoto Univ.* **13**, 497–512.

Yin, G. G. and Zhang, Q. (1998). *Continuous-time Markov chains and applications.* New York: Springer-Verlag.

Index

WILEY SERIES IN PROBABILITY AND STATISTICS
ESTABLISHED BY WALTER A. SHEWHART AND SAMUEL S. WILKS

Editors
Peter Bloomfield, Noel A. C. Cressie, Nicholas I. Fisher, Iain M. Johnstone,
J. B. Kadane, Louise M. Ryan, David W. Scott, Adrian F. M. Smith,
Jozef L. Teugels
Editors Emeriti: *Vic Barnett, Ralph A. Bradley, J. Stuart Hunter,*
David G. Kendall

Wiley Series in Probability and Statistics is well established and authoritative. It covers many topics of current research interest in both pure and applied statistics and probability theory. Written by leading statisticians and institutions, the titles span both state-of-the-art developments in the field and classical methods.

Reflecting the wide range of current research in statistics, the series encompasses applied, methodological and theoretical statistics, ranging from applications and new techniques made possible by advances in computerized practice to rigorous treatment of theoretical approaches.

This series provides essential and invaluable reading for all statisticians, whether in academia, industry, government, or research.

ABRAHAM and LEDOLTER · Statistical Methods for Forecasting
AGRESTI · Analysis of Ordinal Categorical Data
AGRESTI · An Introduction to Categorical Data Analysis
AGRESTI · Categorical Data Analysis
ANDĚL · Mathematics of Chance
ANDERSON · An Introduction to Multivariate Statistical Analysis, *Second Edition*
*ANDERSON · The Statistical Analysis of Time Series
ANDERSON, AUQUIER, HAUCK, OAKES, VANDAELE, and WEISBERG ·
    Statistical Methods for Comparative Studies
ANDERSON and LOYNES · The Teaching of Practical Statistics
ARMITAGE and DAVID (editors) · Advances in Biometry
ARNOLD, BALAKRISHNAN, and NAGARAJA · A First Course in Order Statistics
ARNOLD, BALAKRISHNAN, and NAGARAJA · Records
*ARTHANARI and DODGE · Mathematical Programming in Statistics
*BAILEY · The Elements of Stochastic Processes with Applications to the Natural
    Sciences
BARNETT · Comparative Statistical Inference, *Third Edition*
BARNETT and LEWIS · Outliers in Statistical Data, *Third Edition*
BARTOSZYNSKI and NIEWIADOMSKA-BUGAJ · Probability and Statistical Inference
BASILEVSKY · Statistical Factor Analysis and Related Methods: Theory and
    Applications
BASU and RIGDON · Statistical Methods for the Reliability of Repairable Systems
BATES and WATTS · Nonlinear Regression Analysis and Its Applications
BECHHOFER, SANTNER, and GOLDSMAN · Design and Analysis of Experiments for
    Statistical Selection, Screening, and Multiple Comparisons
BELSLEY · Conditioning Diagnostics: Collinearity and Weak Data in Regression
BELSLEY, KUH, and WELSCH · Regression Diagnostics: Identifying Influential
    Data and Sources of Collinearity
BENDAT and PIERSOL · Random Data: Analysis and Measurement Procedures,
    Third Edition

*Now available in a lower priced paperback edition in the Wiley Classics Library.

BERRY, CHALONER, and GEWEKE · Bayesian Analysis in Statistics and
  Econometrics: Essays in Honor of Arnold Zellner
BERNARDO and SMITH · Bayesian Theory
BHAT · Elements of Applied Stochastic Processes, *Second Edition*
BHATTACHARYA and JOHNSON · Statistical Concepts and Methods
BHATTACHARYA and WAYMIRE · Stochastic Processes with Applications
BILLINGSLEY · Convergence of Probability Measures, *Second Edition*
BILLINGSLEY · Probability and Measure, *Second Edition*
BIRKES and DODGE · Alternative Methods of Regression
BLISCHKE AND MURTHY · Reliability: Modeling, Prediction, and Optimization
BLOOMFIELD · Fourier Analysis of Time Series: An Introduction, *Second Edition*
BOLLEN · Structural Equations with Latent Variables
BOROVKOV · Ergodicity and Stability of Stochastic Processes
BOULEAU · Numerical Methods for Stochastic Processes
BOX · Bayesian Inference in Statistical Analysis
BOX · R. A. Fisher, the Life of a Scientist
BOX and DRAPER · Empirical Model-Building and Response Surfaces
*BOX and DRAPER · Evolutionary Operation: A Statistical Method for Process
  Improvement
BOX, HUNTER, and HUNTER · Statistics for Experimenters: An Introduction to
  Design, Data Analysis, and Model Building
BOX and LUCEÑO · Statistical Control by Monitoring and Feedback Adjustment
BRANDIMARTE · Numerical Methods in Finance: A MATLAB-Based Introduction
BROWN and HOLLANDER · Statistics: A Biomedical Introduction
BUCKLEW · Large Deviation Techniques in Decision, Simulation, and Estimation
CAINES · Linear Stochastic Systems
CAIROLI and DALANG · Sequential Stochastic Optimization
CHATTERJEE and HADI · Sensitivity Analysis in Linear Regression
CHATTERJEE and PRICE · Regression Analysis by Example, *Third Edition*
CHERNICK · Bootstrap Methods: A Practitioner's Guide
CHILÈS and DELFINER · Geostatistics: Modeling Spatial Uncertainty
CHOW and LIU · Design and Analysis of Clinical Trials: Concepts and Methodologies
CLARKE and DISNEY · Probability and Random Processes: A First Course with
  Applications, *Second Edition*
*COCHRAN and COX · Experimental Designs, *Second Edition*
CONGDON · Bayesian Statistical Modelling
CONOVER · Practical Nonparametric Statistics, *Second Edition*
CONSTANTINE · Combinatorial Theory and Statistical Design
COOK · Regression Graphics
COOK and WEISBERG · Applied Regression Including Computing and Graphics
COOK and WEISBERG · An Introduction to Regression Graphics
CORNELL · Experiments with Mixtures, Designs, Models, and the Analysis of Mixture
  Data, *Second Edition*
COVER and THOMAS · Elements of Information Theory
COX · A Handbook of Introductory Statistical Methods
*COX · Planning of Experiments
CRESSIE · Statistics for Spatial Data, *Revised Edition*
CSÖRGÖ and HORVÁTH · Limit Theorems in Change Point Analysis
DANIEL · Applications of Statistics to Industrial Experimentation
DANIEL · Biostatistics: A Foundation for Analysis in the Health Sciences, *Sixth Edition*
*DANIEL · Fitting Equations to Data: Computer Analysis of Multifactor Data,
  Second Edition
DAVID · Order Statistics, *Second Edition*

*Now available in a lower priced paperback edition in the Wiley Classics Library.

*Now available in a lower priced paperback edition in the Wiley Classics Library.

*Now available in a lower priced paperback edition in the Wiley Classics Library.

*Now available in a lower priced paperback edition in the Wiley Classics Library.

*Now available in a lower priced paperback edition in the Wiley Classics Library.